GENETICS AND FISH BREEDING

GENETICS AND FISH BREEDING

Anthony Reid
Ian Wordsworth

Kruger Brentt
Publishers
2023

Kruger Brentt Publishers UK. LTD.
Company Number 9728962

Regd. Office: 68 St Margarets Road, Edgware, Middlesex HA8 9UU

ISBN: 9781787150652

For information on all our publications visit our website at http://krugerbrentt.com/

PREFACE

Fish genetic breeding is a process that remoulds heritable traits to obtain neotype and improved varieties. For the purpose of genetic improvement, researchers can select for desirable genetic traits, integrate a suite of traits from different donors, or alter the innate genetic traits of a species. These improved varieties have, in many cases, facilitated the development of the aquaculture industry by lowering costs and increasing both quality and yield. The potential of fish and shellfish production to feed a growing global population could be significantly enhanced through advances in genetics and biotechnology.

The present book contains 12 chapters covering all related disciplines. These chapters include introduction to animal breeding, introduction of DNA in animal breeding, organization of breeding activities, results of animal breeding, negative effects of animal breeding, basics of animal breeding, challenges for animal breeding, genetic models, genetic diversity and inbreeding, inheritance of monogenic, selection and mating, crossbreeding. Related terminology is given at the end for ready reference. It is written in a lucid style with the most recent advances, and with theoretical discussions for easy understanding of the subject. Well labelled illustrations, diagrams, tables and figures amply support the text wherever necessary. It provides basic principles of Genetics and Fish Breeding in a simple and easy language. The book is also supported with a comprehensive glossary and index in the end. This book would be an invaluable asset for the scientists, students and teachers working in the field of fishery, aquaculture and zoology.

We are grateful to all those persons as well as various books, manuals, periodicals, magazines, journals etc. that helped in the preparation of this book. In spite of great care taken by me and publisher team, some errors may have escaped from our notice. We shall appreciate if such errors are brought to our notice. There will be always being some scope for further improvement. Further queries, constructive suggestions and criticisms for the improvement of the book are always welcome and shall be thankfully acknowledged.

Anthony Reid
Ian Wordsworth

CONTENTS

1

SCOPE OF APPLIED FISH GENETICS

The several areas of aquaculture where biotechnology can be applied include production of fish of high production of fish seedling. Fishes appear to have better scope for research in genetics and breeding. Chromosomal engineering as spermiation occurs in fish eggs prior to the extrusion of the second polar body. There is a hope for activation of the egg with irradiated sperm, retention of the second polar body and inhibition of the spindle formation during the first cleaving. The main vision of aquaculture biotechnology is to achieve improvements of aquacultural stock, preservation of genetic resources, disease diagnosis and control of microbial/ microalgal genetic engineering.

SELECTION

Fish selection is based on the planned utilization of genetic variation inherent in all living creatures. An understanding of the nature of genetic variation and the laws by which it is governed is indispensable for the formulation of an effective programme of selection work.

In comparison to the breeding of domestic animals and agricultural plants, fish culture is a young science. Though in some eastern Asiatic countries, particularly in China, fish culture has been practised for more than two thousand years, until quite recently, their method of culture was based entirely on culture of young fish caught in natural waters. The domestication of fish and the creation of breeds differing from their wild parents in higher productive properties was actually started only a few centuries ago. The only exception is the decorative goldfish (*Carassius auratus*); some goldfish varieties have a thousand-year-long history.

The number of domesticated fish species is not great. Besides carp and goldfish, they include *Crucian* carp (*C. auratus gibelio*), tench (*Tinca tinca*), rainbow trout (*Salmo gairdneri*), lake trout (Salvelinus namaycush), char (*S. alpinus*), gourami (*Osphronemus goramy*), several centrarchids, tilapia and several other fish which are bred in some countries of southern and southeastern Asia. Recent years have seen a sharp increase in the domestication of these. This has been brought about by a swift rise in importance of pond fish for the total balancing of fish production. Attempts to reproduce many new species of various fresh water families in ponds and tanks have proved successful.

Domestication of these fish requires selection to increase their productive properties and adaptation to new conditions of existence greatly differing from natural environments.

Among the principal aims to be accomplished by fish selection are the following:

- To increase the growth rate, by better utilization of food (physiological decrease of food expenditure per unit of growth increment);
- To increase the growth rate, by fuller utilization of natural food in ponds and higher consumption of feed mixtures;
- To increase resistance to oxygen deficiency, to high or low temperature, to higher salinity or to other deviations from the normal environmental conditions;
- To improve resistance to infectious diseases and to infestation with parasites (to develop new breeds resistant to particular diseases);
- To improve the nutritive properties of fish (to increase the calorie content, to decrease the proportional weight of non-edible parts, to decrease the bone content, to increase or decrease the fat content, etc.)

THE ROLE OF GENETICS IN AQUACULTURE

Modern high-yielding plants and animals used in agriculture are derived from 'wild' stocks but their productivity has been enormously increased by a combination of genetic and husbandry methods. Although husbandry has been considerably improved in aquaculture there has been very little genetic improvement. This is partly because of the much shorter history of aquaculture, which for most species is less than a century, and partly because rearing aquatic organisms, which frequently have complicated life cycles, is intrinsically more difficult than rearing land plants and animals. All farmed aquatic animals, with the possible exception of carp, are genetically indistinguishable from the wild populations from which they were captured but there is considerable scope for genetic improvement and for the application of specific genetic manipulations which will significantly improve the productivity of aquaculture.

The role that the application of genetic principles can play in the sustainable use and conservation of living aquatic resources is being increasingly appreciated by resource managers, policy makers and the international community. Gene and genotype frequency data can provide information about:

- Species identification,
- Population stock structure,
- Hybridization,
- Gene flow,

- Inbreeding, and
- Genetic impacts of fishing and aquaculture.

Often, data from other sources, e.g. tag-recapture and morphometrics, cannot provide such information or are extremely difficult to collect. Genetic data and other types of data should be used together to understand better the dynamics of aquatic populations both in culture and in the wild.

Many groups of fishes and invertebrates show similar genetic structuring because of geological changes, current patterns, among other things. For many marine species the differences among sub-groups is not large because of spawning behavior, migrations and not as accurate homing to natal areas as salmon. Genetic differences have also been used to give endangered species protection to valuable groups of fishes for example in the USA under the Endangered Species Act. Application of genetic improvement in aquaculture and culture-based fisheries is progressing rapidly and population genetics can also be applied to the study of cultured groups of fish in order to choose appropriate breeds to crossbreed and to monitor inbreeding.

Fish Propagation

Introduction

1. Several methods exist for the breeding or propagation of cultivated fish. Choosing among them depends on the reproduction biology of the species, on local environmental conditions and on the facilities available. These methods may be grouped into three categories: natural propagation; semi-natural propagation; artificial propagation.
2. For natural propagation, males and females are placed together in a breeding area such as a small pond or an enclosure where they spawn naturally. This method is usually used, for example, to produce tilapias cheaply.
3. The successful breeding of certain species may require some environmental manipulation such as the inflow of new water and a sudden rise of the pond water level for the African catfish Clarias, the presence of grassy vegetation as egg collectors nests for common carp, or the presence of artificial nests for American and European catfishes, *Ictalurus* and *Silurus*.
4. For semi-natural propagation, the fish (usually the females only) are first given one injection of chemicals, such as a pituitary gland extract, which will trigger spawning. Males and females are then placed together in a specially prepared breeding area such as a small grassy pond or an enclosure where spawning takes place. The fertilized eggs are usually collected and reared under improved conditions, either natural or artificial.
5. For artificial propagation, the females are given one or more injections of chemicals which regulate the final ripening of dormant eggs in the ovaries. As soon as the eggs are ripe, they are stripped from the females. The males are usually also

injected. Eggs are artificially fertilized with sperm obtained from the males and reared under controlled conditions.

6. Each of these propagation methods is controlled by a series of environmental factors, as shown in the chart below for common carp, for example.

INTENSIVE FISH CULTIVATION

Fish farming is commonly described as being extensive, semi-intensive or intensive. These terms are not sharply defined but are generally related to the level of inputs of feed and/or fertiliser and to stocking density.

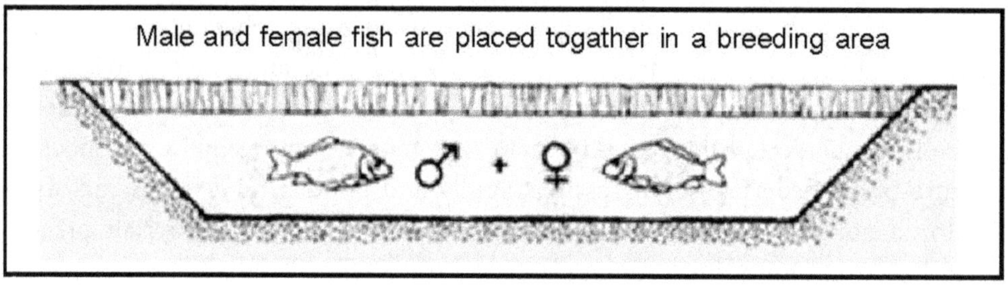

Fig. 1. Natural Propagation

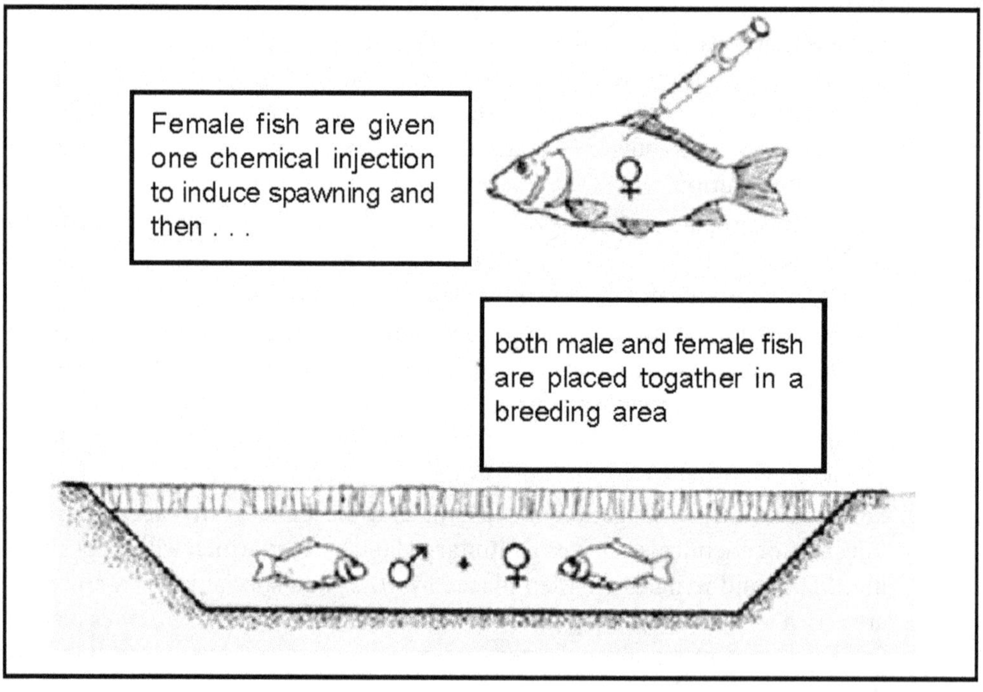

Fig. 2. Semi-Natural Propagation

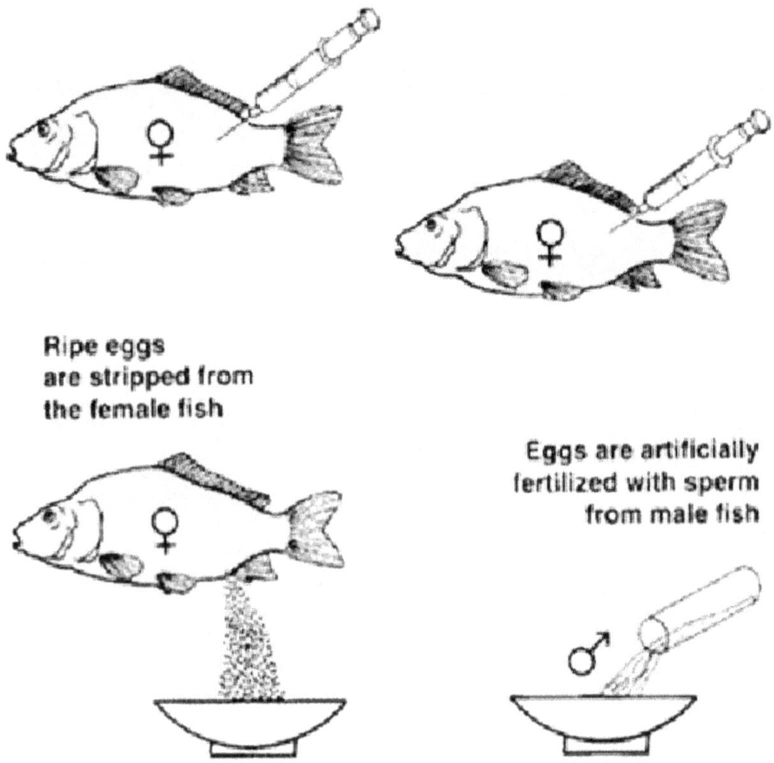

Fig. 3. Semi-Natural Propagation

Fish have been farmed extensively, for many centuries, by fostering a natural environment with little or no inputs. Semi-intensive farming, with a greater use of inputs, developed from this. The end of the 20th century saw the development of intensive fish farming, high input systems with high stocking densities. As with farmed mammals and birds, stocking densities are a key concern for animal welfare.

The three families of fish species with the highest aquaculture production tonnages are carps, tilapias and salmonids, which account for respectively 61%, 9% and 6% of total global farmed fish production tonnage. Carp are often farmed with different species of carp stocked together (polyculture systems) to better exploit the different ecological niches of the pond ecosystem. Carp and tilapia farming can be extensive, semi-intensive or intensive while farming of salmonids (mostly Atlantic salmon and rainbow trout) is generally intensive. Catfish families together comprise 8% of total global farmed fish production tonnage and are often farmed very intensively.

Semi-intensive Fish Farming

In semi-intensive fish farming, the fish still obtain significant nutrition from the food web within their pond, but they are also given supplementary feed. This means the fish can grow faster and/or to a larger size or at a greater stocking density. The feed may be of vegetable origin or may include fish, fish oil and/or fishmeal.

Grass carp is the farmed fish species with the highest global production tonnage and is farmed, mainly in China, in both in semi-intensive pond systems and intensively in cages. In semi-intensive farming, the grass carp are stocked in ponds or pens with other carp species (either as a major or secondary species), where aquatic weeds and terrestrial grasses form the main feed. Commercial feeds, such as pellets, and vegetable by-products may be used instead to save labour costs, or at specific times of the year when growth of water grasses and algae is reduced. The total stocking density is 750-3000 fish per hectare (i.e. 3-13 square metres per fish).

Intensive Fish Farming

In intensive farming, the fish are kept at too high a stocking density to obtain significant amounts of feed from their environment. Instead the fish are dependent on the feed provided and water must be replenished at a high rate to maintain oxygen levels and remove waste. The levels of feed inputs and management of the water affect the stocking density of the fish that can be supported.

Intensive Aquaculture in Existing Tanks and Ponds

Introduction

It is estimated that the country has presently 24.1 lakh hectares under freshwater aquaculture. The NFDB proposes to bring about one-third (33 %) of this area i.e. about 8.0 lakh hectares under intensive aquaculture in the next six years. Generally, the Board will not provide subsidy for the programme. However, in areas where potential for freshwater aquaculture exists but the activity has not picked up due to various reasons, 25 percent of the estimated 8.0 lakh hectare will be covered with subsidy during a span of six years.

Eligibility Criteria

The following criteria will be used to select farmers for subsidy to cover 25 percent of the estimated 8.0 lakh hectares for intensive aquaculture in existing ponds and tanks:

- Past performance of the farmer in undertaking freshwater aquaculture.
- Willingness/ entrepreneurship of the farmer to take up fish culture on scientific lines.
- Assurance / Willingness of the farmer to invest working capital money from 2nd year onwards obtained from the sale of the first crop duly utilizing the NFDB assistance towards the first year input costs.
- Bank's consent to provide the loan/ commitment of the farmer to invest 80% of the unit cost on his own.

Unit Cost (Investment Cost)

The unit cost for renovation/ repair of one hectare (water spread area) existing fresh water tank/ pond has been estimated as Rs 30,000/ha-. Farmers who wish to avail bank loan or who wish to invest on their own be provided with a subsidy not exceeding 20 percent of the unit cost (Rs.6,000/ha) for all farmers except SCs/STs & NE States for whom it will be 25% of the unit cost (7500/ha).

Intensive Aquaculture in New Ponds and Tanks

Introduction

The country has large potential for development of freshwater aquaculture. Due to various reasons, this potential has remained untapped. Besides enhancing production and productivity from the existing tanks and ponds, the NFDB also envisages to tap additional areas under freshwater aquaculture, which will not only add to the fish production of the country but will also generate additional employment, especially in the rural and interior areas. It is proposed to bring in an additional area of 50,000 ha under intensive aquaculture in the country in a span of 5-6 years. To promote this development, subsidy would be extended to cover 50 percent of the area i.e. 25,000 hectares in the next 5-6 years.

Eligibility Criteria

The following criteria will be used to select farmers that would enable coverage of 50 percent of the estimated 50,000 hectares for subsidy to take up intensive aquaculture in new ponds and tanks:

- Past performance of the farmer in freshwater aquaculture including record of his/ her training in the said activity.
- Willingness/ entrepreneurship of the farmer to take up fish culture on scientific lines.
- Clear title of the land.
- If pond/ tank taken on lease, a minimum lease period of 10 years would be preferred.
- Assurance / Willingness of the farmer to invest working capital money from 2nd year onwards obtained from the sale of the first crop duly utilizing the NFDB assistance towards the first year input costs.
- Bank's consent to provide the loan / commitment of the farmer to invest 80% of the unit cost on his own.

Unit Cost (Investment Cost)

The unit cost for development of one hectare (water spread area) new pond for aquaculture in plain areas has been estimated as Rs 2,00,000/ha. Farmers who wish to avail bank loan or who have invested on their own shall be provided with a subsidy of 20 percent of the unit cost i.e. Rs 40,000/ha for all farmers except SCs/STs and & States for whom it will be 25% of the unit cost (Rs.50,000/ha).

ORNAMENTAL FISH BREEDING

Introduction

Ornamental fish farming is a very important commercial component of aquaculture. To maximize profits with the Ornamental fish farming business, you would have to learn the process of hatching Ornamental fish eggs and then groom them to maturity. Ornamental fish farming has been an interesting activity for many, and which provide not only aesthetic pleasure but also financial openings. About 600 Ornamental fish species have been reported worldwide from several aquatic environments. Ornamental fishes are also called living jewels. There are more than 30,000 fish species reported around the world and these Ornamental fishes belong to about 800 fish species. Most of the Ornamental fish species survive in freshwater.

Ornamental fishes can be well-defined as attractive colorful fishes of peaceful nature that are kept as pets in the confined spaces of an aquarium or a garden pool with the main purpose of enjoying their beauty. It is popular as an easy and stress relieving hobby. This culture is an excellent business opportunity in India since there is strong demand from domestic and export markets. Generally, the Ornamental fish farming business is indeed a big farming business because it needs moderate capital and technical skills.

Benefits of Raising Ornamental Fish

Ornamental fishes are peaceful, generally tiny, and available in several attractive colors, and capable of living in confined spaces like aquarium condition or garden pool. They are called 'living jewel' due to their color, shape, behavior, and origin. Ornamental fish farming is a multibillion-dollar industry in the world. Ornamental fish farming is one of many promising job alternatives since it is easy to start. Ornamental fishes are easy to raise and can be sold at a good price.

- **Easy to raise** – Raising Ornamental fish is probably the easiest type of business, considering that it is not hard to raise them. As long as you are knowledgeable about their basic biology, you can easily raise Ornamental fishes in the comfort of your backyard.
- **Does not need a lot of manpower** – Ornamental fish farming does not require hard physical labor. You need a little aquarium unit and you can set up your own fish farming business. The only time you need some physical work is when you want to move the fish to other tanks, but that also does not require a lot of physical extortion. If you do need help, you can hire someone for a one day work and it will require a couple of hours of work only.
- **High demand for Ornamental fishes** – Ornamental fishes are in high demand in many places around the world. Also, an aquarium with Ornamental fishes can be a nice addition to a room as a decoration. Not only it looks pretty,

but also watching fishes swimming around can be relaxing too. So, making the aquarium good ornaments in many places.

Breeding of Ornamental Fishe

In nature, fishes are stimulated by changes in environmental factors which are difficult to extend in captivity. However, such conditions, to some extent, can be stimulated in the aquaria by providing specific biological requirement for spawning. Some fishes breed naturally with ease while others have to be bred by artificial techniques such as hypophysation and hand stripping.

The native aquarium fishes, which have high demand overseas, are harvested from their natural sources (i.e., from the wild). However, overharvesting of these priced native fish species may lead (in the long run) to a depletion of the stock and many such species have been pushed to the brink of extinction.

Hence, captive breeding of such economic ornamental fishes is very important not only for a sustainable trade of these species but also to conserve the natural stock. Breeding of popular exotic species is also undertaken to meet the high demand of the local aquarium fish market.

The various aspects of breeding comprise selecting the breeding pair, conditioning the brood fishes, observing spawning and raising its young. According to the breeding habits of fish, the breeding tanks have to be prepared.

Fishes either produce egg or live young. Egg layers comprises egg-scatterers, egg depositors, egg buriers (bubble-nest builders), mouth brooders, nest builders, etc. In live-bearing fishes live youngs are released by the females:

a. **Identification of sexes:** In the case of live- bearers identification of sexes is essential. The mature males of egg-layers develop 'pale-tubercles' or small 'bumps'-like white spots or cysts on the head, gill cover and pectoral fin which make the male rough to touch. They ooze milt when pressure is applied on the belly. The mature female has a 'gravid spot' on the vent, and the belly is comparatively swollen.

b. **Selection of breeding pair:** A breeding pair is selected that are already showing courting behaviour, when they show interest to each other. The fishes are brought to condition through feeding with good quality food particularly the live food. In case of egg-layers, male fishes often show too much interest in females and they need to be separated after spawning.

c. **Egg-layers or oviparous:** Fertilisation is external in oviparous fishes. The fry of these fishes are small and helpless whey they hatch. The egg-layers depending on the type of eggs laid, may be of two types — (i) Adhesive eggs which have the ability to attach to any substratum and (ii) Non-adhesive eggs.

Another method of dividing the egg layers of ornamental fishes is according to their spawning habit. They are of the following types:

Egg scatterers

The egg scatterers, such as Zebra Danio, Goldfish, etc., show little or no parental care. Fishes like Siamese Fighter Fish have a tendency to eat up their own eggs when they are released. So the eggs or the fishes should be removed to another tank after spawning gets over. In the wild, the eggs probably get hidden in the muddy water or are swept away by the water current.

In an aquarium, the eggs are provided with some egg saving measures, such as:

1. A layer of marble chips or pebbles are provided at the bottom, in which the eggs may lie concealed.
2. In case of fishes which releases adhesive eggs (Ex. common carp), thick bunches of plants or mops of nylon wool may be provided in the breeding tank.
3. For further safeguard of the eggs, a net may be kept submerged in the water, below the breeders, through which the eggs may pass through to the tank bottom.

Important Ornamental Fishes

Usually, exotic fish dominate the Indian market and more than 250 varieties have been recorded from the domestic market. More than 200 freshwater fish species are bred in different parts of India by Ornamental fish breeders. The egg layers lay adhesive or non-adhesive eggs on the aquarium plants or glass walls. Some parents show parental care and destroy their eggs so different breeding setups are needed. In contrast, livebearer fish release young in batches and are very easy to breed compared to egg-layers. Some common exotic live-bearers such as guppy, molly, swordtail, platy and egg layers like goldfish, red-tailed black shark, and red finned shark, etc. dominate the market. Native Ornamental fish include honey gourami, rosy barb, zebrafish, glassfish, and loach, etc.

Categories of Ornamental Fish Species

There are mainly two categories of Ornamental fish are being marketed. They are exotic Ornamental fish and native fish of India, which have Ornamental value for coloration or behavior. Exotic Ornamental fish dominate the domestic market. Already 288 exotic fish varieties have been recorded in the Indian market. More than 200 species of these freshwater fish varieties are bred in different parts of India and others still have to be imported as fry. Based on availability, and climatic conditions the Ornamental fish farmers are engaged in breeding common exotic livebearers and egg layers. Then, the egg layers lay sticky or non-sticky eggs on the glass wall. Some parents show parental care and some destroy their eggs so different breeding setups are required. Livebearers release young in batches and are very easy to breed.

Different Types of Fishes to Raise in an Ornamental Fish Farm

Selecting fish breeders is one of the important steps in fish farming. First, you need to prepare a fish breeding tank specifically used to breed these fishes.

- **Ornamental Fish Breeding of Koi** – This type of fish is perfect for aquariums and pools. Normally, it is a cold water type of fish but now they can adapt to various types of climates due to their ability to survive almost everywhere. They are bright-colored fishes that can be raised easily and this fish can be trained to eat directly from the hand.
- **Ornamental Fish Breeding of Puffer** – The Puffer fishes have thorn-like parts around the body and will blow their body when feeling threatened. The fishes are known as toxic to keep them safe from predators, so you need to be extra careful when raising them.
- **Ornamental Fish Breeding of Tropical Fish** – These fishes originate from tropical environments, but also adaptable in freshwater. They have bright colors and various shapes. Most people prefer tropical fishes compared to other types of Ornamental fishes due to their colorful appearance.
- **Ornamental Fish Breeding** of Angelfish – This is another tropical fish type or freshwater preferred by a lot of people. Although it is considered as predator fish, Angelfish can survive in aquariums and will adapt well.
- **Ornamental Fish Breeding of Parrot Fish** – This fish has a mouth that resembles those of parrot beaks, hence the name Parrot Fish. It has colorful skin that ranges from a combination of blue, yellow color and sometimes with strikes of red.
- **Ornamental Fish Breeding of Arowana** – This freshwater has an elongated body and can easily grow long and big. There are a lot of Arowana species in nature, but only 10 types are kept as pets.

Another method of dividing the egg layers of ornamental fishes is according to their spawning habit. They are of the following types:

(i) Egg scatterers: The egg scatterers, such as Zebra Danio, Goldfish, etc., show little or no parental care. Fishes like Siamese Fighter Fish have a tendency to eat up their own eggs when they are released. So the eggs or the fishes should be removed to another tank after spawning gets over. In the wild, the eggs probably get hidden in the muddy water or are swept away by the water current.

Goldfish Breeding

Although goldfish (*Carassius aural us*) has a lifespan of 10 years, the male and female mature in the first and second years, respectively. The selected breeders are segregated sex-wise. At the time of breeding two males and one female are released into the breeding aquarium, which has been erected for the purpose.

For collection of the eggs, artificial spawning nests are made of small branches of aquatic plants tied on a circular rope, as shown in Fig. 6.90. At the time of mating, the males chases the female and pre-spawning intimation takes place. The eggs laid by the females are fertilised by the milt of the males. The fertilised eggs being adhesive and demersal, stick to the spawning nest.

This nest with attached eggs is then taken out of the breeding tank and transferred to the nearby hatching tank. In a period of three days hatching takes place and the hatchlings initially subsists on the available yolk.

After two days (when the yolk has become exhausted), they are fed with mashed particles of hard-boiled egg yolk seived through a muslin cloth or on small sized zoo-plankton and nauplii of artemia. It has been seen that enhanced feeding takes place only at a high temperature of about 35°C.

Egg depositors

The egg depositors, such as Rasbora doniconius, Hemichronis bimaculatus, Aequidens sp. etc., show highest degree of parental care. They are the most important group among the egg-layers and unlike the egg-scatterers these fishes do not eat their own eggs.

They select sites such as filter tube, glass wall or plant leaf for laying the eggs. The Cichlid fishes deposit their eggs in a hole dug at the bottom. Before spawning the egg depositors forcibly evict any other fish from the surrounding area by chasing them away.

Egg Burriers

The egg burriers, such as Aplocheilus panchax, are annual fishes, as they survive only for a year in the wild and never live to see the fate of their offsprings. In the wild, the adult fishes spawn before the onset of summer. They bury their eggs in the mud bed so that they can survive the forthcoming dry season when the bed dries up. With the onset of monsoon, the eggs hatch out.

In the aquarium, however, the condition is different as the water never dries out and most of the fishes live longer than one year. The eggs are laid in dense planted environment inside the aquarium. As the eggs are capable of remaining viable, hatching is possible even after some weeks or even months.

Mouth Brooders

The female tilapia, being a mouth brooder, picks up the fertilized eggs into its mouth and incubates them in its throat. Even after hatching, the female tilapia carries the spawn and young ones in its mouth. During this incubation period of two weeks, the female does not take any food.

Nest Builders

The gouramis (*Colisa lalia, C. fasciata,* etc.) and their relatives are the most popular among the nest builders. The male generally builds the nest and also guards the eggs from any intruders including the female.

Bubble-nest Builder

The bubble-nest is made by the male next to a floating leaf or among the floating plants. The males of Siamese fighters (Betta splendens), gourami, etc. forms the nest by frequently rising to the surface, gulping atmospheric air and enveloping it with mucus formed in the mouth during spawning to form a bubble. Several such bubbles adhere to one another to form a raft or float.

The male then drives the female near to the raft, flexes around her and fertilises few eggs released by the female. The fertilised eggs sink slowly to the bottom. Both fishes then drift downwards. The male picks up the eggs and blows them into the nest.

The courting and mating behaviour is repeated until the female is empty of eggs. The male, at this stage, becomes aggressive even towards the female and, if not removed, will certainly kill it.

The male guards the eggs until it hatches. Some of the hatched fry fall out and spiral away to the bottom. They are picked up by the male and blown back into the group. After five days when the fry starts to swin, the male fish is then removed. The fry is then fed with infusoria or finely sieved egg yolk for a few days, after which they are given freshly hatched brine shrimp, Daphnia and chopped worms.

Live Bearers of Ovoviviparous

The Live-Bearers due to their beautiful and bright colouration find a special place for the ornamental fish hobbyists. They also have an easy breeding habit. The live-bearers directly release the young ones. The eggs of these fishes are fertilised internally in the belly pouch of the female.

After one copulation, the female collects the sperms which is enough to fertilise 2 to 3 batches of eggs. The pregnant females are then separated in a brood cage, through which the young can escape when born. This is done because the live-bearers are known to be cannibalistic and feed on their young ones. The breeding tanks may be provided with plants for the safety of the young ones.

In India the live-bearing fishes are generally of exotic origin. Some of the common live-bearers are Guppy (*Poecilia reticulatua*), Molly (*Mollienisia latipinna*), Sword tail (*Xiphophorus helleri*), Platy (X. *maculates*), etc.

Feeding of Fry

Food should not be given to the new brood until they are able to consume it, because the unconsumed food would decay and pollute the tank. The fry of live-bearers are able to feed immediately, but the young ones of egg-laying fishes would often remain immobile for several hours while they consume the remainder of their yolk sac.

Commercially available fry food can be had in aquarium shops and is available in liquid or powdered form. The size of the particles should be such that the fry can eat it. Generally brine shrimp, microworms and Infusoria can be fed to new fry. Infusoria are a collection of ciliated protozoans that can be produced in the laboratory.

Boiling water is poured over chopped dry hay or green lettuce leaves in ajar. The mixture is kept in a warm, dark place for a few days. A healthy broth of Infusoria will be produced floating in the Sun. Small quantities of the infusoria can be added, several times a day for feed of the new fry.

Health Care

Water exchange, is a must for maintaining water quality conducive for the fish health. Only healthy fish can withstand transportation and fetch good price. Chemicals / antibiotics, Methylene Blue, Methylene Yellow, Malachite Green, Ampicilin, Vitamins, Potassium permanganate, Copper sulphate etc can also be used for preventing / treating diseases.

Ornamental Fishes

Aquarium fishes are mainly grouped into two categories, viz., Oviparous (egg - layers) and Viviparous (live-bearers). Further, the fresh water ornamental fish varieties can be broadly grouped into Tropical and Cold water species also. Management of these two categories are different in nature. According to water tolerance fishes are hard water tolerant, soft water tolerant species and those with wide tolerance. The fishes and the details of grouping is given below.

Barbs

Barbs, members of the Cyprinidae family, comprise more than 400 species in their entire range, which extends from Africa, Asia and Europe to Central China, the Philippines and the East Indies. About 150 species of barbs are of ornamental value, comprising mostly of wild caught varieties. *Barbus tetrazona* (Tiger barb), *B. conchonius* (Rosy barb) and *Puntius titteye* (cherry barb) are the most common barbs in the aquarium industry. *P. denisonii*, a wild caught variety of barb is mainly found in Kerala and Karnataka.

Fig. 4 Barb fish

Loaches

Loaches, commonly known as Botia, are belonging to the family Cyprinidae. There are around 40 species of loaches known today, belonging to the genus *Botia*. They are natives of Thailand, India, Pakistan, China, Bangladesh and some Indonesian Islands. They are probably the most diverse group of fish in the hobby, both in pattern and behaviour.

Fig. 5. Loaches Fish

Danios

Danios, another member of the Cyprinidae family, are small and lively fish which are native to the Indian peninsula, Sri Lanka, Pakistan, Thailand, Myanmar, Malaysia and Indonesia. They are found in a variety of habitats from boulder-strewn mountain torrents to small pools in dry zone streams. There are more than 12 species reported today of which *Danio malabaricus* (Pearl Danio), *D. albolineata* and *Brachydanio rerio* (zebra fish) are common in the hobbyist market. *Brachydanio rerio* is mainly found in Kerala and Karnataka.

Fig. 6. Danios Fish

Freshwater Sharks

Freshwater sharks are different types of minnows, belonging to four genera, Balantiocheilos, Epalzeorhynchos, Labeo and Luciosoma, under the Cyprinidae family. They are commonly referred to as freshwater sharks in the hobby market. These minnows are widely distributed over Southeast Asia, the Malay Archipelago, Indonesia, and parts of the Middle East and Africa. Genus Epalzeorhyncho have four types of fresh water sharks, *E. bicolor*, E. frenatus, E. kalopterus, and E. munense. The genus, Labeo has over a hundred species, of which two species, Labeo chrysophekadion and L. cyclorhynchus are popular in the aquarium trade. The genus, Luciosoma has five species, of which one, L. siplopleura, is popular in the hobby. Bala shark, as popularly known, is a short form of its Latin name, Balantiocheilus melanopterus, are native of Malaysia, Thailand, Cambodia and Indonesia (Sumatra and Borneo).

Fig. 7. Freshwater sharks

Gourami

Gourami, the labyrinth fish, is a native of western and southern Africa, northeastern, southeastern, south and Southeast Asia. *Colisa lalia*, the Dwarf Gourami, occurs naturally in freshwater ponds, streams and paddy fields of northeastern India and Bangladesh, and is most popular among the Gouramis. Important varieties include The Honey Gourami (*C. chuna*), Indian Gourami (*C. fasciata*), Thick-lipped Gourami (*C. labiosa*), three spot gourami (*Trichopodus trichopterus*), Pearl Gourami (*T. leeri*), Snakeskin (*T. pectoralis*), Moonlight Gourami (*T. microlepis*) and the Kissing Gourami (*Helostoma temmincki*).

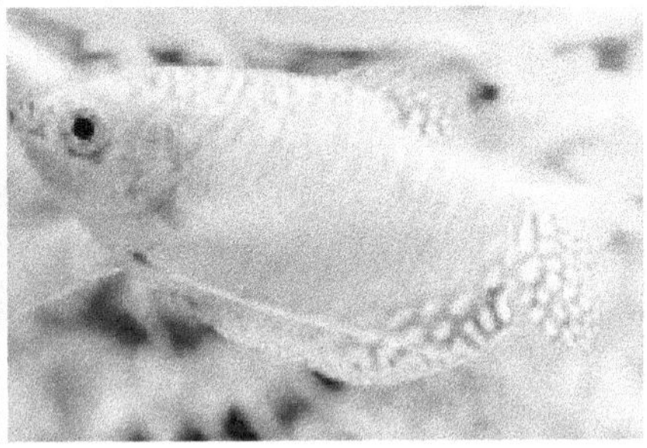

Fig. 8. Gourami Fish

Glassfish

Glassfish, belonging to the family Chandidae or Ambassidae, comprise mostly of fish with transparent bodies. They originate from Asia, especially Pakistan, India, Nepal, Bangladesh, Myanmar and Thailand. There are four species commonly referred to as glassfish in the trade: *Chanda ranga*, *C. lala*, *C. nama* and *C. baculis*.

Fig. 9. Glassfish

Gold Fish

The gold fish (*Carassius auratus*) is considered to be the most popular and attractive pet fish among all ornamental fishes, due to its many variations such as colour, fin shape, size and body structure. Though similar in appearance to carp (*Cyprinus carpio*), gold fish lack barbels and a dark spot at the base of each scale. The goldfish is the most common aquarium fish and one of the oldest and best-known fish in the industry.

Platy

Fig. 11. Platy

The Platy (*Xiphophorus maculatus*) belonging to the family Poeciliidae is native to the east coast of Central America and Southern Mexico. It is a live bearer and most popular among aquarium fishes. Several different colour variations have been developed so far such as red, yellow, orange, blue and white. Three species are commonly available; they are Southern Platy (*Xiphophorus maculatus*), Variatus Platy (*X. variatus*) and Swordtail Platy (*X. xiphidium*).

Oscar

Astronotus ocellatus is a member from the cichlid family, known under a variety of names such as Oscar, tiger Oscar or marble cichlid or the velvet cichlid. They are native of South America particularly Peru, Colombia, Brazil and French Guiana.

Fig. 12. Oscar fish

FISHING

Fishing is the activity of trying to catch fish. Fish are often caught as wildlife from the natural environments, but may also be caught from stocked bodies of water such as ponds, canals, park wetlands and reservoirs. Fishing techniques include hand-gathering, spearing, netting, angling, shooting and trapping, as well as more destructive and often illegal techniques such as electrocution, blasting and poisoning.

The term fishing broadly includes catching aquatic animals other than fish, such as crustaceans (shrimp/lobsters/crabs), shellfish, cephalopods (octopus/squid) and echinoderms (starfish/sea urchins). The term is not normally applied to harvesting fish raised in controlled cultivations (fish farming). Nor is it normally applied to hunting aquatic mammals, where terms like whaling and sealing are used instead.

Fishing has been an important part of human culture since hunter-gatherer times, and is one of the few food production activities that have persisted from prehistory into modern age, surviving both the Neolithic Revolution and successive Industrial Revolutions. In addition to being caught to be eaten for food, fish are caught as recreational pastimes. Fishing tournaments are held, and caught fish are sometimes kept long-term as preserved or living trophies. When bioblitzes occur, fish are typically caught, identified, and then released.

According to the United Nations FAO statistics, the total number of commercial fishers and fish farmers is estimated to be 38 million. Fishing industries and aquaculture provide direct and indirect employment to over 500 million people in developing countries. In 2005, the worldwide *per capita* consumption of fish captured from wild fisheries was 14.4 kilograms (32 lb), with an additional 7.4 kilograms (16 lb) harvested from fish farms.

Different Types of Fishing

You can catch a fish in many ways. You can catch it with hands; you can impale it on a spear, catch it on a hook, trap it, and throw an explosive on it (not recommended and even illegal). We had time to invent many possible ways of fishing because we have been fishing and eating fish for 40,000 years.

Here are some methods and types of fishing, some more, some less popular:

- **Noodling** is fishing with hands practiced in South of the United States. Fishermen catch catfish by sticking hand into a catfish hole where this fish lives.
- **Flounder tramping** is a method of fishing practiced in Scottish village of Palnackie on every August. People compete in catching the flounder (which is a species of flatfish) by stamping on them.
- **Spearfishing** is fishing with ordinary spears or with their variants like harpoons, tridents, arrows, Hawaiian slings, and spearguns.
- **Netting** is method of fishing which uses fishing nets. There are many types of nets for different uses and different fish. Cast net (or throw net) is a smaller round net with weights on it edges. Gillnet is placed in water vertically (using combination of weights and floats) and catches fish which try to pass through it. Trawl net is large, conical a dragged by ship.

- **Angling** is fishing with a hook (angle), line and rod. Hook has bait on it and is sometimes weighted with a sinker.
- **Fly fishing** uses artificial flies as lures with specially constructed fly rods and fly lines. Artificial flies are usually hand made in variety of shapes.
- **Bottom fishing** is angling with heavy weight at the bottom of water. It can be done from boats and from the land and its aim is catching the fish that lives at the bottom.
- **Ice fishing** is angling through the hole in the ice during the time of a year when a body of water is frozen.
- **Fishing from float tube** is fishing from a small inflatable boat (very small). It is used from fly fishing and when fishermen use it they don't scare the fish with splashing.
- **Kite fishing** is using kites to carry line and fishhook to the places that are not easily reachable. It was invented in China and it is still used on New Guinea and other Pacific Islands.
- **Trapping** is fishing done with traps. Basket weir fish traps are woven from branches and made so a fish can enter it but not exit. Lobster traps are similar to basket traps but are smaller and have more compartments. Fishing weir is a large trap made of logs and fences, placed in water to fore fish to enter parts of trap from which it can not exit.
- **Cormorant fishing** was a practice in China and Japan since 14th century. It uses trained cormorants with ring on their neck which prevents the bird to swallow the fish when it catches it. Instead it brings the bird to the fisherman.
- **Electrofishing** is method used in freshwater by fisheries scientists. Electricity is used to stun fish, check fish population and then return it into water unharmed.

Threats to Habitat

Habitat destruction occurs when the conditions necessary for plants and animals to survive are significantly compromised or eliminated. Most areas of the world's oceans are experiencing habitat loss. But coastal areas, with their closeness to human population centers, have suffered disproportionately and mainly from manmade stresses. Habitat loss here has far-reaching impacts on the entire ocean's biodiversity. These critical areas, which include estuaries, swamps, marshes, and wetlands, serve as breeding grounds or nurseries for nearly all marine species.

Causes of Ocean Habitat Loss

Climate and Seas Changes

Scientists explain that when there is higher concentration of carbon dioxide in the air, the water absorbs much of it. The water gets contaminated and the level of heat in the water rises beyond the expectations of life. When the ocean temperatures rise; there are other associated adverse impacts that are experienced in the aquatic environment.

Climate Change-related Heat

The heat melts the icecaps and as a result, there is a resultant rise in the ocean level. The melted ice caps and glaciers contaminate the water and this threatens the life of the aquatic plants and animals. The increase in temperature has another aeration negative impact. It obviously limits the concentration or the solubility of oxygen in the water. Plants and animals will then suffocate. The changes that take place in the sea due to climate change-related heat are also responsible for the marine habitat loss and destruction as they may contaminate the water or even alter the water temperatures.

Pollution

There are some human activities which are responsible for the changes in the marine ecological conditions and such may mean alteration or destruction and loss to the marine habitat. Such can entail water pollution, air pollution, and land pollution which intoxicate the environment and makes the water contaminated. Thermal pollution as a result of industrial activities also destroys marine habits. The end result is the depletion of marine animal and plant species.

Unsustainable Fishing

It is always important that the fishing activity has the ability to support the environmental conservation attempts. Too much aggressive fishing activity can damage the marine habitat as it leads to the loss of many fish and aquatic species.

Lack of Protection From the Governments

Where governments have failed to acknowledge the importance of conserving the natural habitat, there is higher probability that negative impacts will be made. In the case of marine habitats, when there are no policies regulating access to such areas, sea plants and animals will shoulder the burden of irresponsible acts.

Shipping Impact

In many cases, countries rely on the shipping means of transport for bulky cargo. But the effect of the process does not always augur well with the well-being of the sea habitat. Marine habitat is therefore destroyed by the oil spills among other associated hazards which poison the water.

Eutrophication

Eutrophication can be understood as the process where the fertilizers from land are washed off into the waters. These fertilizers always contain phosphate and nitrate ions and when they come into contact with the waters; they form basic and acidic compounds thereby leading to dense growth of algal blooms, which in return create intolerable environment that cannot support aquatic life.

Development of Coastal Places

Due to the existence of water transport, many countries develop a consequential drive to develop their coastal regions. But in the process, less or no concentration is rendered on the need to conserve the natural marine habitat and all that it contains.

Effects of Marine Habitat Loss and Destruction

Low Oxygen Concentration

When the marine habitat is destroyed and is at the brink of vanishing, there various effects that would be likely. The first major impact that acts as the determinant and influences the outcome of the others is that oxygen concentration in the water gets to the lower level, to the extent that it can barely support aquatic life.

This can be as a result of water or air pollution which even makes the water to be contaminated. The resultant situation of such destruction is that most plants and animals disappear through death.

Migration of Marine Animals

Some animals may be forced to migrate on sensing that there is that element of environmental hostility. It means that the country or the people that border this particular water body will be obviously deprived of the opportunity to freely gain from the benefits of the aquatic plants and animals.

Food Reduction

Because there is some degree of dependence between the terrestrial and aquatic lives; the former will be hit by the depletion of the latter. A significant example that can be cited in this particular scenario is that humans depend on some sea fish such as octopus, star fish, salmon and much more for food. When they are depleted, humans will definitely suffer from that negative impact.

Extinction of Animal and Plant Species

Animals such as whale, shark and many others depend on other aquatic ones such as seals for food. Extinction of one means the extinction of the others in the food chain. The whole consequence of the marine habitat loss and destruction is that it leads to death and migration of animals. Some plants also die and become extinct due to the extreme ecological conditions.

Rapid Land Conversion Rate

Where marine habitat loss and destruction is characterized by the disappearance or the decrease in water masses, there is quick land conversion rate. This may benefit humans for settlement but other benefits drawn from the aquatic life will surely vanish.

Loss of Coastal Natural Beauty

The beauty of the coastal places is dictated by the variety of plants and animals that are existing. This implies that when the natural habitat is encroached or even destroyed, such species or animals and plants become extinct. The extinction takes away the beauty of the coastal areas because its natural aspect is taken away.

Loss of Revenue to the Governments

Coastal regions are major tourist attraction sites. Not only do the sandy beaches attract people but also the picturesque view of the aquatic plants and animals play crucial roles. Whenever such animals and plants are rendered extinct, the beauty of these places is taken away and this implies that the countries have to contend with decrease in tourism activities and as such, the loss in revenue.

Solutions of Marine Habitat Loss and Destruction

Controlling Pollution

The first measure that would be useful in combating this environmental threat would be the step to control environmental pollution. Water, air and soil pollution are responsible for the adverse climatic changes. Such changes result into glaciation, melting of the icebergs and in the process, the marine habitat is threatened by degradation.

Environmental Conservation Policies

Human discipline would be quite necessary in ensuring that the problem of marine destruction is solved. How would this happen? Perhaps, various environmental conservation bodies such as UNEP can be in liaison with the various governments so that laws seeking to control industrialization and waste disposal are ratified. Eventually, global warming shall have been controlled and the threat posed to the aquatic lives will have minimized.

Controlled Fishing

There are other associated activities that can be used constructively. This may entail the controlled or moderated fishing which is sustainable. Overfishing tampers with the aquatic ecology and food chain. It can be done in a way that other plants and animals are not rendered extinct in the process. Overfishing can render certain animal species extinct due to the deprivation of food. Fishing regulations can therefore help in giving guidance.

Civic Education

Humans sometimes need to be educated on how they can respect the sanctity of the laws of nature. This encompasses the duty to treat the plants and animals with care. Civic education can be important in the marine areas as it would equip the people with the necessary skills to help them avoid the destruction of marine habitat. This activity can be conducted in small groups.

Restricting Over Aggressive Coastal Developments

Governments can only use the relevant coastal locations to boost other activities such as trade. This move will be very helpful in limiting the spread of urbanization and coastal developments which pose serious threat to the marine habitat. Other policies can be formulated which support and give guidance on what needs to be done during coastal urbanization to preserve the marine habitats.

Limiting Land Conversion Drive

When uncontrolled, human settlements can be a nuisance to the existing ecosystem. In the coastal regions, it can get over ambitious and this connotes doom to the marine habitat. As a consequence, plants and animals will be endangered. Such human activities can be moderated by limiting land conversion drive.

Using Designate Routes Only During Shipping

Countries that are using water transport can go ahead and control the activities. This can be achieved through the strict use of designate routes in order to limit the spread of danger that can be caused by the ships. It can reduce the oil spill disasters and other heat related dangers.

Strategies for fish biodiversity conservation

It is essential to prevent the further decline of our fish germplasm resources by devising all the possible measures of conservation and rehabilitation. The conservation policy should promote the management practices that maintain integrity of aquatic ecosystems, prevent endangerment and enhance recovery of the threatened species.

Researchers have identified five principal elements or tasks in the recovery programmes such as

- Habitat management
- Habitat development and maintenance
- Native fish stocking
- Non-native fish and sport-fishing
- Research data management and monitoring

The irreparable harm caused to fish and habitats need be compensated through afforestation, eco-restoration, soil conservation, complete ban on deforestation particularly in the fragile mountains and strict implementation of Indian Fisheries Act 1897 (modified in 1956) along with the following measures would positively help in restoration of the threatened fish fauna.

2

SEXUALITY AND REPRODUCTION

Nearly all fish reproduce by sexual reproduction - the fusion of sperm produced from testes and eggs produced from ovaries. Most species have separate female and male sexes, but there are also many species that are hermaphrodites, meaning that an individual has both testes and ovaries. Generally, a hermaphrodite can only produce one type of gamete, sperm or eggs, at a time, so their gametes must be cross-fertilized by another member of the species. Many hermaphroditic fish switch sexes, alternating between producing sperm or eggs during different reproductive cycles. There are a few hermaphroditic species, such as hamlets and salmon, that produce both eggs and sperm.

A more unusual form of reproduction found in a few fish species is parthenogenesis, which is where the female egg develops into a new individual without fusing with a sperm cell. In parthenogenesis, the offspring are identical copies of their mothers. Until fairly recently, this type of reproduction had only been shown in bony fish. Scientists confirmed that a pup born to an isolated female hammerhead shark, similar to the one shown in Figure above, was the result of parthenogenesis.

Fig. 1. A bonnethead shark (a species within the hammerhead shark genus). A member of this shark species was recently shown to be capable of a form of a sexual reproduction called parthenogenesis.

Fertilization is external in most fish species. Large quantities of eggs and sperm are simultaneously released into the water by females and males. This process of egg release is termed spawning.. During spawning, members of a species will come together to form an enormous group, and all release their eggs and sperm at the same time. This enhances fertilization and ensures the survival of some fertilized eggs because even if predators are in the neighborhood, they are not likely to be able to consume all of the embryos.

A disadvantage of spawning is that there is no parental care for the young. There are a few species of fish that release their eggs but maintain them nearby, sometimes generating a nest, and guard them until they hatch. There are also a number of species called mouthbrooders, including jawfish, sea catfish, and cardinalfish, that brood their eggs in their mouths. In many of these species, it is actually the male that broods the eggs in his mouth until hatching.

Since most fish species undergo external fertilization, it only makes sense that, for the vast majority of fish species (>95%), the eggs develop outside of the mother's body. This mode of development is called oviparity and is used by many species that undergo internal fertilization as well. In these cases, the eggs are released into the water for external development only after fertilization has taken place. In contrast, most shark and ray species develop their eggs inside of their bodies but without direct nourishment from the mother, a system called ovoviviparity.

Developing ovoviviparous embryos are nourished internally from a yolk sac, and they are delivered as live young. There are also some species of sharks, such as lemon sharks, that do provide maternal nourishment for their internally developing eggs. This is called viviparity, and nutrients are usually supplied through an organ called the placenta that connects to the mother's blood supply.

Once the eggs of oviparous fish have hatched, the offspring emerge as larvae that are very different from the adult form into which they will develop. The larvae usually possess a large yolk sac for nourishment. At the end of this stage they undergo a process of transformation into a juvenile form that is essentially a sexually immature version of the adult form. This transformation is called metamorphosis.

SEX DETERMINATION

- The sexuality of fish has great significance in fish breeding because there are differences in growth rate, behaviour pattern, breeding time, body colour, shape, or size between male and female in each cultured species.
- The expression of sexuality in fishes is greatly diversified, including various types of sex chromosomal mechanisms, gynogenetic reproduction, differentiated and undifferentiated gonochorism, synchronous, protandrous and protogynous hermaphroditism, species specific external diversified ethological sexes.
- However, methods of sex determination are known for relatively few species of fishes.

- From simple hermaphrodites to species with marked sexual dimorphism and complex mating and parental care behaviour reported in fishes imply an evolutionary adaptive radiation of sex control.
- Sexuality in fishes is under low grade of differentiation. The authors studying fish chromosomes were unable to find heterochromosomes in them for quite a long time. This is explained by the fact that many fishes possess a primitive mechanism of sex determination as compared to higher vertebrates.

PHYSIOLOGICAL SEX AND GONADAL SEX

Physiological Sex

Physiological sex is formed through the biochemical process of onto genesis under the control of genetic sex.

Gonadal Sex

The basis of the physiological sex depends upon the type of primary sex organ. This basic sex is known as the gonadal sex which is further classified into two groups, gonochorism and hermaphroditism.

GONOCHORISM

Gonochorism is the basic mode of sexuality and the majority fish species express it. Separate male and female individuals are involved by the existence of either testes or ovaries in one individual fish. Most cultured fish species have this gonochoristic type of sexuality. In the gonochoristic fishes, there are two types, undifferentiated and differentiated species. About 90% of Osteichthyes have a distinct female with ovaries and distinct male with testes. This is known as gonochorism. Following the early stages of sex determination, a gonochoristic fish develops either testes or ovaries, and after puberty they reach maturation.

Sex reversal in hermaphroditic fish species has been reviewed by researchers; therefore, our review will focus on gonochoristic fish that represent most of the teleost species.

Gonochoristic fish display an amazing diversity of sex determining systems, ranging from a strict genetic control (genetic sex determination, GSD) to an environmental control (environmental sex determination, ESD) or the joint action of both factors. GSD was originally reported in medaka but has been shown to be widespread within gonochoristic teleost species. In most fishes, sex is determined at fertilization by sex chromosomes with a monofactorial system often based on a XX/XY male heterogamety or less frequently a ZZ/ZW male homogamety.

Sex in some fishes as well as in a few other animal species relies upon a polyfactorial system where the sum of various independent loci determines the sexual phenotype. Contrary to mammals and birds, where the sex chromosome pair can be identified by its heteromorphy with a decayed 'heterogametic sex chromosome', in most of the poikilotherm vertebrates including fish, sex chromosomes are homomorphic and poorly differentiated.

Because they cannot be directly identified in fish through karyotype analysis, indirect approaches have been developed to search for the sex chromosomes of various species in order to analyze or control sex determination. More recently, molecular sex-linked markers have been identified in some model fish and aquaculture species. All-male or all-female populations (also known as monosex) have been produced for various aquaculture species to benefit from the best sex-linked dimorphisms (e.g., growth, age at puberty).

The sexual phenotype in fish is a plastic trait that can be easily manipulated by hormonal treatments (induction of sex reversal) during critical stages. Complete and functional sex reversal was produced by oral administration of androgens to masculinize or estrogens to feminize medaka fry. The use of Y-linked color genes brought evidence that part of the phenotypic males from the masculinization groups were neomales (genetic females masculinized by the treatment = mXX).

Similarly, in the feminization process, part of the phenotypic females were neofemales (genetic males feminized by the estrogens = fXY). By mating mXX with classic fXX females, all-female progenies have been obtained (mXX × fXX → fXX). Conversely, by mating fXY with mXY, viable and fertile YY males (mYY) were obtained (mXY × fXY → 1 fXX + 2 mXY + 1 mYY), and these YY males when mated with classic fXX breeders sired only male progenies (mYY × fXX → mXY).

Following this pioneer work, numerous studies have challenged the hypothesis that estrogens and androgens could be the natural 'gynoinducers' and 'androinducers', respectively. These studies mainly relied upon in vivo treatments using steroids, steroid enzyme inhibitors, or steroid receptor antagonists, in vitro steroid metabolism, steroid enzyme immunodetection, and gene expressions, concluding that aromatase and estrogens are the ones which play a key role in fish sex differentiation.

All the hormonal sex reversal treatments that have since been applied to numerous fish species are based upon these pioneer studies. As reported by researchers, this plasticity of sex differentiation can also lead to xenobiotic-induced sex reversal in fish due to pollution. Since these endocrine disrupting effects have been recently reviewed.

Identification of an ESD system is more recent in gonochoristic fish species. ESD implies that the environment experienced by the offspring during early ontogeny determines the sex of a progeny. Temperature was the first environmental factor demonstrated to be able to determine the sex of an Atlantic silverside *Menidia menidia* progeny through a temperature-dependent sex determination (TSD). Fifteen years later, a temperature-induced sex reversal (TISR) was first evidenced in the Nile tilapia *Oreochromis niloticus*, with high temperatures being able to masculinize sensitive progenies despite the fact that sex in this species is usually determined by major and/or minor genetic factors.

TSD was then found in the pejerrey *Odontesthes bonariensis* [Strüssman et al., 1996]. Together, these studies strongly stimulated research on TISR/TSD in various other fish species. EISR/ESR (environmentally induced sex reversal/environmental sex reversal) and especially TISR have been obtained in an increasing number of fish species under controlled conditions. This included species with (i) a male heterogametic GSD system, such as the Japanese flounder *Paralichthys olivaceus*, the medaka *Oryzias latipes*, and the Southern flounder *P. lethostigma*; (ii) a female heterogametic GSD system, such as the blue tilapia *O. aureus*, the turbot *Scophthalmus maximus*, or the half-smooth tongue sole *Cynoglossus semilaevis*; or (iii) a polygenic GSD system, such as the seabass *Dicentrarchus labrax* or domestic strains of the zebrafish *Danio rerio*.

SEXUAL CHARACTERS IN FISHES

This is grouped into three categories.

Monomorphism

No external characters to distinguish the sexes even when they are sexually matured. This includes most of the pelagic fishes like sardine, seer fish, carangids, *etc.*

Sexual Dimorphism

In many species, it is possible to determine the sex from their external body features. This phenomenon of differentiation of male and female sexes by external characters is called "sexual dimorphism".

a. Permanent dimorphism

b. Temporary dimorphism

In permanent dimorphism, sexes can be distinguished after the onset of sexual maturity including the colour, e.g.: fighter fish (*Betta splendeus).*

In temporary dimorphism sexes can be distinguished only during spawning season. During other times, sexes cannot be distinguished, e.g.: common carp (*Cyprinus carpio)*

In respect of species which do not exhibit sexual dimorphism, the separation of sexes mostly rely on internal examination and observation of the gonads.

Sexual Polymorphism

In this, both the sexes can be distinguished by more than one character. e.g: Salmon.

Sexual differentiation can be made by observing the gonads only after attaining the maturity. There are two types of characteristic:

1. **Primary characters:** the characters which are actually associated with reproductive process; in case of males – testis and ducts, in case of female – ovaries and ducts. This can be found out by dissecting the fish.
2. **Secondary characters :** These are more useful because the fish need not be sacrificed or killed. These characters occur in mature fishes. The secondary

characteristics which have no relation with reproductive process but serve as additional structures for spawning. e.g: Claspers, Gonopodium, Papillae etc.

However in certain species of finfishes, variations occur in the morphology of fish:

i. **Body shape**: Females are heavier and larger in size compared to the males because of the ovaries.

   Genital papillae: It is a small tube in cloacal aperture and which distinguishes male from females e.g. darters, lampreys etc.

   Pearl organ (Nuptial tubercles): These are horny short structures seen on the snout, cheek (head region) only in males. Once spawning is over nuptial tubercles will disappear, e.g.: common carp, minnows.

ii. **Fins** : Generally fins are larger in males than the females. In some fishes, pectoral fins can be used to distinguish between males and females. In males, they are rough and grainy in nature (Indian major carps). In some fishes, the caudal fin can be used to distinguish, e.g.: male sword tail has lower lobe much longer.

iii. **Coloration**: Most male fishes are brightly coloured and more intense when compared to females. This is common in most of the aquarium fishes, e.g. parrot fishes.

   In **Bow fish** (*Amia* sp.), the juvenile develop a coloured circular spot in their caudal fin of both sexes but when they attain maturity, it disappears in females and it becomes very intense in males.

   Accessory sexual characters: It Includes modification of anal fin to an organ called gonopodium (in males) which helps in the transfer of sperms during maturity, e.g.: mosquito fish, guppies.

   Pelvic fins: The pelvic fins are modified into claspers in males and serve as claspers in many elasmobranches.

   Female accessory sexual characteristic is seen in the form of egg laying tube or ovipositor, e.g.: Asiatic lump sucker.

iv. **Head characters**: In chimaerids, especially the males develop a spiny stout retractile knob like structures called frontal claspers. This sort of structure is also seen in forehead brooders. In salmons, males develop knob like hook and this is called as leype, seen at the tip of both the jaws.

v. **Size** : Deep sea male angler fish parasitic on the body of female. Fishes which have parental care, the secondary sexual characters are more pronounced. Sexual dimorphism is least pronounced in case of fishes which don't exhibit parental care.

SECONDARY SEXUAL CHARACTERISTICS

Secondary sexual characteristics (SSCs) are important features that have evolved in many fish species because of inter-individual competition for mates. SSCs are crucial not only for sexual selection, but also for other components of the reproductive process and parental care. Externally, they are especially clear in males (for instance, tubercles, fatpad, anal finnage, colouration) but are also externally present in the females (for instance, ovipositor). These characters are under hormonal control and as such there has been much interest in incorporating them as measures in fish test methods to assess the potential endocrine activity of chemicals.

Here we describe the external SSCs in typical laboratory test species for endocrine testing - fathead minnow (*Pimephales promelas*), Japanese medaka (*Oryzias latipes*), zebrafish (*Danio rerio*) and the three-spined stickleback (*Gasterosteus aculeatus* L.). We also provide some examples and discuss the utility of SSC responses to the endocrine activity of chemicals in the field and the laboratory. This paper is not aimed to provide a comprehensive review of SSCs in fish but presents a view on the assessment of SSCs in regulatory testing. Due to the current regulatory importance of establishing an endocrine mode-of-action for chemicals, we also consider other, non-endocrine factors that may lead to SSC responses in fish. We conclude with recommendations for how the assessment of SSCs in fish could be usefully incorporated into the endocrine hazard and risk assessment of chemicals.

REPRODUCTION IN FISHES

The methods of reproduction in fishes are varied, but most fishes lay a large number of small eggs, fertilized and scattered outside of the body. The eggs of pelagic fishes usually remain suspended in the open water. Many shore and freshwater fishes lay eggs on the bottom or among plants. Some have adhesive eggs. The mortality of the young and especially of the eggs is very high, and often only a few individuals grow to maturity out of hundreds, thousands, and in some cases millions of eggs laid.

Males produce sperm, usually as a milky white substance called milt, in two (sometimes one) testes within the body cavity. In bony fishes a sperm duct leads from each testis to a urogenital opening behind the vent or anus. In sharks and rays and in cyclostomes the duct leads to a cloaca. Sometimes the pelvic fins are modified to help transmit the milt to the eggs at the female's vent or on the substrate where the female has placed them. Sometimes accessory organs are used to fertilize females internally-for example, the claspers of many sharks and rays.

In the females the eggs are formed in two ovaries (sometimes only one) and pass from the ovaries to the urogenital opening and to the outside. In some fishes the eggs are fertilized internally but shed before development takes place. Members of about a dozen families each of bony fishes (teleosts) and sharks bear live young. Many skates and rays also bear live young. In some bony fishes the eggs simply develop within the female, the young emerging when the eggs hatch (ovoviviparous).

Others develop within the ovary and are nourished by ovarian tissues after hatching (viviparous). There are also other methods utilized by fishes to nourish young within the female. In all live-bearers the young are born at a relatively large size and are few in number. In one family of primarily marine fishes, the surfperches from the Pacific coast of North America, Japan, and Korea, the males of at least one species appear to be born sexually mature, although they are not fully grown.

Some fishes are hermaphroditic, an individual producing both sperm and eggs, usually at different stages of its life. Self-fertilization, however, is probably rare. Successful reproduction and in many cases defense of the eggs and young is assured by rather stereotyped but often elaborate courtship and parental behaviour, either by the male or the female or both.

Some fishes prepare nests by hollowing out depressions in the sand bottom (cichlids, for example), build nests with plant materials and sticky threads excreted by the kidneys (sticklebacks), or blow a cluster of mucus-covered bubbles at the water surface (gouramis). The eggs are laid in these structures. Some varieties of cichlids and catfishes incubate eggs in their mouths.

Some fishes, such as salmon, undergo long migrations from the ocean and up large rivers to spawn in gravel beds where they themselves hatched (anadromous fishes). Others undertake shorter migrations from lakes into streams or in other ways enter for spawning habitats that they do not ordinarily occupy.

Types of reproduction

Generally two types of reproduction are seen, these are:

1. Sexual reproduction
2. Asexual reproduction

Sexual reproduction

A from of reproduction that involves the fusion of two reproductive cells (ova and sperm) in the process of fertilization normally especially in animals, it requires two parents, one male and the other female. The female of the species produces an egg which is fertilized by sperm from the male. Depending on the species, fertilization takes place either within the female's body or externally, after the female lays her eggs and the male releases his sperm close by. When the egg is fertilized the genes of both parents are combined to produce zygote which develops into a new individual. Nearly all animals reproduce sexually.

Bisexual

Which is the prevalent kind; sperm and eggs develop in separate male and female individuals.

Hermaphrodites

Both sexes are in one individual and, as among certain errands and a dozen or more other families. Self-fertilization or true functional hermaphrodites sexists. Hermaphroditic sex glands are known for several species, including some trout relatives (salmonoids), perches (Perca), walleyes (Stizostedion). Darters (Etheostoma). And some of the black basses (Micropterus).

Asexual Reproduction

Reproduction in which new individuals are produced from a single parent without the formation of gametes. Asexual reproduction enables animals to reproduce without a partner. It occurs chiefly in lower animals, microorganisms, and plants. Many invertebrates reproduce asexually, including coral and starfish.

Coral grow small buds that break off to become separate organisms, whilst starfish are able to generate an entirely new being from a fragment of their original body.

Other Types

Without above classification, another three types of reproduction are also possible.

Parthenogenesis

Parthenogenesis is the development of young without fertilization, and a condition that has been called parthenogenesis (more properly gynogenesis) occurs in a tropical fish, the live-bearing Amazon molly (*Poecilia formosa*), and is also known in poeciliopsis. Mating with a male is required, but the sperm serves only one of its two functions, that of inciting the eggs to develop; it does not take any part in heredity. The resultant young are always females, with no trace of parental characters.

Sexual Reproduction

Sexual reproduction is a process of biological reproduction by which organisms give rise to descendants that have a combination of genetic material contributed by two different gametes, usually i.e. two different individuals, male and female. A gamete is a mature reproductive or sex cell. Sexual reproduction results in increasing genetic diversity, since the union of these gametes produces an organism that is not genetically identical to the parent(s).

Sexual reproduction is characterized by two processes: meiosis, involving the halving of the number of chromosomes to produce gametes; and fertilization, involving the fusion of two gametes and the restoration of the original number of chromosomes. During meiosis, the chromosomes of each pair usually cross over to achieve genetic recombination. Once fertilization takes place, the organism can grow by mitosis.

While typically sexual reproduction is thought of in terms of two different organisms contributing gametes, it also includes self-fertilization, whereby one organism may have "male" and "female" parts, and produce different gametes that fuse.

Sexual reproduction is the primary method of reproduction for the vast majority of visible organisms, including almost all animals and plants. The origin of sex and the prominence of sexual reproduction are major puzzles in modern biology.

Overview: Gametes, Meiosis, Fertilization, and Mitosis

In the first stage of sexual reproduction, 'meiosis', a special type of that takes place in gonads prior to the a production of gametes, the number of chromosomes is reduced from a diploid number (2n) to a haploid number (n). During 'fertilization', haploid gametes come together to form a diploid zygote and the original number of chromosomes (2n) is restored in the offspring.

Sexual reproduction involves the fusion or fertilization of gametes from two different sources or organisms.

Typically, a gamete or reproductive cell is haploid, while the somatic or body cell of the organism is diploid. A diploid cell has a paired set of chromosomes. Haploid means that the cell has a single set of unpaired chromosomes, or one half the number of chromosomes of a somatic cell. In diploid organisms, sexual reproduction involves alternating haploid (n) and diploid (2n) phases, with fusion of haploid cells to produce a diploid organism. Some organisms, however, exhibit polyploidy, whereby there are more than two homologous sets of chromosomes.

Meiosis and mitosis are two diverse type of cell divisions. Mitosis occurs in somatic (body) cells. The resultant number of cells in mitosis is twice the number of original cells. The number of chromosomes in the daughter cells is the same as that of the parent cell. Meiosis occurs in reproductive or sex cells and results in gametes. It results in daughter cells with half the number of chromosomes present in their parent cell. Essentially, a diploid cell duplicates it's chromosomes, then undergoes two divisions (tetroid to diploid to haploid), in the process forming four haploid cells. This process occurs in two phases, meiosis I and meiosis II.

Fertilization involves the fusion of haploid gametes to give a diploid organism, which can then grow by mitosis. Thus, in sexual reproduction, each of two parent organisms contributes half of the offspring's genetic makeup by creating haploid gametes that fuse to form a diploid organism. For most organisms, a gamete that is produced may have one of two different forms. In these anisogamous species, the two sexes are referred to as male, producing sperm or microspores as gametes, and female, producing ova or megaspores as gametes.

In isogamous species, the gametes are similar or identical in form, but may have separable properties and may be given other names. For example, in the green alga, Chlamydomonas reinhardtii, there are so-called "plus" and "minus" gametes. A few types of organisms, such as ciliates, have more than two kinds of gametes.

Sexually reproducing organisms have two sets of genes (called alleles) for every trait. Offspring inherit one allele for each trait from each parent, thereby ensuring that offspring have a combination of the parents' genes. Having two copies of every gene, only one of which is expressed, allows deleterious alleles to be masked.

As with the other vertebrates, sexual reproduction is the overwhelming dominant form of reproduction. However, there are several genera of fish that practice true or incomplete parthenogenesis, where the embryo develops without fertilization by a male.

Although vertebrates in general have distinct male and female types, there are fish species that are both males and females gonads (hermaphrodites), either at the same time or sequentially. For example, the amenone fish spend the first part of their lives as males and later become females, and the parrot fish is first female and then male. Some members of the Serranidae (sea basses) are simultaneous hermaphrodites, such as the Serranus and their immediate relatives, Hypoplectrus (the synchronoous hermaphroditic hamlets).

Fertilization may be external or internal. In the yellow perch, eggs are produced by ovaries in the female and sperm is produced by the testes, and they are released through an opening into the environment, and fertilization takes place in the water. In some live bearers, such as guppies and swordtails, females receive sperm during mating and fertilization is internal.

Other behaviors related to sexual reproduction include some species, such as the stickleback, built nests from plants, sticks, and shells, and many species that migrate to spawn.

Asexual Reproduction

Asexual reproduction is a form of reproduction in which an organism gives rise to a genetically-similar or identical copy of itself without a contribution of genetic material from another individual. It does not involve meiosis, ploidy reduction, or fertilization, and only one parent is involved genetically. A more stringent definition is agamogenesis, which refers to reproduction without the fusion of gametes.

Asexual reproduction is the primary form of reproduction for single-celled organisms such the archaea, bacteria, and protists. However, while all prokaryotes reproduce asexually (without the formation and fusion of gametes), there also exist mechanisms for lateral gene transfer, such as conjugation, transformation, and transduction, whereby genetic material is exchanged between organisms. Biological processes involving lateral gene transfer sometimes are likened to sexual reproduction. The reproductive variances in bacteria and protists also may be symbolized by + and - signs (rather than being called male and female), and referred to as "mating strains" or "reproductive types" or similar appellations.

Many plants and fungi reproduce asexually as well, and asexual reproduction has been cited in some animals, including bdelloid rotifers, which only are known to reproduce asexually, and various animals that exhibit parthenogenesis under certain conditions. In parthenogenesis, such as found in some invertebrates and vertebrates, an embryo is produced without fertilization by a male.

Generally, parthenogenesis is considered a form of asexual reproduction because it does not involve fusion of gametes of opposite sexes, nor any exchange of genetic material from two different sources however, some authorities classify parthenogenesis as sexual reproduction on the basis that it involves gametes or does not produce an offspring genetically identical to the parent (such as a female domestic turkey producing male offspring).

A wide spectrum of mechanisms may be exhibited. For example, many plants alternate between sexual and asexual reproduction, and the freshwater crustacean Daphnia reproduces by parthenogenesis in the spring to rapidly populate ponds, then switches to sexual reproduction as the intensity of competition and predation increases. Many protists and fungi alternate between sexual and asexual reproduction.

A lack of sexual reproduction is relatively rare among multicellular organisms, which exhibit the characteristics of being male or female. Biological explanations for this phenomenon are not completely settled. Current hypotheses suggest that, while asexual reproduction may have short term benefits when rapid population growth is important or in stable environments, sexual reproduction offers a net advantage by allowing more rapid generation of genetic diversity, allowing adaptation to changing environments.

HERMAPHRODITISM

A hermaphrodite is defined as any individual organism that possesses both male and female reproductive organs during their life span. The main advantage of hermaphroditism is the assurance of a reproductive partner. Although hermaphroditism is quite common in invertebrates and plants, it is an exceedingly rare occurrence in vertebrates. Hermaphroditism in mammals and birds are almost always a pathological condition (often leading to infertility). Only in Perciforms (fish) does hermaphroditism occur naturally and in high frequency.

Hermaphrodites are divided into two main categories: synchronous hermaphrodites, and sequential hermaphrodites. In the synchronous hermaphrodites, organisms possess both active male and active female reproductive organs at the same time. In sequential hermaphrodites, both male and female reproductive organs may be present, but only one is active and viable at any given time.

Synchronous, or simultaneous hermaphrodites in reef fish are relatively atypical. A few Serranids (sea basses, e.g. *Serranus* sp.) and Hamlets are known synchronous hermaphrodites. During mating, one individual will lay eggs while another fertilizes the eggs, after which both will reverse roles and perform fertilization again. Synchronous hermaphrodites do not fertilize themselves; Self-fertilization does not promote genetic diversity, and can amplify genetic defects from parent to offspring. The interesting fact is most synchronous hermaphrodites form monogamous pairs.

Sequential hermaphrodites are so named because they are capable of transforming from one sex to another. Theses transformations entail a full conversion of gonads from one sex to another. The gonads of sequential hermaphrodites possess the genetic information to produce both male and female reproductive organs, but only the dominant gene is expressed at any giventime. Different cues - varying from species to species - may induce sex changes.

Sequential hermaphrodites are further categorized into two main categories: protogynous and protandrous. Protandrous hermaphrodites are those that develop into males first, then possibly to females. Protogynous hermaphrodites are the exact opposite, with juveniles first developing female reproductive organs that may possibly change into male reproductive organs in select circumstances. It should be noted that hermaphrodites do not necessarily have to change sexes, but by definition, are capable of this feat.

Protandrous hermaphrodites are the rarer of the two types. Pomacentrids (damselfish) are the most famous of these hermaphrodites. For example, clownfish of the genus Amphiprion live in communities that consist of one dominant female specimen and several smaller male (or asexual juvenile) specimens. If the female should be removed, a male will convert to a female, insuring a reproductive partner for the community.

Protogynous hermaphrodites are most often haremic fish. These fish form monoandric harems comprising of 1 male overseeing numerous females for life. The two primary responsibilities of the male are to defend its territory against other conspecific males, and to court and fertilize females of its territory. If the male should die (either of natural causes or conflict-related mortality), the dominant female of the harem will undergo a sex change from female to male. This sex change may take as little as 5 days. The new male will then resume the full responsibilities of the previous male until he should die. Protogynous hermaphrodites that form harems include the wrasses of the genus Cirrhilabrus and Paracheinlinus, Dwarf Angelfish of the genus Centropyge, and Anthias (e.g. Pseudoanthias sp.).

Some protogynous fish do not form harems, but may form pairs. The dottybacks (*Pseudochromis* sp.) are presumably protogynous hermaphrodites that fall under this category.

There are also sequential hermaphrodites that waver between sexes with no discernable order. The sexes of these fish are often determined by the ratio of sexes in an immediate community. These types of hermaphrodites include numerous gobies.

MODES OF REPRODUCTION AND DEVELOPMENT

The ability to produce new living individual is a basic characteristics of all plants and animals. All reliable evidence indicates that new life comes only from pre-existing life; this is the process of biogenesis or reproduction. The different modes of reproduction can be classified according to the environment in which the embryos develop, and the sources of the nutrients supporting embryonic growth. These are mentioned into the following term:

Oviparity

Oviparity or egg laying refers to the situation where the development or the fertilized of egg occurs outside the body of the female. The young hatch when the egg envelope, shell or capsule is broken. Oviparous fish may be further categorized as being either ovuliparous or zygoparous.

ovuliparity refers to the release of ova from the reproductive tract of the female followed by fertilization or activation in the external environment. Thus, all organisms that have external fertilization and this includes most teleost are said to be ovuliparous.

Zygoparity refer to the oviparous condition in which the zygotes (i.e. fertilized ova products of fusion between the eggs and sperm) are retained within the body of the female for a short period of time before being released into the environment. Obviously, zygoparous species display internal fertilization with their being a transfer of male sperm to the reproductive tract of the female. zygoparous reproduction characterizes all skates some sharks and a small number of teleost.

Irrespective of weather the fertilization of the eggs occurs internally or the egg yolk provides externally the nutrient for the developing embryos of oviparous species.

Oviparity in Fishes

Amongst fish which reproduce by ovipartity, there are several groups.

Egg Scatterers

Fish which breed in this way either spawn in pairs or in groups. Males and females release milt (the sperm and spermatic fluid of the male) and eggs into the water at the same time. These mix together, fertilising the eggs. The fertilised eggs are broadcast (or spread) into the plankton column and float away in the current or sink to the bottom. No parental care is given, so large amounts of eggs are produced. It is easy to produce many eggs and because they are in the water, they don't dry out - necessary oxygen and nutrients aren't scarce.

When theoffspring settle out of the plankton, they might be in totally new environments: this gives the young a chance to survive across a wide area. The main disadvantage of this method is that the fish must go through a larval stage before they transform into adults. In this larval stage, they are very vulnerable while they try to find food and avoid predators. Also, they may not find a suitable environment when they settle out of the plankton column. The survival rate for individual eggs is very low, so the parent has to produce millions of eggs.

Egg depositors

These fish either lay eggs on a flat surface, like a stone or plant leaf or may even place them individually among fine leaved plants. The parents usually form pairs and guard the eggs and fry (young fish) from all danger. The Cichlids such as Koi are the best known species for this. Some Catfish and Rainbow fish are also egg depositors.

Nest Builders

Many fish species build nests. These might be a simple pit dug into gravel (trout do this) or an elaborate bubble nest. When they are ready to spawn, the fish may construct a nest by blowing bubbles, and they often use vegetation to anchor the nest. The male will keep the nest intact and keep a close eye on the eggs. The Gouramis, Anabantids and some catfish are the most common of this type.

Mouthbreeders

These are particularly odd, since eggs are fertilised externally, but raised internally. The females usually lay their eggs on a flat surface where they are then fertilised by the male. After fertilization the female picks up the eggs and incubates them in her mouth. Broods tend to be small, since by the time the fry are released by their mothers they are well formed and suffer minimal losses. The best known mouthbreeders are the African lake Cichlids.

Egg buriers

The annual Killifish reproduce in this way. As the pools they live in dry out, the fish spawn, pressing their eggs into the mud. The pools eventually dry out completely, killing the adults, but the eggs remain safe in the dried mud. When it rains and the pool refills the eggs hatch and the cycle is repeated. Killifish eggs can survive for many years in dried out mud.

Ovoviviparity

In the ovoviviparous species the eggs usually develop within the uterus (modified oviduct) of the female. Fertilization of the egg occurs internally and the eggs are retained until hatching or beyond. The developing embryos do not receive any supply of foodstuffs from the female but must rely on the yolk of the egg for nutrition. This form of nutrition is known as lecthotrophy. The developing embryos must, however, rely on the female for supply of oxygen. During pregnancy the wall of the oviduct becomes enlarged and richly supplied with blood vessels.

Viviparity

Viviparous species have eggs, which develop either within the uterus (i.e. several species of shark), or within the ovary (i.e. teleosts). The developing embryos receive some nutrient supply from the female in edition to that provided as yolk in the egg. This form of nutrition is known as matrotrophy. The additional nutrient can be supplied in various forms. in some species of female secrets a nutrient-rich fluid which is taken up by the developing young has a form of 'soup'. In other species some form of 'placenta' may develop allowing a more direct transfer of nutrient from the blood of the female to the developing embryo.

A distinction is sometimes made between viviparous animals in which the embryos derives nourishment from the mother's tissues as in most mammals and animals that are called ovoviviparous in which the embryo is nourished entirely by food stored in the egg, the hatching before being laid.

Viviparity in fishes

Fishes provide a key to understanding vertebrate viviparity in as much as the first viviparous vertebrates were fishes and they display the most diverse maternal fetal relationships of all the live bearing vertebrates. Viviparity occurs in three major fish groups: the chondrichthyans, the teleosts, and the actinistians. Although widespread, it represents the dominant mode of reproduction only among the sharks and rays. Approximately 420 of the estimated 600-700 species of chondrichthyan fishes are viviparous in contrast, only 510 of the estimated 20,000 species of teleost fishes are viviparous. The coelacanth, the only extant species of actinistian, is viviparous.

Semelparity and Iteroparity

Semelparity and Iteroparity refer to the reproductive strategy of an organism. A species is considered semelparous if it is characterized by a single reproductive episode before death, and iteroparous if it is characterized by multiple reproductive cycles over the course of its lifetime. Some plant scientists use the parallel terms monocarpy and polycarpy.

In truly semelparous species, death after reproduction is part of an overall strategy that includes putting all available resources into maximizing reproduction, at the expense of future life. In any iteroparous population there will be some individuals who die between their first and second reproductive episodes, but unless this is part of a syndrome of programmed death after reproduction, this would not be called semelparity.

Semelparity

The word semelparity comes from the Latin semel, once, and pario, to beget. It is also known as "big bang" reproduction, since the single reproductive event of semelparous organisms is usually large, as well as fatal. A classic example of a semelparous organism is Pacific salmon (Oncorhynchus spp.), which lives for many years in the ocean before swimming to the freshwater stream of its birth, for laying eggs, and dying. Other semelparous animals include many insects, including some species of butterflies, cicadas, and mayflies, some molluscs such as squid and octopus, and many arachnids. Semelparity is much rarer in vertebrates, but in addition to salmon, examples include smelt, capelin, and a few lizards, amphibians, and didelphid and dasyurid marsupial mammals.

Iteroparity

The term iteroparity comes from the Latin *itero*, to repeat, and *pario*, to beget. An example of an iteroparous organism is a human-though many people may choose only to have one child, humans are biologically capable of having offspring many times over the course of their lives. Iteroparous vertebrates include all birds, most reptiles, virtually all mammals, and most fish. Among invertebrates, most mollusca and many insects (for example, mosquitoes and cockroaches) are iteroparous. Most perennial plants are iteroparous.

Fecundity

Fecundity, the number of eggs ripened by female fish during a spawning season, or event, varies from a few dozen in some continuously reproducing livebearing fishes to millions in some species that spawn pelagic eggs on an annual basis. In general, fecundity varies among species inversely with the amount of "care" given to the individual progeny:

viviparous fishes have lower fecundity than ovoviviparous fishes, which in turn have lower fecundity than oviparous fishes, and nest builders have lower fecundity than pelagic spawners. Within species fecundity is positively related to size; generally it is close to a function of the cube of fish length. Although fish eggs range from about 0.5 to 20 mm in diameter, the size of the adult limits the fecundity in smaller species. In most species with high fecundity, several batches of eggs are usually produced at intervals of a few days to weeks during the spawning season.

Some general reasons for determining fecundity is that it is useful for making total population estimates, it is useful in studies of population dynamics or productivity, and it is useful for characterizing specific populations, subpopulations, and/or stocks of fishes.

- **Absolute fecundity :** is the number of ripe eggs produced by a female in one spawning season or year (this is the usual meaning when the general term "fecundity" is used, although on occasion it might also mean the number of eggs produced in a lifetime).
- **Relative fecundity :** is the number of eggs produced in a season per unit somatic weight of the fish (*i.e.*, eggs/gram), and is useful if it is shown that the fecundity of a fish is proportional to its weight, which is not uncommon.
- **Population fecundity :** is the number of eggs spawned by the population in one season, is the sum of the fecundities of all females, and is usually expressed as the product of the expected fecundity of an average female, the number of breeding females in the population (an example of why classifying maturity stages may be useful).

Spawning

Spawning refers to the release of unfertilized planktonic eggs by female fish, which is the reproductive pattern for most marine fishes. The eggs are fertilized shortly after release by males. Some fishes also deposit unfertilized Fish Reproduction eggs in nests where they are fertilized and develop. Fishes with internal fertilization release free-swimming larvae, or juveniles. The ripening of eggs and spawning are controlled by hormones, nutrition of the female, and external (ecological) factors). Usually maturation and spawning are controlled by a combination of endogenous and exogenous controls and are not governed by any specific factor.

CRYOPRESERVATION

Introduction

Cryopreservation is a long-term storage technique with very low temperatures to preserve the structurally intact living cells and tissues for extended period of time at a relatively low cost. Cryopreservation is to preserve and store the viable biological samples in a frozen state over extended periods of time. A very important part research in cryopreservation is to reveal the underlying physical and biological responses of the cell and cause of cryoinjury, especially those associated with the phase change of water in extracellular and intracellular environments. From the original slow-cooling study, another cryopreservation approach has moved to easier and more efficient technique-vitrification, Cryoprotective agents has to gain access to all the parts of the system.

Cryopreservation considers the effects of freezing and thawing. Therefore, the diffusion and osmosis processes have important effects during the introduction of cryoprotective agents, the addition or removal of cryoprotectants, the cooling process, and during thawing. These phenomena are amenable to the experimental design and analysis. Thus, reliable methods can be developed for preserving a very wide range of cells and some tissues. These methods have found widespread applications in biology, biomedical technology and conservation.

Germplasm cryopreservation includes storage of the sperm, eggs and embryos and contributes directly to animal breeding programmes. Germplasm cryopreservation also assist the *ex situ* conservation for preserving the genomes of threatened and endangered species. The establishment of germplasm banks using cryopreservation can contribute to conservation and extant populations in the future. Since the first successful cryopreservation of bull semen, cryopreserved bull semen has been used to propagate the rare and endangered species using assisted reproduction techniques.

Every year, more than 25 million cows are artificially inseminated with frozen-thawed bull semen and many bovine calves have been produced using the transfer of cryopreserved embryos into cow. Tissues, cultured cell lines, DNA and serum samples could be frozen and store in cryogene bank. For example, mice and sheep have been generated from frozen-thawed pieces of ovary that have been replaced in a female and stimulated to ovulation. The principle of testicular cell freezing and transplantation has been demonstrated and is currently used for human male infertility. Significant efforts are being made on non-mammalian species using cryobiology techniques.

In fish aquaculture, the successful cryopreservation of gametes and embryos could offer new commercial possibilities, allowing the unlimited production of fry and potentially healthier and better conditioned fish as required. Cryopreservation of reproductive products of many aquatic species has been successfully achieved.

Cryopreservation of aquatic sperm is relatively common in the breeding and management of fish species, including salmonid, cyprinids, silurids, and Acipenseridae (família) is well documented. However, cryopreservation of embryos and oocytes of aquatic species have not been successful, except for eastern oyster eggs (*Crassostrea virginica*)), larvae of eastern oyster and larvae of the sea urchin.

Cryopreservation technology applied to the preservation of fish gametes in aquaculture plays an important role in seed production, genetic management of broodstock and conservation of aquatic resources. Fish germplasm also plays a significant role in human genomic studies because its relatively small size of the genome makes it easier for sequencing and ideal models for studying the human disease. This would help in identifying the roles for human genes from fish mutations and also in fish models for genes identified by human disease. Aquatic species preservation would assist the development, protection and distribution of research lines and would offer benefits for restoration of endangered species.

Sperm

In 1949, researchers successfully cryopreserved the avian spermatozoa using glycerol as a cryoprotectant. Thereafter, cryopreservation of male gamete became possible. Researchers applied a similar approach for fish gametes and reported success with Atlantic herring spermatozoa, achieving approximately 80% cellular motility after thawing. Since then, cryopreservation of fish sperm has been studied and has been successful in more than 200 species and techniques of sperm management have been established for freshwater and marine fish species, including carp, salmonids, catfish, cichlids, medakas, white-fish, pike, milkfish, grouper, cod, and zebrafish.

Many studies on cryopreservation of fish sperm have been carried out on economically important freshwater species and attempts to cryopreserve sperm from the marine fish species tended to be more successful when compared with those obtained from the freshwater fish. Although freshwater fish sperm are generally more difficult to cryopreserve, the fertilization rates obtained from the cryopreserved marine fish sperm are similar to those obtained with mammalian species. Controlled-rate slow cooling in cryopreservation has been mainly used for fish sperm. Common carp has been studied using frozen-thawed sperm with 95% fertilization and hatching rate.

Salmonid species spermatozoa have been successfully cryopreserved. Another well studied cryopreserved group is cyprinids and some of these cyprinid fishes are widely farmed throughout Asia and Europe. A fertilization and hatching rate of 95% using the frozen-thawed sperm has been reported for the common carp and these results are not significantly different from fresh sperm.

Tilapias are among the exotic freshwater fishes that have been successfully established for fish farming in Taiwan; they have been cryopreserved successfully and produced 40-80% motility with cryoprotectant DMSO. The sperm of more than 30 marine fish species have been cryopreserved successfully. Generally, high survival and fertilization capacity has been obtained in frozen-thawed spermatozoa when compared to freshwater species.

Successful cryopreservation of the sperm of aquatic invertebrate has been carried out for sea urchin, oyster, starfish, abalone and coral . Dimethyl sulfoxide has also been reported as a successful cryoprotectant for sperm cryopreservation; the concentration range used was 5 to 30% for these species. Various levels of motility, ranging from <5% to 95%, have been reported for the cryopreserved aquatic invertebrate sperm.

Embryos

Cryopreservation of embryos has become an integral part of assisted reproduction. Successful cryopreservation of embryos is important because the biodiversity of both the paternal and maternal genomes will be preserved. While cryopreservation techniques have been largely established for the mammalian embryos, successful cryopreservation of intact fish embryos has not yet been achieved. Factors limiting fish embryo cryopreservation include their multicompartmental biological systems, high chilling sensitivity, low membrane permeability and their large size, which gives a low surface area to volume ratio.

The effect of such low ratio is a reduction in the rate at which water and cryoprotectants can move into and out of the embryo during cryopreservation (Mazur 1984). Fish embryos are osmoregulators; they are released into the external medium and activated. Then the vitelline envelope separates from the plasma membrane and forms chorion. Studies on the chorion permeability of zebra fish embryos clearly showed that it was permeable to electrolytes and a range of cryoprotectant, including propane-1,2-diol, methanol, DMSO, ethylene.

The chorion structure plays a crucial role as flexible filter for the transport of some materials and protects against the microorganisms. Studies on zebra fish embryos have shown that the water permeability of the plasma membrane at different developmental stages remained relatively stable. The permeability to methanol (cryoprotectant) appeared to decrease during embryo development. This also indicated that there was a gradual reduction in the permeability following the fertilization in zebra fish embryos, as opposed to the generally held belief that the membrane permeability of fish embryos reduced rapidly to minimum shortly after the fertilization.

The studies on the kinetics of subzero chilling injury in *Drosophila* embryos and chilling sensitivity of zebra fish embryos have demonstrated that chilling injury plays an important role in reduction of embryo survival during the exposure to subzero temperatures. Chilling sensitivity has been shown for many species and has been analyzed in fish embryos, including brown trout (*Salmo trutta f. fario*), rainbow trout (*Oncorhynchus mykiss*), carp (*Cyprinus carpio*) (Dinnyes et al. 1998), fathead minnows (*Pimephales promelas*), goldfish (*Carassius auratus*) and zebrafish (*Danio rerio*). These studies demonstrated that the later stages (after 50% epiboly) were less sensitive to chilling, but chilling sensitivity increased significantly as the temperature fell below zero.

The high chilling sensitivity of fish embryos, especially at early stages, their complex membrane structure and large yolk are the main obstacles to achieve successful cryopreservation of these embryos. Chilling injury in embryos has been linked to the inhibition of metabolic and enzymatic processes from low temperatures injuries which could be detrimental in the embryonic development such as fish embryos.

Cryoprotectant toxicity follows a similar pattern to chilling sensitivity with later stages being less sensitive to cryoprotectant. Several studies have determined membrane permeability for zebra fish embryos and membrane permeability to water and most cryoprotectants has been shown to be low. Studies on the cryopreservation of zebra fish embryos demonstrated 8% embryo survival in 2M methanol at -25 °C; however, no embryo survival was observed when frozen to -30 °C or below.

Cryopreservation studies on the embryos and larvae have been conducted on marine invertebrate such as oysters, sea urchins, polycheate worms, coral and penaeid shrimp species. However, survivals of most of these species has been inadequate in maintaining the structure and activity of embryos and larvae after freezing to cryogenic temperatures. Embryonic and larval development of marine invertebrates after cryopreservation often showed abnormalities in structure and colour. The problems with invertebrate embryo cryopreservation associated with those identified with the fish embryos are their low membrane permeability and high chilling sensitivity. Although cryopreservation of the embryos has not been fully achieved, considerable progress has been made in understanding the conditions required for fish embryo cryopreservation and this would undoubtedly assist the successful protocol design in the future.

Oocytes

Oocyte cryopreservation is potentially the best way to preserve the female fertility. Cryopreservation of fish oocyte has been studied which offers several advantages such as the smaller sizes range, much lower water content in oocytes and absence of a fully developed chorion that the permeability to water and solutes in oocyte is higher than embryo.

Fish embryos are too large to apply traditional cryopreservation protocol. Immature oocytes can be an alternative for the mature eggs because of their smaller size. However, there is no practical technique available to induce the small oocyte to mature *in vitro*. A technique to obtain the mature eggs from the late stage oocytes is available. Thus, the combination of this technique and their cryopreservation could be a breakthrough.

However, at present, late stage oocytes cannot be successfully cryopreserved because their size is still not small enough to result in much lower surface area to volume ratio. These reduce the rate at which water and cryoprotectant move into and out of oocytes during the cryopreservation. Developing the methods for cryopreservation of oocytes requires the screening of potential cryoprotectant treatments, evaluation of tolerance to chilling, determination of the appropriate rate of freezing to cryogenic temperatures and rate of thawing. Viability assessment methods of oocytes with trypan blue (TB), fluorescein diacetate (FDA) + propidium iodide (PI) and adenosine triphosphate (ATP) content assay have been developed for quick assessment of viability. A functional test based on *in vitro* maturation, followed by germinal vesicle breakdown (GVBD) has also been shown effective for late stage III oocyte.

The permeability of the zebra fish oocyte membrane to water and cryoprotectants has been studied and membrane permeability was shown to decrease with the temperature and permeability was generally lower than those obtained from sea urchin eggs but higher than the immature medaka oocyte. Studies on zebra fish oocyte chilling sensitivity showed that those oocytes were very sensitive to chilling and their survival decreased with decreasing temperature.

Chilling sensitivity in zebra fish oocytes was thought to be due to lipid phase transition of the oocyte membrane. The phase transition in zebra fish oocytes showed that chilling injury could occur when oocytes were exposed to temperatures between 12 to 22°C above the water freezing temperatures. Cryopreservation of late stage zebra fish oocytes has been studied using the controlled slow cooling and an optimum cryoprotective medium and cooling rate identified. Researchers showed that although the oocyte viability obtained immediately after freeze-thawing was relatively high with 88% using TB staining; oocyte viability decreased to 29.5% after 2 h incubation at 22 °C. The study also showed that the ATP level in the oocytes decreased significantly after thawing and all oocytes became translucent. Cryopreservation of early stage zebra fish oocytes using the controlled slow freezing has been reported by researchers.

The results suggested that 4M methanol in KCl buffer was identified as the optimum cryoprotective medium. Although results obtained after the cryopreservation using trypan blue and FDA+PI staining were promising with 69% and 54%, especially with stage II ovarian follicles, the ADP/ATP ratio assay showed that the energy system of these follicles had been compromised. Apparently the ADP/ATP ratio could be a valuable measure of cellular injury after post-thaw incubation period as it reflected the metabolic and energy status of population as well as indicating some measure of the potential for repair. Furthermore, *in vitro* culture method is effective for assessing early stage zebra fish oocytes growth competence *in vitro.* The early stage zebra fish oocytes can be cultured *in vitro* for 24 h, stage I and II oocytes can grow to the sizes of early stage II and stage III oocytes after hCG treatment. and also can be used for other teleost species.

Studies on the cryopreservation of invertebrate oocytes and eggs over the past several decades have been extraordinarily difficult to achieve. However, it was found that intracellular crystallization occurred in the starfish oocytes at relatively high temperature that was very close to the temperature of extracellular ice formation. In order to avoid this problem, researchers successfully cryopreserved starfish oocytes using ultra-rapid freezing technique, called vitrification. High chilling sensitivity and low membrane permeability of zebra fish oocytes are major obstacles to the development of a successful protocol for their cryopreservation as chilling sensitivity or cold shock can hinder slow cooling processes. Vitrification may be another option to achieve successful cryopreservation for the oocytes.

Blastomeres

Blastomeres are the cells produced as the result of cell division and cleavage in the fertilized egg. They are totipotent and pluripotent (depending on the stage of embryonic development) having the ability to differentiate into any of the three germ layers or entire organism. They are different from the muscle cells, blood cells or nerves cells. Although cultured somatic-cells from fish have been cryopreserved successfully, their value is limited because of loss of development potential. Cryopreservation of blastomeres can maintain the genetic diversity of both, nuclear genome and mitochondrial DNA.

Blastomeres from the early embryos of fish still retain pluripotency and their cryopreservation may be a promising approach to preserve the genotypes of zygotes and reconstitution of the organism. Indeed, there are several reports of germ-line chimeras created using the transplantation of blastomeres into goldfish, zebra fish, medaka and rainbow trout embryos. Researchers transplanted the frozen-thaw blastomeres into goldfish embryos and the blastomeres differentiated into primordial germ cells. This report demonstrated that germ-line cells from the cryopreserved blastomeres could develop into mature gametes of chimeric fish because the blastomeres were not damaged by cryopreservation. Therefore, the cryopreservation techniques are very important.

Cryopreservation of blastomeres has been successful in several fish species. In the first reported studies, researchers used a two-step freezing procedure, with ice-seeding at -6°C, and cooling to -25°C, followed by immersion in liquid nitrogen. The survival rate of 84.8% was obtained after cryopreservation of 50% epiboly zebra fish blastomeres. However, the results obtained from a very small sample size and freezing rates were not controlled, rather tubes were allowed to equilibrate in the cooled alcohol baths. Researchers demonstrated the effect of cryopreservation on zebra fish blastomeres survival using the controlled slow cooling method. It was shown that DMSO was the most toxic to zebra fish blastomeres.

However, DMSO was the best cryoprotectant in terms of survival of zebra fish blastomeres. Therefore, it is possible that the cryoprotective effect of DMSO may be greater than its toxicity effect. Although the survival rate in Lin's results progressed from 25% to 70%, it was still lower than that obtained by using two step methods. The comparisons between these studies must take into consideration the different methodology. Vitrification of zebra fish blastomeres was studied more recently and the highest blastomere survival was 93.4%. Cryopreservation of blastomeres was also carried out in rainbow trout, carp and medaka after post-thawing. Rainbow trout blastomeres have been cryopreserved using the controlled slow freezing procedures with a survival of 95%.

It has been reported that the controlled slow freezing protocol adopted for rainbow trout was successfully applied to carp blastomeres with survivals of 94% and 96%. Lower survival rates of cryopreserved blastomeres using controlled slow freezing have also been reported for other fish species such as whiting (20%), medaka (34%), pejerrey (67%) and chum salmon (59%).

Conclusion

Cryopreservation of gametes and embryos are already routinely applied in the mammalian. Cryopreserved sperm, oocytes and embryos are used for artificial insemination and embryo transfer in the livestock industry. Cryopreservation also has enormous applications in the artificial propagation of widely diverse aquatic organism. Sperm and embryonic cells cryopreservaiton has been successful in a number of teleosts and invertebrate species.

However, cryopreservation of embryos and oocytes remain a major challenge. The practical application of cryopreservation in the aquatic species needs more vigorous research efforts in this area and the efforts may be prioritized on endangered, economical value and representative species from various aquatic habitats. Cryopreservation of gametes, embryos and embryonic cells has become of immense value in aquatic biotechnologies which provide an important tool for protecting the endangered species, genetic diversity in aquatic species. The establishment of cryobanks to utilize the cryopreservation worldwide would be a significant and promising task in the future.

MENDAL'S LAW OF INHERITANCE

- Gregor Johann Mendel(1822-1884) initiated a series of hybridization experiments in 1857 by experimental crossing of pea varieties, *Pisum sativum*.
- After 8 years of research in 1865, he delivered first lecture on pea experiments to Brunn Natural Science Society.
- In1866, his paper "*Experiments onplant hybridization*" was published in volume IV of Proceeding of Natural Science Society. Based on his hybridization experiments, Mendel proposed some basic genetic principles, rules of inheritance. But his remarkable piece of work was unattended and not appreciated up to 1900.
- In 1900 three botanists simultaneously, CarlCorrens (Germany), Eric Von Tshermak (Austria) and HugoDevries (Holland)reported the rules governing the transmission of traits from parents to offspring independently, which gave the credit for the discovery to Mendel.
- After death Mendel received recognition and honour as "Father of Genetics". This marked the beginning of modern genetic research. Mendel's experimentation is now recognized as a classic example of carefully planned and executed scientific research and his paper as an excellent illustration of scientific genius.

Mendel also considered the results of inheritance of one, two, and more distinct traits. Although Mendel knew nothing about chromosomes, he deduced the behavior of the two genes for each trait, and the behavior is exactly described by the way in which chromosomes are distributed during meiosis. For a single gene pair, the gametes carry only one of the two chromosome pairs. The diploid complement is restored at fertilization (Fig. 2).

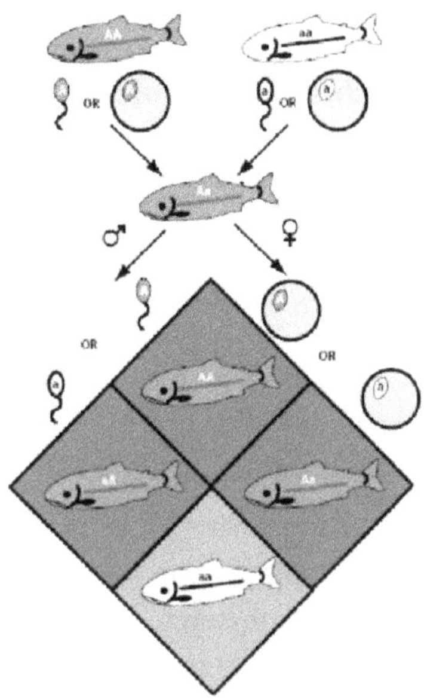

Fig. 2. Mendel's first law. The two alleles at a locus separate during gamete formation. The diploid state is restored at fertilization. Note that a recessive trait (albinism here) is not expressed unless it is homozygous. In the F1 generation, both males and females are heterozygous. The products in the second generation are shown in the grid as the intersection between possible sperm and eggs. Second generation individuals will be expected to have a phenotypic ratio of 3 dominant: 1 recessive. Note that there are two ways to obtain heterozygotes in the second generation—the sperm can contribute the A and the egg the a or vice versa.

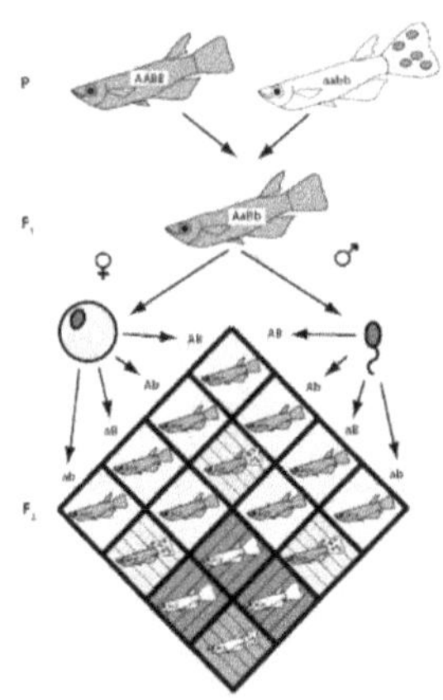

Fig. 3. Mendel's second law. Inheritance of alleles at one locus is not influenced by the inheritance of alleles at other loci. Here we follow two traits, one for coloration—pigmented (A) or albino (a)—and a second trait for tail spotting—no spotting (B) is dominant to spotting (b). Gametes that have four different composite genotypes (AB, Ab, aB, and ab) can be produced by the F1 double heterozygotes (AaBb). The grid shows all possible ways in which the different possible eggs and sperm could combine. Note

that in the second generation of these crosses, the phenotypic ratio is 9 that are dominant for both traits: 3 dominant for the first and recessive for the second: 3 recessive for the first and dominant for the second: 1 recessive for both. Of course these frequencies will only be accurate if many progeny are produced because the inheritance of alleles follows laws of probability.

If two pairs of genes are carried by different chromosomes, they will behave independently. That is, inheritance of one trait is in no way connected with inheritance of the second trait on the other chromosome. The idea is similar to tossing a coin and rolling a die. Whether the coin lands with heads or tails up has nothing to do with which face of the die is up. The easiest way to visualize this process is to use a matrix to follow the possible results after two generations of matings (Fig. 3).

Sexual reproduction and diploidy provide a mechanism that generates huge amounts of genetic variation. For example, pink salmon have 26 pairs of chromosomes. If each chromosome pair carried only a single heterozygous locus (and of course each usually has many, many more, some exceeding 1,000), 2^{26} different (haploid) gametes cold be produced from the combination of alleles at the different loci.

The number of diploid individuals that could be generated is $(2^{26}) \times (2^{26})$. That is more than 67 million different gametes and more than 4 quadrillion (15 zeros) genotypes. Those numbers result from just 26 variable loci. In reality, we will never see all of the genotypes that are possible because we will never see 4 quadrillion pink salmon.

The phenotypes produced by alleles at a locus usually result from the expression of the information carried in the instructions. That is, the cake that results from the recipe. Some alleles have faulty information— for example, if the recipe instructed you to clean up and discard everything before the cake batter would be poured into the pan and baked, no cake would be produced by that faulty recipe. If you had two recipes, however, and one recipe was complete, you would still successfully produce a cake if you followed both recipes, and the result (phenotype) would be "successful cake."

If both were copies of the faulty recipe, however, the phenotype would be "no cake." Many recessive alleles carry defective information. Albinism results when an individual carries two defective alleles that specify one of the enzymes involved in the biosynthetic pathway for the pigment melanin. The recessive wrinkled peas that Mendel studied resulted from defective instructions for an enzyme that synthesized starch. Starchy peas retain water and remain plump; without the starch, the peas shrivel and are wrinkled. Most recessive metabolic diseases (like cystic fibrosis and phenylketonuria) also result from defective instructions for key enzymes.

Codominant and Partial Dominant Phenotypes

Even though Mendel's laws are most often explained from traits that are expressed as dominant or recessive phenotypes, the expression of all loci is not restricted to those modes. At some loci, both alleles are expressed; this is referred to as codominant expression. This is like the effect of wearing bowling shoes—ordinarily, one shoe is designed for sliding (the left shoe for right-handed bowlers) and the other for breaking. The function of both shoes is expressed in the complete phenotype of the bowling action.

The standard example for codominance is the blood type AB, the expression of which results from the chemical groups attached to the outside of blood cells. The group attached as a result of expression of the A allele differs from that attached as a result of expression of the B allele. Type A blood can result from a genotype homozygous for the A allele (AA) or from a genotype that involves another allele at that locus, the o allele which is recessive and attaches no chemical group. The genotype oo produces type O blood, in which neither of the chemical groups is attached to the red blood cell. Similarly, type B blood can result from either BB or Bo genotypes.

Another mode of inheritance is partial dominance. An example of this is carnation color. Red carnations are homozygous for the R allele (RR), white carnations are homozygous for the r allele (rr), and pink carnations are heterozygous (Rr). Expression of the R allele apparently produces red pigment, but the amount produced by a single R allele is insufficient to turn the carnation red; red requires the combined expression of both R alleles.

Sex-linked and Nuclear Genes Versus Mitochondrial Genes

Most organisms also carry some genes that are not diploid and do not obey Mendel's laws. In some species, the chromosomes that determine the sex of an individual are a special pair of chromosomes. Just like all of the normal chromosomes, sex chromosomes pair during meiosis and are sorted. However, the two sex chromosomes often do not carry the same set of genes (loci) in many species (including humans). As a result, an individual may carry only a single copy of a gene (allele) on a sex chromosome.

For example, human females have two X chromosomes; and human males have just one X chromosome but also carry one Y chromosome. The human Y chromosome carries very few genes, so males carry only a single allele at each locus (one copy of each gene) that is on their X chromosome. And they inherited that allele from their mothers. Traits that are coded for by genes that are on the X chromosome are referred to as sex-linked traits.

Two sex-linked traits in humans are hemophilia (the gene specifies instructions for a blood clotting factor) and color blindness. If a male inherits a defective gene for one of these traits, he will be a hemophiliac or be color blind because he has only that one copy of the allele, the one on the single X chromosome he received from his mother. The incidence of sex-linked phenotypes is higher in human males than in females. Relatively few genes are sex-linked; geneticists have little specific knowledge about sex-linked traits in fishes.

Simple Gene Effect

Mendel postulated that genes (characteristics) are inherited as pairs of alleles (traits) that behave in a dominant and recessive pattern. Alleles segregate into gametes such that each gamete is equally likely to receive either one of the two alleles present in a diploid individual. In addition, genes are assorted into gametes independently of one another. That is, alleles are generally not more likely to segregate into a gamete with a particular allele of another gene.

A dihybrid cross demonstrates independent assortment when the genes in question are on different chromosomes or distant from each other on the same chromosome. For crosses involving more than two genes, use the forked line or probability methods to predict offspring genotypes and phenotypes rather than a Punnett square.

Although chromosomes sort independently into gametes during meiosis, Mendel's law of independent assortment refers to genes, not chromosomes, and a single chromosome may carry more than 1,000 genes. When genes are located in close proximity on the same chromosome, their alleles tend to be inherited together. This results in offspring ratios that violate Mendel's law of independent assortment. However, recombination serves to exchange genetic material on homologous chromosomes such that maternal and paternal alleles may be recombined on the same chromosome.

This is why alleles on a given chromosome are not always inherited together. Recombination is a random event occurring anywhere on a chromosome. Therefore, genes that are far apart on the same chromosome are likely to still assort independently because of recombination events that occurred in the intervening chromosomal space.

Whether or not they are sorting independently, genes may interact at the level of gene products such that the expression of an allele for one gene masks or modifies the expression of an allele for a different gene. This is called epistasis.

GENE-ENVIRONMENT INTERACTIONS

Environmental exposure to nutritional, chemical and physical factors can alter the DNA methylation patterns, leading to imprinting disorders and congenital birth defects. Deficiencies of four essential dietary sources of methyl groups (choline,methionine, vitamin B 12 and folic acid) have been associatedwith an increased risk for a number of diseases. It has further been shown in animal experiments that nutritional supplements to the mother can alter gene expression in her offspring without altering the genes themselves. Increasing nutritional supplementation of folic acid to pregnant women reduces the risk of spina bifida and other neural tube defects in the offspring. Environmental chemicals, such as certain metals and other xenobiotics may have a probable role in abnormal methylation processes. The toxic effects exhibited by metals such as Ni, As, and Cd, as well as by Zn deficiency, may lead to cancer, atherosclerosis, birth defects, neurological disturbances, and pancreatic lesions.

Mendel's Experiments

- The essence of his experiments was to cross breed plants that had discrete, non-overlapping characteristics and then to observe the distribution of these characteristics in the next several generations.
- He studied parental lines (P) of chosen plant. For 2 years he conducted experiments to get pure breed. He got pure line for each character of the 7 pairs by selfing for several generation. He got genetically pure tall and dwarf plants.
- The two pure lines were crossed and both strains were used as male and female parents.
- He recorded the results of F_1 and F_1 were selfed. He recorded the results of subsequent generation.
- He found that Tall, Round or Smooth seeds as dominant. Dwarf, wrinkled as recessive. Cotyledon colour, Yellow as dominant; Green as recessive.
- The expression of recessive character in F_1 is masked. But, when F_1 were selfed, the recessive character reappeared in F_2 offsprings in a definite ratio of 3:1. Among 4 plants of F_2 , 3 displayed dominant and 1 recessive characters. Mendel observed 787 tall (T) and 277 dwarf (t) plants for a ratio of 2.84 to 1. This 3:1 F_2 ratio was observed to occur in the monohybrid crosses for each of the 7 pairs of characters. When recessive F_2 were selfed in F_3 they produced only offsprings with recessive character.

Dihybrid Cross

Later on, Mendel tried to find out how different phenotypic traits behave in relation to each other in the inheritance from generation to generation. For this he crossed two varieties of pea plants differing in two pairs of contrasting characters. *When a cross is made between two parents differing in two characters, the cross is known as dihybrid cross.* When round yellow plants were crossed with wrinkled green plants all the F_1 were Round Yellow. When F_1 were selfed the F_2 offspring produced were in the ratio of 9:3:3:1.

The irregularity of F_2 offspring ratio was explained by his second law, Law of independent assortment or recombination of genes. It states that "alleles for different traits are distributed to sex cells independently of one another. The allele Y was associated with allele R in parent but it does not always remain associated with it and, also becomes associated with alleler.

- Mendel's genes which underwent independent assortment are located on different chromosomes which recombine freely in meiosis.

Mutagenesis and Genomic Resources

In vertebrates, genomic functions are often safeguarded by functional redundancy. This is especially true in teleost fish, in which the genome seems to have been duplicated at the base of the clade's radiation. Thus, a mutagenesis screen in one species might not suffice to uncover all potential gene functions. The medaka was chosen to complement the zebrafish and other vertebrate species in systematic mutagenesis screens for determinants of embryonic patterning and other aspects of development, organogenesis and behaviour because it is evolutionarily closer to the genomic model system *Fugu* than to zebrafish and its genome is only half the size of that of zebrafish.

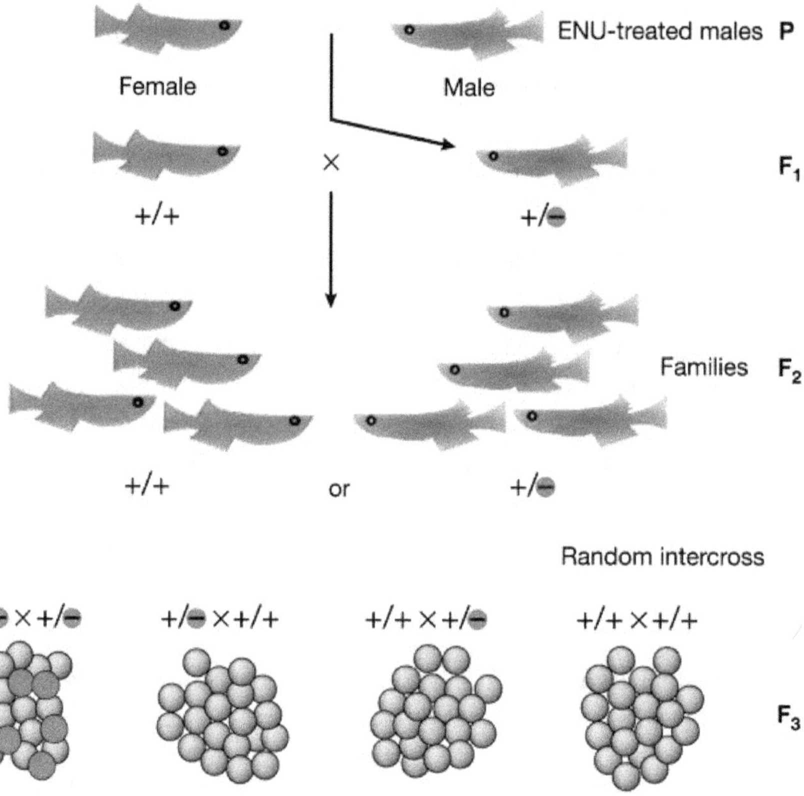

Fig. 4. Example of an F3 mutagenesis screen scheme. Males were mutagenized with Nethyl- N-nitrosourea (ENU) and crossed with wild-type females. Mutations induced by the treatment are eventually uncovered in the F3 generation after random intercrosses within the F2 families. Carriers for mutant genes are highlighted in red.

The availability of different inbred strains and an established genetic map are specific advantages in using this species. M. Furutaniseiki (Kyoto, Japan) gave an update on the status of a continuing international large-scale mutagenesis project in which more than 12 groups are participating. The classical F_3 screen (study of mutants in the third generation after mutagenesis; Fig. 4), uses *in situ* hybridization and antibody staining in addition to morphological characterization, and the frequent occurrence of a temperaturesensitive phenotype is an additional benefit.

The international team has so far analysed 628 F_2 families, obtained 1,505 embryonic lethal mutations and identified 352 of those causing specific embryonic patterning defects.

In the mutants, many specific structures and organs are affected, including the early axis, central nervous system, eye, sensory neural network, developing gonads and thymus. Whereas several phenotypes are strikingly similar to those of known zebrafish mutants, many medaka mutants exhibit novel phenotypes to those that so far have been uncovered in zebrafish. Many medaka mutants have already been identified. Thus, genetic and physical mapping approaches are of crucial importance for isolating the corresponding genes.

Physical maps based on bacterial artificial chromosome (BAC) clones are key tools for both the structural and functional characterization of individual genes. In addition to providing templates for the genomic sequencing of complex genomes, BACs can be used for the positional cloning of genes that cause particular mutant phenotypes, for the rescue of mutants once candidate genes have been mapped to them, and for the establishment of links between physical, genetic and cytogenetic maps through fluorescence *in situ* hybridization studies.

Another strategy, the whole-genome shotgun (WGS) approach, is complementary to the BAC approach in terms of whole-genome sequence analysis. With WGS, it is straightforward to produce a large body of sequence information, but assembly can be a major hurdle in repeat-rich regions of the genome. In contrast, BAC sequencing is hampered by the necessity to generate, and perform quality control on, multiple shotgun libraries, but reduces the assembly problem by reducing complexity. Proposals to integrate these strategies include: first, large-scale preparation of BAC end-sequences with which to put contiguous WGS sequences (contigs) into the context of larger scaffolds, and second, the sequencing of BACs at low coverage, followed by the detection of matching sequences from a pool of WGS reads by using appropriate software. H. Himmelbauer (Berlin, Germany) reported on progress towards the generation and completion of a BAC contig map of the medaka genome, based on his previous experience and on technology developed for a yeast artificial chromosome/BAC map of the mouse. The medaka BAC map will encompass segments of all the available BAC libraries (northern strain HNI and southern strains CAB and Hd-rR) and will therefore permit comparisons across strains and genotypes. A total of 10,000 probes are now being mapped to BACs, representing a 15-fold coverage of the medaka genome. Probes are being derived from medaka genes, expressed sequence tags (ESTs) are being generated and genetically mapped in the University of Tokyo Medaka Genetic Mapping Project and pairs of end probes are being generated from CAB and Hd-rR BACs and sequenced at Keio Medical School.

Additionally, information on synteny (the location of genes on the same chromosome or chromosome segment) from the recently completed *Fugu* and *Tetraodon* genome sequences is being used in the rapid generation of contigs across large intervals. This project is expected to provide a deeper understanding of the structure and large-scale organization of the medaka genome and of the genes encoded therein.

Genetic linkage maps have been used for various kinds of analysis, such as positional cloning, quantitative trait locus analysis, comparative vertebrate genomics and the detection of radiation-induced DNA mutations. H. Mitani (Kashiwa City, Japan) reported that, in an effort to expand the existing map, the group at the University of Tokyo has derived 2,400 pairs of mapping primers based on ESTs that have a significant degree of similarity to previously identified genes from other species. About 800 of these were polymorphic between the mapping strains and thus were incorporated into the existing genetic map. Studies of the zebrafish genome had determined that the conservation of synteny between the zebrafish and human genomes is extensive.

Given the evolutionary relationships between humans, zebrafish and medaka, this synteny was also expected to be conserved between the medaka and human chromosomes, and this has now been confirmed by a comparison of linkage relationships between orthologous genes in these three organisms. Although some conserved syntenies are extensive, even conserved groups of genes can have changes in gene order, apparently reflecting the relatively frequent occurrence of inversions and other intra-chromosomal rearrangements. In most analysed medaka and zebrafish linkage groups, the distribution of duplicated genes suggests that two chromosomes in the medaka and the zebrafish are derived from each ancestral chromosome. This and extra-Hox clusters in the teleost lineage reported previously support the argument that a 'whole-genome duplication' was the basis of the teleost radiation.

Two laboratories are working on large-scale whole-mount *in situ* screens to complement the mapping approaches discussed above. So far, randomly picked cDNA clones from libraries generated from different developmental stages have been analysed by sequencing and whole-mount *in situ* analysis by both laboratories, but with a slightly different emphasis: whereas the Bourrat group focuses on genes expressed in the proliferative area of the optic tectum by screening brain-specific libraries of corresponding stages, the Wittbrodt group is interested in early patterning markers and genes expressed in the developing and differentiating eye.

Insights from the Outside

Autonomous retrotransposable elements certainly have a major role in the evolution of eukaryotic genomes, and might have been involved in generating the high genetic diversity observed in fish. Such elements can duplicate and integrate into new genomic locations in which they might disrupt genes, recombine ectopically to produce genomic rearrangements, modify the expression patterns of flanking genes, and duplicate different types of non-autonomous sequence. Retrotransposable elements can be used as molecular tools for genome analysis, but can also be a source of problems during the assembly of genome projects. J.-N. Volff (Würzburg, Germany) reported that recent analyses, particularly those involving medaka, have strongly suggested that fish genomes contain an amazing diversity of active retrotransposons that is not observed in mammals.

Retrotransposable elements atypical of the vertebrate lineage were first identified in the medaka. These include the first vertebrate retrotransposons found to encode a restriction-enzyme-like endonuclease or an endonuclease related to certain mobile DNA introns. Numerous other elements, including retrotransposons with or without long terminal repeats and short interspersed nuclear elements (SINEs), are also present in the genome of medaka. Thus, the medaka has the potential for development into an outstanding model for functional and evolutionary studies of retrotransposable elements that are absent from the genomes of higher vertebrates.

M. Nonaka (Tokyo, Japan) discussed evolutionary aspects that have emerged from his extensive study of the structure of the medaka's major histocompatibility complex (MHC). This is a genomic region that harbours several genes essential for adaptive immune recognition (class I antigen presentation) and is specific to jawed vertebrates. The MHC region serves as a model for the survey of the evolution of the vertebrate genome.

In contrast to its centralized structure in cartilaginous fish (Chondrichthyes), amphibians, birds and mammals, in the medaka and other teleosts the MHC genes are dispersed over several chromosomal loci, suggesting extensive genomic reorganization in this lineage. However, several genes involved directly in class I antigen presentation form a tight cluster in the medaka genome, and although the products of these genes are structurally unrelated, they are intimately linked at a structural level, indicating that adaptive processes have maintained the linkage between these genes throughout vertebrate evolution.

Although the genes that determine the development of the male or female sex are known in *Caenorhabditis elegans*, *Drosophila melanogaster* and most mammals, the identity of the sex-determining factors in several species remains unknown. M. Schartl (Würzburg, Germany) reported the results of the joint efforts of his group, together with the teams of N. Shimizu and A. Shima, in cloning the medaka sex-determining gene. They found that the Y-chromosome-specific region spans only about 280 kilobases, a region that has been entirely sequenced, and that it contains a duplicated copy of the autosomal *DMRT1* gene, named *DMRT1Y*. This is the only functional gene on this chromosome segment and it maps precisely to the male sex-determining locus. The gene is expressed during male embryonic and larval development and in the Sertoli cells of the adult testes. These features make *DMRT1Y* a highly likely candidate for the medaka male sex-determining gene.

COLOR IN PLATY FISH (*XIPHOPHORUS* SPP.)

Platys are one of the several livebearing species of freshwater fish popular with aquarium enthusiasts. Grouped together with their close cousins, the swordtails, platys are part of the genus known as *Xiphophorus*. They are hardy, compatible with other fish, and are extremely easy to breed. They also come in a variety of attractive colors and forms.

Although platys comprise only a couple of species within the *Xiphophorus* genus, there are considerable variations in color and even fin type within these species. Furthermore, the species interbreed so readily that many specimens sold in the trade are mixed hybrids. However, they all do well in similar conditions, so unless your goal is to breed a pure line, this doesn't really matter. Choose your variety based on whatever appearance you like.

Platy fish come in a wide range of patterns and colors. Most platy fish are peaceful and get along well with fish that have similar temperaments.

Colors, Patterns, Fins, and Sex Differences

Platy fish have flattened, short bodies with fan-shaped tails and triangular heads. Platy fish are sexually dimorphic. Males are smaller than females and have a gonopodium. They become dull when stressed or sick.

Color Variations

Platys are one of the more colorful freshwater fish, with base colors that range from pale yellows all the way to deep black and many shades in between. As with many fish species, there is also an albino variant. Within each of the color groups, there are many shades, along with a number of names used to describe them. It is common for platys to exhibit several colors as well as various shades all in the same fish. Colors seen in platys include:

- Black, half-black, and black finned (wag tail)
- Blue
- Brown
- Gold/yellow: referred to as gold, golden, marigold, sunburst, sunset
- Green
- Red: referred to as blood red, brick red, coral red, velvet red

Wagtail Pattern

When the rays of the caudal and dorsal fins are black, the pattern is said to be a wagtail. Wagtails can occur with virtually any body color and can be combined with other color patterns. For instance, the painted or variegated pattern may be combined with the wagtail trait. Red or gold are the most commonly seen colors of wagtails, but the variation can also exhibit blue, gold, or green colors. The wagtail color pattern is also frequently seen in another species within the genus, the swordtails.

Variegated Pattern

Dark splotches of various sizes and shapes all over the body is a common color pattern. Usually called the variegated pattern, it's also referred to as *painted*, as it resembles the dabbling of an artist's brush. Variegated patterns may be combined with any base color, as well as with other color patterns or tail variations. The highly popular Mickey Mouse platy is a type of variegated color pattern.

Salt-and-Pepper Pattern

In this slightly different twist on the variegated pattern, there are a number of dark or light spots (rather than blotches) that are liberally sprinkled over the body. As with other color patterns, the salt-and-pepper type may be combined with several different color and fin variations.

Tuxedo Pattern

The tuxedo variation refers to a dual pattern in which the posterior portion of the fish is black while the anterior portion is another color. Red and gold are commonly seen colors in tuxedo patterns and the combination can be extremely attractive. As with other types of color patterns, the tuxedo trait is often combined with additional color variations, such as the comet or twin bar pattern.

Rainbow Pattern

A number of colors in a single fish is known as a rainbow color variation. Often these fish display colors in true rainbow fashion, ranging from dark to lighter colors, ending with a black tail. An attractive iridescent rainbow color pattern is also available, sold under the name *neon*.

Comet or Twin Bar Pattern

The comet or twin bar trait is another color variation that is often combined with other color patterns. In this variation, the caudal fin is edged in black on both of the outside margins. The twin black bars makes the tail fin stand out noticeably.

Hifin Variation

There are two common fin variations in platys—the *hifin* is the most frequently seen trait. In hifins, the dorsal fin is elongated, sometimes quite significantly. As with other traits, the hifin variation can be found in fish with a variety of different color patterns.

Unfortunately, any tail variation in which all or part of the tail is elongated lends itself to being nipped. Elongated fins are also more prone to disease when the fish is stressed, or if the water conditions are not optimal. Owners of fish with these variations should monitor their health closely and address problems quickly.

Seen less often than the hifin, this variation is easily spotted. In the pintail, the center portion of the tail fin is elongated, jutting out like a pin. Sometimes the fish is mistaken for a swordtail, which is actually a different species within the same genus.

3

HEREDITY AND CYTOGENETICS OF FISH REPRODUCTION

INTRODUCTION TO GENETICS

Genetics is a branch of the biology involved with the study heredity, its biological process, the study of genes, genome, cell cycle, heredity, inherits genes and lot more.

The exploration of the working and major codes of variation and heredity is termed as Genetics. The groundwork on which heredity stands is known as inheritance. It is defined as the procedure by which characteristics are handed down from one generation to the other. Gregor Johann Mendel is known as the "Father of Modern Genetics" for his discoveries on the basic principles of **heredity**.

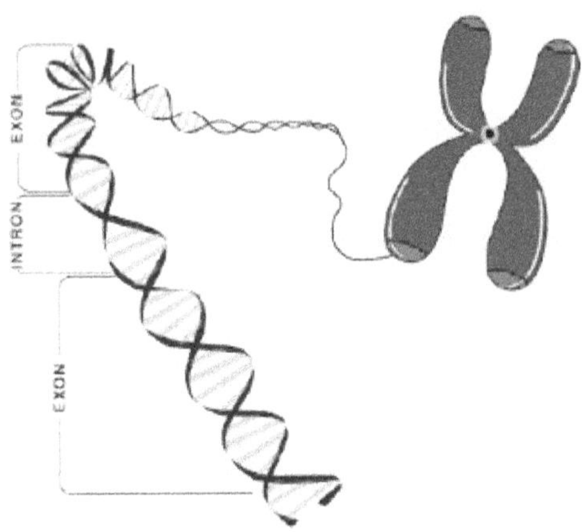

Fig.1. Genes

Variation, as the name suggests is the amount of dissimilarity that exists in between children and their parentages. It can be deter-mined to keep in view the behaviouristic, cytological, physiological, and morphological characters of individuals fitting into similar species.

Some of the major reasons that variation are

- Genetic/Chromosomal rearrangement.
- Mutated genes due to the influence of the ecosystem.
- Crossing over.

Let us have a detailed look at genetics notes to learn about genes and the principle of inheritance.

LAW OF INHERITANCE BY GREGOR MENDEL

Garden Pea (*Pisum aativum*) was the plant that Mendel experimented on for 7 years to get to the point to propose the laws of inheritance in live creatures. Mendel carefully chose seven distinct characteristics of Pisum Sativum for the investigation concerning hybridization. Mendel used true-breeding lines i.e. those that go through constant self-pollination and display steady characteristic heritance.

Augustinian monk Gregor Mendel developed the science of genetics. Mendel performed his experiments in the 1860s and 1870s, but the scientific community did not accept his work until early in the twentieth century. Because the principles established by Mendel form the basis for genetics, the science is often referred to as *Mendelian genetics.* It is also called *classical genetics* to distinguish it from another branch of biology known as molecular genetics.

Mendel believed that factors pass from parents to their offspring, but he did not know of the existence of DNA. Modern scientists accept that **genes** are composed of segments of DNA molecules that control discrete hereditary characteristics.

Most complex organisms have cells that are diploid. **Diploid** cells have a double set of chromosomes, one from each parent. For example, human cells have a double set of chromosomes consisting of 23 pairs, or a total of 46 chromosomes. In a diploid cell, there are two genes for each characteristic. In preparation for sexual reproduction, the diploid number of chromosomes is reduced to a haploid number. That is, diploid cells are reduced to cells that have a single set of chromosomes. These haploid cells are **gametes,** or sex cells, and they are formed through meiosis . When gametes come together in sexual reproduction, the diploid condition is reestablished.

The offspring of sexual reproduction obtain one gene of each type from each parent. The different forms of a gene are called **alleles.** In humans, for instance, there are two alleles for earlobe construction. One allele is for earlobes that are attached, while the other allele is for earlobes that hang free. The type of earlobe a person has is determined by the alleles inherited from the parents.

The set of all genes that specify an organism's traits is known as the organism's **genome.** The genome for a human cell consists of about 20,000 genes. The gene composition of a living organism is its **genotype.** For a person's earlobe shape, the genotype may consist of two alleles for attached earlobes, or two alleles for free earlobes, or one allele for attached earlobes and one allele for free earlobes.

The expression of the genes is referred to as the **phenotype** of a living thing. If a person has attached earlobes, the phenotype is "attached earlobes." If the person has free earlobes, the phenotype is "free earlobes." Even though three genotypes for earlobe shape are possible, only two phenotypes (attached earlobes and free earlobes) are possible.

The two paired alleles in an organism's genotype may be identical, or they may be different. An organism's condition is said to be **homozygous** when two identical alleles are present for a particular characteristic. In contrast, the condition is said to be **heterozygous** when two different alleles are present for a particular characteristic. In a homozygous individual, the alleles express themselves. In a heterozygous individual, the alleles may interact with one another, and in many cases, only one allele is expressed.

When one allele expresses itself and the other does not, the one expressing itself is the **dominant** allele. The "overshadowed" allele is the **recessive** allele. In humans, the allele for free earlobes is the dominant allele. If this allele is present with the allele for attached earlobes, the allele for free earlobes expresses itself, and the phenotype of the individual is "free earlobes." Dominant alleles always express themselves, while recessive alleles express themselves only when two recessive alleles exist together in an individual. Thus, a person having free earlobes can have one dominant allele or two dominant alleles, while a person having attached earlobes must have two recessive alleles.

INTEGRATIVE MOLECULAR CYTOGENETICS

Introduction

Investigating 50% of the total biodiversity is a hard and exciting task for most ichthyologists. Fish are represented by 32,900 species; more than 20,000 are marine and 8,000 are living in Neotropical continental waters.

Cytogenetics, which is the study of chromosomes, is becoming an important biodiversity-detection tool also used to measure biodiversity evolutionary aspects. Cytogenetics also enables the development of evolutionary, taxonomic, and phylogenetic inferences resulting from the support provided by the conservation of Neotropical fish species. Firstly, this scientific field took steps towards describing and defining the chromosomal morphology of the species according to usual cytogenetic analysis methods such as the conventional staining. The first marks in the chromosome of fish were identified by classic chromosome techniques, C-banding, and the detection of nucleolar organizing regions (Ag-NORs). The identification marks revealed much information about the evolutionary processes within this group such as chromosome rearrangements, structural and/or numeric polymorphisms, and sexual chromosome systems and variations associated with the geographic distribution of some species and/or populations.

Although these techniques have enabled good insights into the understanding of chromosome diversity in fish, the access to the genome was limited, mainly in many families that have quite stable conservative karyotypes and banding patterns, a fact that has hindered the detection of their most diverse genetic divergence levels.

The methodological advances on chromosome analyses drastically grew in the last decades; they showed refinement and more accurate resolution power, due to molecular cytogenetics by fluorescence *in situ* hybridization (FISH). Such technique strongly improved the transition from classic to molecular cytogenetics. It allowed integrating the quantitative and qualitative data from DNA sequences and their physical location in chromosomes and genomes. The FISH technique enables identifying the DNA sequences in the studied cytological material (target DNA), no matter if they are chromosomes or interphase nuclei fixed on the surface of the slide.

Many changes in this technique have been adapted and improved in cytogenetics laboratories, due to advances in microscopy and bioinformatics. However, FISH principles and procedure steps (marking, hybridizing, and detecting) have remained the same. Its greater resolution power plays an important role in precisely featuring the chromosome structures of many fish species. Different technique types, based on the herein referred methodology, such as genome *in situ* hybridization (GISH), *in situ* simultaneous location of different chromosome regions by Multicolor FISH (m-FISH), and identification of specific regions by chromatin fiber extended (Fiber-FISH) or spectral karyotype (SKY), are known and applied to distinct means.

The mapping of repeating DNA has been the main use of the FISH applied to fish. Among these sequences, it is possible to highlight the chromosome locations of multigene families such as 18S and 5S ribosomal DNA, histones, telomeric and centromeric sequences, and transposable elements. The isolation and application of specific satellite DNA probes have also been the routine method in intraspecific characterization as well as in the understanding of evolution sequences among related species. These probes help detecting chromosomal homeologies by identifying syntenic groups kept or rearranged during karyotype divergence among analyzed species. These studies aim to understand the structural nature and the likely origin of B or supernumerary chromosomes and to draw the origin and evolution of sexual chromosomes and their behavior in the meiotic cells.

FISH potentialities are highly broad. The technique enables the better grounding of hypotheses heading towards structural, taxonomic, population, and/or phylogenetic aspects. It happens along with the easiness and availability of prospecting probes through different strategies. These data are crucial for understanding the chromosomal dynamics and evolution and how it may be linked to speciation processes and macroevolutionary events. Throughout the data presentation, the current study aims to assess the contributions provided by the herein described technique and to integrate fish genome.

Cryptic Species Detection and Population Structuring in Marine Environments

Marine biogeography studies the history of marine taxa geographic distribution and it aims to set endemism areas as well as broad global distribution patterns. The combined action of certain parameters such as the pelagic larval period duration, the individual's power to disperse in adulthood, and the actions resulting from ocean currents have shaped the chromosomal diversity of species, although it is difficult to indicate one single determining factor of cladogenetic events.

There is an exuberant diversity of shapes and wide body size variety in the Perciformes order, which holds more than 10,000 species. This morphological diversity contrasts with the high karyotype stability found in most representatives of this order. They are acrocentric diploid and have fundamental number chromosomes, simple ribosomal sites, and little heterochromatin mostly concentrated in the pericentromeric regions. However, the stronger chromosomal dynamism is shaped by pericentric inversions and, in smaller proportion, by centric fusions, just as it happens in some families, mainly in those that live in coral reefs. These evidences enable discriminating populations in the Brazilian coast by classic banding, although such procedure is not always resolute.

Studies focused on population cytogenetics are incremented by the use of *in situ* hybridization. FISH was used to discriminate so far nondiagnosed cryptic karyotypes by conventional techniques. Indeed, its use detected different fishery stocks of the circumtropical species *Caranx lugubris* in São Pedro and in São Paulo Archipelagos (1100 km away from the Brazilian coast).

Although they apparently present homogeneous karyotypes regarding chromosomal morphology, C-banding, and Ag-NORs, these morphotypes diverged on the frequency of 5S ribosomal gene sites, thus corroborating the distinct morphological patterns. These data suggest that the possible contact among stocks from other oceanic regions may take place in these oceanic islands.

This same analytical approach also shows population fragmentation in *Bathygobius soporator* within the Northeast region of Brazil as well as revealing remarkable chromosomal differentiations among coastal populations and populations in Rocas Atoll, regions approximately 267 km geographically distant from each other, a fact that points towards new species in the area.

The use of available molecular phylogenies enables associating chromosomal patterns and preestablished topologies. It allows follow-ups in the history of these characters. The interaction among DNA sequences and chromosomes enables polarizing the chromosomal changes, identifying the consistency of phylogenetic signs or the possible homoplasies and evolutionary trends, and diversifying mechanisms in the group.

Drawing the history of these characters seems to be a big challenge in Perciformes, since the established chromosomal conservatism strongly features many families. Actually, this trend is found in the Gerreidaefamily, which shows most of its representatives with acrocentric chromosomes. It is possible to identify the independent evolution of these genes in both genera, *Diapterus* and *Eucinostomus* (Fig. 2), by crossing the history of the characters and the information from the 18S and 5S ribosomal genes. Whereas the 18S rDNA sites present higher location variability in *Diapterus*, this same gene, in *Eucinostomus*, appeared to be quite conservative and located in the same pair of different species of this genus. On the other hand, inverse pattern is found in 5S rDNA sites, in which the loci are located in different pairs in representatives of the genus *Eucinostomus*, whereas, in *Diapterus*, this same genus appeared to be conservative in the analyzed species.

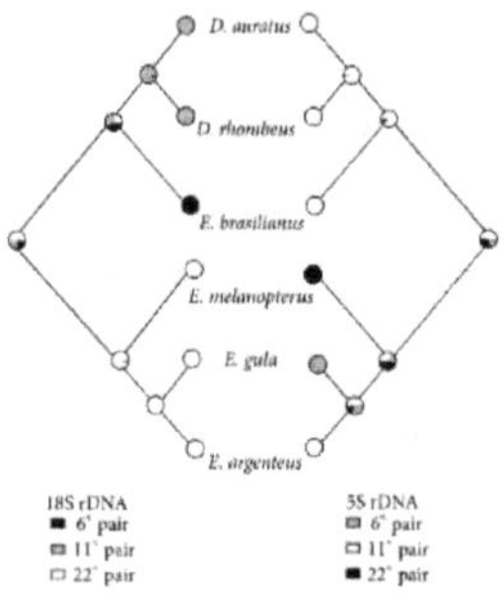

Fig. 2. Reconstruction of ancestral characters of 18S and 5S rDNA in six species from the Gerreidae family (*Eucinostomus gula, Eucinostomus melanopterus, Eucinostomus argenteus, Diapterus auratus, Diapterus rhombeus*, and *Eugerres brasilianus*), which was gotten by means of the Mesquite software using the MK-1 model. Phylogenetic hypothesis estimated by likelihood analysis based on mtDNA COI sequences.

Intra- and Interspecific Karyoevolutionary Aspects of Freshwater Groups

Freshwater fish offer extremely informative models to investigate geologic background and connectivity among basins during the identification of biogeographic processes. The Neotropical region offers excellent opportunities to the study on diversification mechanisms among freshwater fish. The region holds the biggest biodiversity in the world due to complex ecological and historical processes that deal with isolation and specialization.

More than 1,000, out of the 8,000 estimated species, already have information about their karyotype and demonstrate great diversity in diploid values. It covers from chromosomes in *Pterolebias longipinnis* (Rivulidae) up to in catfish *Corydoras aeneus*.

Continental systems physical subdivisions are often effective in blocking gene flow. Such condition leads to species endemism in some regions, whereas, in others, despite demanding broad contribution, they are morphologically categorized as a single taxon. They show remarkable differences in the chromosomal number as well as showing morphology that highlights the occurrence of complexes of species.

Hoplias malabaricus consists of seven karyomorphs. It provides an exceptional way to understand historical relations in different draining within the Neotropical region. Close relations are found among populations from adjacent basins that share the same karyomorph and they indicate the phylogenetic ancestry among these hydrographic basins.

However, differences in the number of 18S ribosomal sites point towards historical gene flow restrictions among these populations. The mapping analysis of 5S*Hind*III satellite DNA is effective in characterizing allopatric populations of such species in DNA sequence mapping when there is no divergence between pairs carrying the 18S rDNA. It shows clear interpopulation divergences that result from different evolutionary histories triggered by this basin's geological isolation.

The taxonomic relations in the Parodontidae family, which presents stable karyotypes, are the target of controversies. Although the location of 18S ribosomal genes appears to be discriminating between the genera *Parodon* and *Apareiodon*, the first genus possesses more conservative conditions.

The first ones show more conservative conditions located in the terminal region of the long arm of one subtelocentric chromosome pair in all species studied thus far and in *Parodon* it reveals higher dynamism among the species. The 5S rDNA sites appear to be more conservative in Parodontidae, found in pericentromeric position in a submetacentric pair similar in Anostomidae, sister family.

The mapping of 5S rDNA enables discriminating sister-species such as *Oligosarcus solitarius* and *O. argenteus* (Characiformes). The complementary localization data, in 18S rDNA sites, allow interpopulation differentiations in *O. solitarius*.

Briefly, for instance, the countless FISH applications through 18S and 5S ribosomal genes appear to be resolute in establishing phylogenetic relations in species discrimination as well as in understanding population historical relations in either freshwater or marine environments.

Evolutionary Dynamics of Marine Fish Chromosomes

Some fish groups, mostly in the Perciformes order, show a well-known low karyotypes dynamics. Overall, such condition is due largely to biological and environmental characteristics of the marine biome. Species with high dispersive potential and high population contingents seem to be particularly refractory to chromosomal change fixations in their karyotypes. These conditions are particularly pronounced in marine species. The absence of geographic barriers and the attention given to the other two conditions are more easily found.

Karyotype evolution rates in Perciformes can be classified in three diversification levels, that is, low, moderate, and high. Some families present bradytelic evolution, which is once more exemplified by taxa as low as 0.094×10^{-2}/m.a. in Haemulidae. There is a second group, which holds families that present moderate karyotype evolution, horotelic, and in which Sparidae and Labridae show rates that, respectively, vary from 2.933×10^{-2}/m.a. to 4.031×10^{2}/m.a. There is one last group composed of families such as Pomacentridae and Gobiidae that present remarkable tachytelic evolution with rates that vary from 6.124×10^{-2}/m.a. to 8.943×10^{-2}/m.a. The karyotype evolution rate varies in the range of one hundred times.

It is possible to find some families of marine Perciformes such as Sciaenidae, Chaetodontidae, Gerreidae, and Lutjanidae among groups that show slow karyotype evolution. These groups present karyotype composed of acrocentric chromosomes and it is a plesiomorphic condition shared by Percomorpha; thus it possibly reaches Acanthopterygii.

Entire families can share the same karyotype, but it is practically unchangeable under the sieve of classic cytogenetic analyses. As for Lutjanidae, the so far analyzed species often present common karyotype (acrocentric chromosomes). This is the same condition found, for instance, in Sciaenidae, Chaetodontidae, Haemulidae, and Gerreidae. Such karyotypes share features such as symmetric chromosomes (small size difference among bigger and smaller karyotype elements), reduced heterochromatic regions, homogeneous heterochromatin, and single ribosomal sites (Ag-NORs). According to an evolutionary perspective, these chromosomal characters significantly hinder the establishment of intraspecific diversity indicators as well as of phylogenetic inferences among the species.

The Ag-NORs sites are effective cytotaxonomic loci in conservative karyotypes. Approximately 330 teleost species from 77 families distributed into 22 orders were already analyzed. NORs present single sites in 72% of the species; besides, they constitute notably diverse regions in comparison to other genes. Thus, in some cases, rDNA may appear limited to identify exclusive differentiations in its positioning and frequency, mostly among species from families that present notable chromosomal conservatism and, therefore, low evolutionary dynamics.

Despite the fact that cytogenetic analyses provide more robust data, their exclusive use is quite limited and it just covers a short percentage (1.3%) of teleost species. The *in situ* DNA sequence mapping enables identifying high syntenic chromosomal conservatism among species from marine fish families. It shows that it can be an extensive condition to have a broad spectrum in one single clade. On the other hand, in some cases, it also enables evidencing the occurrence of extensive evolutionary changes with regard to the dispersion of some repeating sequences, both coding or noncoding.

Multigene families are repeating sequences of coding DNA that belong to a family of related proteins coded by a set of similar genes. These families are formed by duplication events during evolution. The observed differences reflect the mutations that took place throughout time. Histone and ribosomal genes are found among these differences.

Chromosomal rearrangements disperse the multigenic families through the genome, which may be followed by the physical mapping of its sequences. Generally, the multigene families present a considerable number of pseudogenes; thus they show similarities with functional genes in the same family, although they enable functionality due to the acquired mutations.

rRNA genes are among the better known multigene families in fish. In fact, the 18S and 5S ribosomal genes are the highlighted repeating sequences mostly found in fish chromosomal evolutionary studies.

The evolution of 5S rDNA genes is progressively better understood. The molecular variability observed for 5S rDNA gene is due mainly to NTS regions (nontranscribe spacer regions). Retrotransposons and microsatellites also seem to be involved with the high dynamism of their sequences. The 5S rDNA is distributed into one or a few sites, among various fish groups. However, the physical mapping of these sequences by FISH in rare situations has revealed a massive dispersion in most of the chromosomes of some species. Such situations are observed in families with conservative evolutionary patterns such as Pomacanthidae as well as in those with more dynamic patterns such as Gobiidae. In this last family, the species *Ctenogobius smaragdus*, with chromosomes, exhibits 5S hybridization marks in 42 chromosomes of the karyotype (Fig. 3), and it suggests that not all these regions are active.

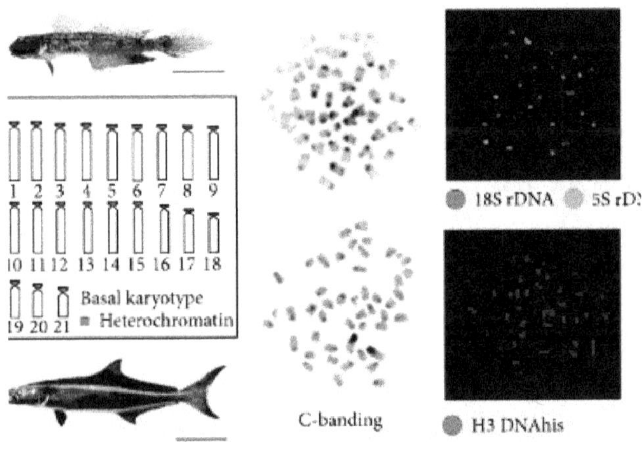

Fig. 3. Basal karyotype pattern in the Perciformes and extensive dispersion events of sequences of the 5S DNAr multigenic families in *Ctenogobius smaragdus* (above) and of DNAhis H3 in *Rachycentron canadum*. (below)

The complementary mapping of the 18S and 5S ribosomes enables the better and more efficient option of specific cytotaxonomic loci in some groups of species with high karyotype conservatism. As for Haemulidae, which is a group with genera that show low evolutionary dynamics (e.g., *Haemulon*, *Pomadasys*, and *Conodon*), the species share a uniform karyotype. Species carrying symmetric karyotypes formed by acrocentric chromosomes with reduced heterochromatic content and, in general, with single ribosomal sites present limitations to identify interspecific diversification patterns by means of classic cytogenetic methods. Indeed, as for these cases, the combined use of mapping by applying the double FISH to 18S and 5S genes is particularly indicated. It enables analyzing these genes' genomic dynamics as well as identifying the interspecific diversity.

With regard to other marine fish groups, the physical map of H1, H2B, H2A, and H3 histone genes shows interesting information about their evolution and it leads to insights into their dynamics in the species' karyotype. These genes play a key role regarding changes in the chromatin structure, in cellular cycle progression, and in gene activity repression. Besides, little is known about their physical positioning in the chromosome of fish. The H1 multigenic family constitutes the histone class with the faster evolutionary rate, whose diversification, mainly by evolution in concert, has been questioned in favor of birth-and-death evolution.

Cytogenetic analyses among populations and *Bathygobius* (Gobiidae) species, which is a group with high evolutionary dynamism, showed that H1 histone genes positioning and frequency are conserved with two sites kept in homologs chromosomes. On the other hand, the H2B-H2A DNAhis and H3 DNAhis genes in *Rachycentron canadum* (Rachycentridae) presented the more diversified condition. Actually, this species of H2B-H2A genes presented multiple sites distributed in up to 6 chromosome pairs in the karyotype, whereas H3 sequences are surprisingly distributed throughout all the chromosome pairs. These data suggest a diversification pattern that matches the birth-and-death evolutionary model. It is followed by the purifying selection found in H2B-H2A histone genes as well as by the variation resulting from the evolution in concert, which involves H3 histone genes. These data reinforce the independent evolution of histone genes in this species.

These genes appear to be colocated with many other repetitive DNA, microsatellites, 18S and 5S rDNA, transposons, and retrotransposons. Distribution analysis of H3 DNAhis site, in five Lutjanidae species, showed that these genes are found in a single site in most of species (*Lutjanus analis, L. synagris,* and *L. alexandrei*). However, they can be found in two loci (*L. jocu*) or extensively dispersed in 44 out of the 48 chromosomes in *Ocyurus chrysurus*. This low diversification pattern, which is highlighted by significant changes in the dispersion of sequences, is associated with the participation of transposable elements in the species' karyotype.

Repetitive DNA represent approximately 50% of the eukaryote genomes. The accumulation of these repetitive sequences is responsible for the variation in the genome size in the eukaryotes. It points out their influence on DNA replication, recombination, and gene expression band on the differentiation of sexual and B chromosomes. Besides, the repetitive DNA are also involved with chromosomal rearrangements such as deletions, duplications, inversions, and reciprocal translocations, thus providing karyotype diversification in many groups. The physical mapping of these sequences in the chromosomes enables accessing countless aspects related to the origin and evolution of sexual and B chromosomes.

The Origin and Differentiation of Sexual and B Chromosomes

Another main element in the karyotype structure of fish regards the presence of sexual chromosomes. Fish, differently from birds and mammals, do not present sexual chromosomes on the basis of their phylogeny. Thus, this character independently and repeatedly emerged in the evolutionary history of this group. Some species present quite differentiated sexual chromosomes, although such occurrence is not frequent.

Sex chromosomes systems with female heterogametic (ZZ/ZW) are found in some species in the genera *Leporinus*, *Parodon*, and *Triportheus*. It is evident, in most of these cases, that sex chromosome differentiation was followed by the heterochromatinization process and by changes in the size of W chromosome. XX/XY systems were described in *Pseudotocinclus tietensis* (Loricariidae) and also in the genus *Hoplias*, respectively. In addition, simple systems (XY and ZW) may undergo rearrangements and originate multiple sexual systems. There is the case of multiple sex chromosomes with female heterogamety (ZZ/ZW_1W_2) reported in *Apareiodon affinis*.

The systems with male heterogamety have the first records in Neotropical fish in the genera *Hoplias*, with multiple sex chromosomes such as $X_1X_1X_2X_2/X_1X_2Y$ and XX/XY_1Y_2, and *Eigenmannia*, with a $X_1X_1X_2X_2/X_1X_2Y$ system.

So far, the chromosomal location of these genes in most fish species is still an open issue, although sex-determining genes may occur even without the presence of morphologically differentiated sex chromosomes. On the other hand, consistent advances have been achieved in the detection of these sexual chromosomes as well as in the origin and differentiation processes, by applying the whole chromosome painting (WCP).

Some examples well depict this situation either in simple or multiple systems. Within simple systems, it is possible to identify the genes differentiation discerning homologous segments among the allosomes, as it can be seen in species in the genus *Triportheus*. In multiple systems, the origin of sex neochromosomes is evidenced by Robertsonian fusions, as it is seen in species in the genus *Harttia*.

Furthermore, the use of chromosome painting also helps solving issues of common and/or independent origin such as the case linked to the Erythrinidae family in which the multiple system $X_1X_1X_2X_2/X_1X_2Y$ found in *Hoplias malabaricus* and *Erythrinus erythrinus* presented independent origins.

Using FISH to investigate sex chromosomes is also a way to determine the location of ribosomal genes related to heterogametic sex chromosomes differentiation. There is an example of it in the W chromosome from the ZW system in genus *Triportheus* and from the XX/XY system in *Hoplias*. Besides, it has been possible to detect the residual interstitial telomeric sequences (ITS) from the centric fusion process in the formation of the XX/XY_1Y_2 multiple sexual chromosome system in *Harttia carvalhoi*.

Heterochromatization processes played an active role in the process of differentiating diverse sexual chromosome systems. The location of repetitive sequences over the sexual chromosomes helps elucidating differentiation processes due to the presence of *Rex* elements spread over the W chromosome in *Semaprochilodus taeniurus* and the presence of microsatellites and different *Rex* in the Z and W chromosomes in *Triportheus trifurcatus*.

Another emblematic matter concerning the karyotype of fish regards the occurrence of B or supernumerary chromosomes, as it is observed in *Apareiodon piracicabae* and *Paraligosarcus pintoi*, *Prochilodus lineatus* and *P. cearensis*, *Curimata modesta*, *Steindachnerina insculpta*, *Schizodon*, among others.

Those are additional chromosomes that do not recombine with those from the standard karyotype complement and follow their own evolutionary path. Among fish, it is possible to highlight the complex of *Astyanax scabripinnis* species as the currently more studied model concerning supernumerary chromosomes distribution, behavior, and origin (Fig. 4).

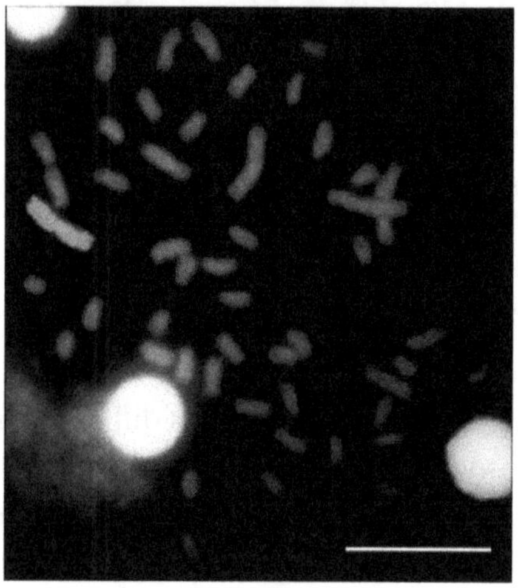

Fig. 4. *Astyanax scabripinnis* metaphase chromosomes subjected to WCP with B chromosome probe amplified by DOP/PCR and marked by nick translation with streptavidin (red). Bar = 10 μm.

The use of probe *As*51 from repetitive DNA over the chromosomes, gotten by means of total DNA cut with the *Kpn*I restriction enzyme, within species from the complex *A. scabripinnis,* identified its association with the 18S ribosomal DNA and the heterochromatic regions located in different autosomes and in the B chromosome. These data reinforce previous hypotheses about the possible intragenic origin of B chromosome in *A. scabripinnis* due to the formation of isochromosome in the standard complement. Actually, recent evidences corroborate B chromosome autopairing in the pachytene of meiotic cells in *A. scabripinnis* by applying FISH with *As*51 and WCP with B chromosome probe, gotten by microdissection and amplification by degenerate nucleotide primed and polymerase chain reaction (DOP-PCR).

GAMETOGENESIS AND FERTILIZATION

Introduction

Gametogenesis and Meiosis Meiosis is a very important component of gametogenesis. Meiosis is a type of division, which occurs only in sex cells. Meiosis has two main functions: a) reduction of chromosome number from the diploid state (2n), typical for somatic cells, to the haploid state (n), typical for gametes, and b) recombination of genetic material. Reduction in chromosome number during meiosis results from two consecutive cell divisions with only one reduplication of chromosomes. The female and male gametes differ to a great extent both morphologically and functionally. Therefore the processes of oogenesis and spermatogenesis have essential differences, although the general scheme of meiosis is universal. Figure 5 presents the basic scheme of chromosome behavior during gametogenesis in animals.

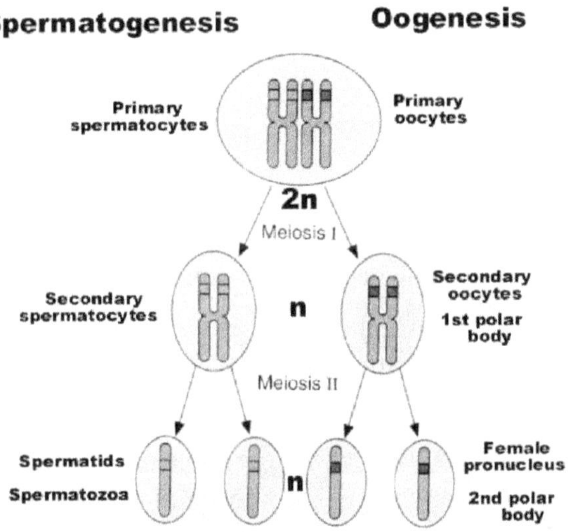

Fig. 5. Scheme of chromosome behavior during gametogenesis in animals.

Oogenesis

The development of female sex cells begins with mitotic divisions of oogonia; as a result their number significantly increases. In interphase before meiosis the replication of DNA occurs; each chromosome reduplicates with formation of two sister chromatids.

Female sex cells coming into meiosis are named oocytes. Until completion of the first meiotic division they are referred as primary oocytes. The process of oocyte development in fish is divided into four basic periods:

- Chromatin-nucleolus stage
- Previtellogenic (small) growth
- Vitellogenic (big) growth
- Maturation

The scheme of development of fish oocytes during chromatin-nucleolus and growth stages is given Fig. 6. This scheme shows the correspondence between the stages of oogenesis and the phases of meiosis. During the chromatin nucleolus stage of oogenesis complex transformations in the structure of chromosomes, caused by conjugation and crossing-over in prophase I, take place.

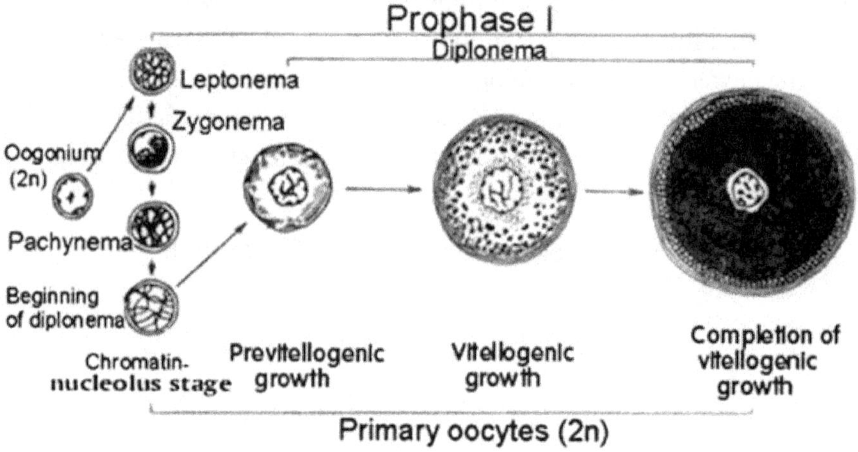

Fig. 6. Development of fish oocytes during chromatin-nucleolus and growth periods

Prophase I is divided to several sub-phases. During the first sub-phase of prophase I, leptonema, thin chromosome threads are evenly distributed all over the nucleus (see Fig. 6).

During this period, preparation for further conjugation of homologous chromosomes occurs. Conjugation of homologous chromosomes (or synapsis) takes place during the next stage – zygonema. For zygonema the specific arrangement of chromosomes is typical, on one side of the nucleus a dense agglomeration of chromosomes is seen, while the other side of the nucleus is filled with single chromosome threads.

This arrangement of chromatin distribution, sometimes called a "bouquet", is observed since conjugated chromosomes attach with their ends to a certain region of the nuclear membrane. At the end of zygonema chromosomes are condensed forming a one clew.

At the beginning of the pachynema, unwrapping of the chromosome clew begins. At the end of this stage thick chromosome threads are evenly distributed throughout the nucleus. At pachynema, homologous chromosomes tightly adjoin to each other; during this period the crossing-over occurs.

At the beginning of diplonema homologous chromosomes begin to separate from each other and the space between them can be observed. By early diplonema the chromatinnucleolus stage of development of the oocytes finishes, meiotic processes block and female sex cells begin to grow.

During the period of previtellogenic and vitellogenic growth, oocytes accumulate a great amount of nutritious substances needed for embryo development. During the growth period, high transcriptic activity of chromosomes, i.e. intensive synthesis and accumulation of ribonucleic acid (RNA), is observed. The storage of large amounts of

RNA is necessary since the genes in fish embryos become transcriptionally active during gastrula stage. Until this period of development all proteins are synthesized on the basis of RNA accumulated previously in oocytes. For the synthesis of large quantities of RNA, numerous side loops, which become regions of active transcription, appear in chromosomes. Due to the appearance of such loops, chromosomes at this stage are named "lampbrush chromosomes".

Numerous nucleoli are located at the edges of the oocyte nuclei during the period of growth; they also play a significant role in the RNA synthesis. Since nucleoli are clearly visible during period of previtellogenic growth this stage is frequently referred as "perinucleolus stage".

During previtellogenic growth the cell size is increased by means of accumulation of homogeneous cytoplasm. Around the oocytes the follicle layer is formed, which is important for transfer of nutritional substances to oocytes. During vitellogenic growth oocytes accumulate yolk (or vitellus).

Numerous vacuoles, oily drops and globules of yolk appear and gradually fill the whole cytoplasm. A multi-layer wall appears, surrounding the growing oocyte. The micropyle - a canal in the membranes through which a spermatozoon penetrates to the surface of the egg, is formed.

During the period of vitellogenic growth oocytes drastically increase their volume and reach a final size. In the fully-grown oocytes the nucleus is located either in the middle of the cell or slightly shifted towards the animal pole, where the micropyle and a small region of cytoplasm free from yolk is located. At this stage the oocyte's large nucleus is frequently called the germinal vesicle (GV).

Oocyte growth may last many years depending on the age when females reach sexual maturity. During this period the oocytes are at the stage of diplonema of prophase I. Microphotographs of some stages in fish oocyte development are shown in Fig. 7.

The general scheme of nuclear transformations during the maturation period in fish oocytes is given in Fig. 8. The period of maturation begins with migration of the nucleus to the animal pole. Later, the nuclear membrane breaks down and chromosomes come out into the cytoplasm - at this period oocyte is at the final stage of prophase I – diakinesis (see Fig. 8). Soon the spindle of the first meiotic division is formed, and chromosomes are placed in its equatorial plane forming the metaphase plate of metaphase I. After that the first meiotic division quickly completes.

During the meiotic process in female sex cells only one egg results from one oogonium. The other products of meiotic divisions are called polar bodies, which later degenerate. In the first meiotic division the homologous chromosomes detach; therefore its products, the first polar body and secondary oocyte have haploid numbers of chromosomes (n), each of which consists of two chromatids.

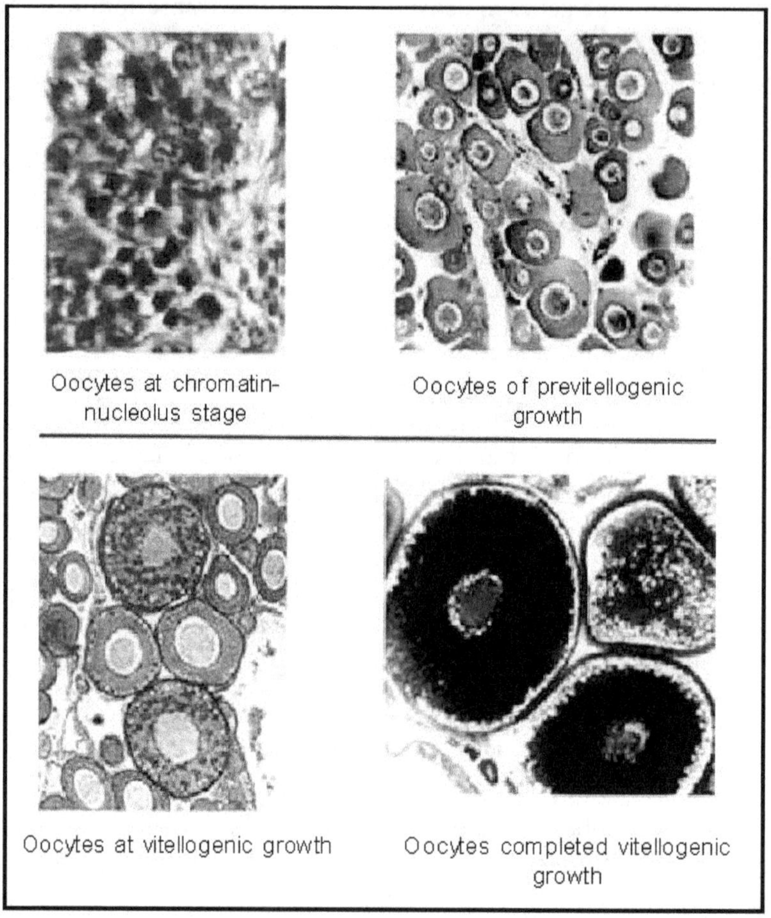

Fig. 7. Microphotographs of common carp oocytes at different stages of development.

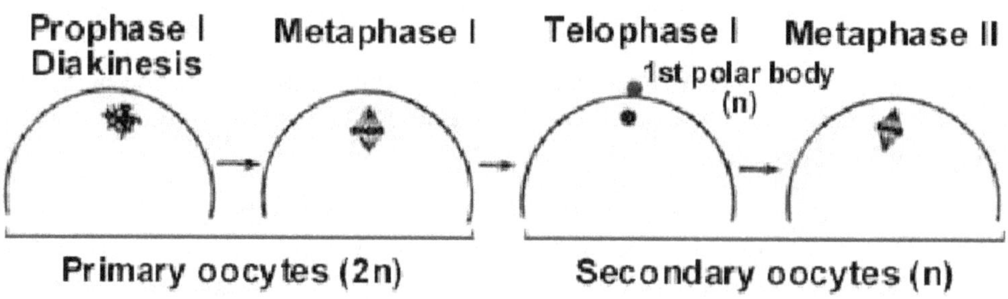

Fig. 8. Nuclear transformations in oocytes during maturation period.

Almost at once after termination of the first meiotic division, the second division begins; it proceeds until metaphase II (see Fig. 8). At this stage meiosis is blocked again, and oocytes ovulate: the follicle layer breaks down and oocytes come out to the ovarian cavity (in most fish) or in the body cavity (sturgeons and salmonids). Microphotograph of metaphase II in ovulated oocyte of silver crucian carp (*Carassius auratus gibelio*, Cyprinidae).

Ovulated oocytes, which are ready for fertilization, are called eggs; they can be spawned by females during natural spawning or stripped during artificial spawning in the hatchery. From a cytogenetic point of view the fish egg is a secondary oocyte at the metaphase of the second meiotic division. Second meiotic division terminates later during the fertilization process.

In contrast with a long growth period, the maturation of oocytes is a quick process. Depending on the temperature, the time period from the hormonal injection until the stripping of eggs lasts only about 12-24 hours.

HYBRIDIZATION

Hybridization has become a common practice in fish culture, though it has not yet attained its rightful place in this branch of economy. The hybridization of fish for practical purpose has demonstrated its economic value. In the U.S.S.R. hybridization has solved the problems of carp culture in the northern regions. New productive forms have been successfully created by crossing species and varieties of Carassius, as well as by crossing carp and Carassius. Research on crossing of herbivorous fishes for their acclimatization in new areas has been initiated. Economic characteristics of a number of hybrids of ganoids have been determined. Experiments have shown that hybridization has extensive possibilities in sturgeon culture.

One may ask why hybrids should be created when excellent non-hybrids forms, sturgeon for example, are available. Hybrids are offered along with the pure species and not as substitutes. Preferences for hybrids are based on inherent characteristics of hybrid heterosis (hybrid vigour), which manifest as intensive growth rate, higher viability, adaptive flexibility, and sometimes early sexual maturation.

Charles Darwin attached great significance to hybridization: "If we do not take it for granted, at least we shall consider it highly probable - the existence of a great law in nature, the law of crossing animals and plants which are not closely related to each other, and this is extremely useful and even necessary".

Natural Hybridization

Study of causes and regularities of hybridization processes occurring in nature reveals possibilities and methods of utilization of hybridization if fish culture. This is particularly true of fresh water fishes, as reproductive isolation is violated more frequently among them than among sea fishes because of less constant hydrological conditions in rivers as compared to the seas. In nature, hybrids occur mostly as single specimens, but sometimes large scale hybridization may occur as a result of violation of reproductive conditions of the species. Abundance of individuals of one species occurring along with less numbers of other species may result in an increase of interspecific crossing. That is why the probability of crossing is high when large numbers of a species are introduced into a new area, while it remains low in the aboriginal species.

Artificial Hybridization

Without detailed cytogenetic analyses of the causes of fertility and sterility of fish hybrids, there are more chances of producing completely fertile hybrids if the species crossed have equal chromosome number, which must be homologous and able to conjugate in the course of gametogenesis. When there is unequal chromosome number and violation of the conjugation process, sterility occurs in various degrees.

Intergeneric Hybridization

Intergeneric hybridization consists of crossing fish belonging to different species. The aim of producing intergeneric hybrids is as follows:

- To unite the useful characteristics of two different species in one of the new forms.
- Producing a new construction in consequence of the heterosis effect.

Intraspecific Hybridization

The aim of intraspecific hybridization is to produce more productive hybrids by crossing different genotypes of females and males having the best results during the previous progeny testing. The phases of intraspecific hybridization are as follows:

- Selection of various genotypes of the same species;
- Inbreeding of the selected lines;
- Polyallel crossing of the different lines;
- Comparative testing of the F1 hybrid generation;
- Propagation of the best hybrid producing parental lines for the fish seed production farms;
- Producing three-line hybrids;
- Producing four-line hybrids.
- After the inbreeding of parental lines, the heterosis effect can be increased. The fastest method of inbreeding is the artificial gynogenesis.
- As a result of intraspecific hybridization, the increasing productivity of fish species of about 15–25 percent can be expected.

ARTIFICIAL GYNOGENESIS AND HORMONAL SEX REVERSION

The main purpose of gynogenesis is to produce inbred individuals having only maternal chromosomes for the further genetic work. For causing gynogenesis artificially, inactivated sperm can be used for fertilization of eggs, with a cold shock treatment after fertilization. For the inactivation of sperm a cobalt 60 gamma ray source or ultra-violet germicide lamp has to be used. The individuals of gynogenetic populations are females with diploid chromosome sets similar to the mother. The technique of artificial gynogenesis was demonstrated on the common carp. The method can be adopted for the other fish species with necessary modifications.

For producing gynogenetic inbred male populations, methyl-testosteron hormone feeding has to be provided for the females at a young age to influence the sex formation in the direction of the male. The technique of sex reversion was demonstrated with gynogenetic common carp. The same method can be used for all fish species which are able to feed with artificial food.

For the purpose of increasing the inbreeding level, the gynogenesis can be applied from generation to generation. It is important to make a proper selection within the gynogenetic population to avoid the undesirable effect of inbreeding depression. The keeping and maintenance of gynogenetic inbred females and males requires more care than the normal ones because of their high breeding value and more sensitiveness to the environmental effects. Therefore, the location, rearing and keeping of this fish in small ponds is necessary, with a good water supply and other necessary conditions and to protect them from poaching. Some outstanding individuals can be kept in an indoor raceway system.

The gynogenetic inbred populations or individuals can be used for improving the species with their useful characteristics, to make inbred lines for intraspecific hybridization, for producing monosex fish populations and also for genetic examinations for particular purposes.

CHROMOSOME MANIPULATION

Chromosome set manipulation is a technique to control the number and combination of a haploid set of chromosomes. This technique is useful for genetic analyses and the improvement of aquaculture traits through sex manipulation, sterilization, inbreeding, cloning and so on. Chromosome manipulations can be carried out easily in fishes due to their external fertilization and embryogenesis. In recent years a great deal of research has been conducted on chromosome set manipulation (polyploidy, gynogenesis and androgenesis) in a number of species. The aim of Chromosome manipulation is to produce sterile progeny for culture; for sex manipulation; and to produce highly inbred or cloned lines for crossbreeding for research purpose. Also producing Haploid, gynogenetic diploid, triploid and tetraploid genotypes that is beneficial for world-wide culture industry.

4

PROCESS OF GYNOGENESIS AND HYBRIDOGENESIS IN FISH

The gynogenesis is defined as the process of development of ovum after penetration of sperm, but in the process the sperm nucleus is unable to form syngamy with the nucleus of ovum. So gynogenetic fish inherit genes only from mother.

The penetration of sperm is responsible for initiation of development alone. Gynogenetic haploid embryos contain a haploid chromosome complement representing a single product of a meiotic division, one quarter of a tetrad embryo do not develop entirely normally.

Gynogenetic diploid embryo contains a diploid chromosome complement, also derived from mother only and are obtained by treatments to prevent either the second division in an egg fertilized with UV-irradiated sperm.

In fish, the gynogenesis is achieved both naturally and artificially. In some fishes, gynogenesis is a natural mode of reproduction, these unisex fish may be often triploid. In a few fishes such as silver crucian carp (Prussian crucian carp of European goldfish) and Poeciliapsis, unisex form present in natural population.

Population, which are produced by natural gynogenesis. Artificial gynogenesis can also be produced under suitable experimental conditions. During artificial gynogenesis, manipulations in event of fertilization are made by destroying DNA of the spermatozoa and to restore diploidy to the ovum with either meiosis or cell cleavage.

In this process the male inheritance is excluded, the offspring obtained in this way possesses inheritance of female characters and they are of only one sex, i.e., female. The unisexual population obtained by artificial gynogenesis are used for fisheries management.

Gynogenesis can also be used to produce clonal lines of tilapia. Hybrid among such inbred lines are expected to exhibit significant heterosis and uniform performance with respect to growth rate and body conformation.

The most important breeding goal of tilapia is to improve growth rate and feed conversion efficiency, which are prime factor regulating profitability of the industry. Sex determination mechanisms, genetically male tilapia have higher growth rates than even hormonally reverse female.

Researchers observed that gible carp, *Carassius gibelio* exhibits gynogenetic reproduction by utilizing the sperm of other species which resulted in the increase population in the lake as gible carp females have little opportunity for mating with gibel carp male.

According to them Ylikiensis roach is the best available sperm donar option for gynogenetically gibel carp of the lake. These authors made artificial fertilization of gibel carp eggs with sperms of other species such as ylikiensis roach (Cr × r), gold fish (Cr. × G) adb common carp (Cr × C), hatching occurred after three days at 14 °C in all groups.

NATURAL GYNOGENESIS OF *CARASSIUS AURATUS GIBELIO*

Genetic analysis of the bisexual form of *C. auratus gibelio*

Investigation of the bisexual form of *C. auratus gibelio* was carried out with the population of the Volma fisheries (in the Byelorussian Soviet Socialist Republic), which was brought from the Amur in 1948. It was found that specimens of the bisexual form of *C. auratus gibelio*, when mated with each other, yielded equal numbers of female and male offspring.

The diploid number of chromosomes of the bisexual form was found to be 94 (by the Japanese scientist Makino in 1939). Maturing of bisexual females is accompanied by the usual meiosis. In the last prophase of the first meiotic division all chromosomes are represented by bivalents. The first division is reducing, and takes place shortly before ovulation ends with discharge of the polar body. As with all fishes studied, the ovum of the bisexual *C. auratus gibelio* is in the metaphase of the second meiotic division after ovulation and contains the haploid number of 47 chromosomes.

Shortly after insemination the second (equational) division comes to an end and the second polar body detaches itself. The remaining female telophasic group and the sperm head turn into female and male pronuclei. When preparing for the first division of segmentation both pronuclei draw closer to each other. This process ends with the formation of the diploid metaphasic plate of the first division of segmentation.

Genetic analysis of unisexual forms of *C. auratus gibelio*

Genetic data testifying to natural gynogenesis of the unisexual form of Crucian carp were obtained by the analysis of progeny of females of unisexual populations and males of other species. In all the crossings tabulated the progeny consisted entirely of females of *C. auratus gibelio*.

Cytological analysis of natural gynogenesis of *C. auratus gibelio*

The exclusion of the male nucleus in the reproduction of the unisexual form of *C. auratus gibelio*, which was quite clearly proved by experimental crossings, was also confirmed by direct cytological observations of the sperm nucleus in the plasma of the ova. The first cytological analysis of the process of fecundation of the unisexual form of *C. auratus gibelio* was carried out by K.A. Golovinskaya in 1954. Later these results were confirmed in experiments with females of the unisexual form from many populations.

Observations showed that the centrosome brought in by the sperm doubles, and each of the two centrosomes forms a pole of the spindle of the first division of segmentation. However, during the whole period of preparation for the first division of segmentation the male nuclear apparatus remains as a dense chromatin formation and does not turn into male pronucleus as in usual fecundation. In the first minutes after insemination, the head of the sperm penetrates into the plasma and gradually comes closer to the female pronucleus. After that it can be detected in form in the zone of one of the poles of the spindle of the first division of segmentation, and later in one of the blastomeres. Lieder managed to trace the head of sperm up to the stage of eight blastomeres. The further destiny of the male chromatin remains unknown. Evidently it is absorbed by the plasma of the ovum.

By analyzing the maturing process of the gynogenetic females in the population of the Volma fishery it was possible to ascertain the cytological mechanism securing the permanent number of chromosomes of the unisexual forms of *C. auratus gibelio*.

Chromosomes in the somatic and sex cells of the Volma unisexual females were found to be 141, triploid set of the basic number, 47. Triploidy has also been confirmed by indirect methods, by determining ploidy by the number of nucleoli in the nuclei of the epithelial cells of the fin border of one-day larvae and by cytometric data (by the size of cells of some tissues).

At present all these indices are widely used for determining ploidy without directly counting the number of chromosomes. Cytometrical investigations have shown that the size of erythrocytes and nuclei of triploid females is on the average 1.4 times greater than of diploid bisexual females.

The process of maturation of unisexual females differs considerably from the usual meiotic process, which is evidently closely connected with triploidy. In the late prophase of the first meiotic division all chromosomes are univalent and no conjugation of chromosomes is observed. In the last period of maturation, which ends with ovulation, it is possible to single out several main stages.

i. Stage 1 - concentration of univalents in a limited area of the cytoplasm of the animal pole and formation of an achromatic, multi-pole figure in the zone of chromosomes. This stage is observed immediately after the nuclear capsule is broken and corresponds to prometaphase 1.

ii. Stage 2 - formation of a three-pole spindle and distribution of chromosomes in three groups at the poles.

iii. Stage 3 - transformation of the three-pole spindle into a bipolar spindle. At this stage, conventionally corresponding to anaphase 1, chromosomes either distribute themselves into two equal groups or in proportion of 1:2, forming haploid and diploid groups.

The above process is abortive and results in no reduction. It ends in uniting all univalents into the triploid metaphasic plate of the only equational division of maturation. It is at this stage that the egg undergoes ovulation. Soon after insemination the division of the polar body takes place. The female triploid complex that remains in the ovum plasma gradually turns into a triploid pronucleus. Further, the process results in forming the triploid plate of the first division of segmentation.

Thus the stable number of chromosomes of the unisexual form of *C. auratus gibelio* of the Volma population is secured by the exclusion of the reduction division. Due to the exclusion of crossing-over and reduction which during the usual process of meiosis cause a genetic recombination, the progeny of each gynogenetic female from the Volma fishery is a genotypically uniform clone.

Females from the unisexual population of the Yakot fishery (the Moscow region) have proved to be triploid with ameiotic maturation. It may be presumed that the cytological peculiarities of Volma gynogenetic females are characteristic of other populations of the unisexual form of *C. auratus gibelio*.

Experiments to obtain unisexual *C. auratus gibelio* by parthenogenesis gave no positive results. Unfertilized eggs when submerged in water began irregular divisions and soon perished. In order to develop normally the eggs of unisexual females should be inseminated. No case of parthenogenesis of these fish has been observed. Suspected cases of parthenogenesis of some species of fish require reappraisal.

A comparison of the cytology of gynogenesis in *C. auratus gibelio* with data on parthenogenesis of other animals shows that natural gynogenesis of *C. auratus gibelio* is a typical case of triploid apomixis. It should be noted that polyploidy and, first of all, triploidy is quite a common phenomenon under natural parthenogenesis. As in other animals, triploidy of *C. auratus gibelio* is accompanied by the loss of ability for usual sexual reproduction.

The origin of triploidy in gynogenetic females of *C. auratus gibelio* is still not clear. It can be presumed that initially the species had a diploid gynogenetic form, and triploidy came about later as a result of crossing diploid gynogenetic females which produce unreduced gametes with males of the bisexual form of their own species or related species. This assumption is based on the results of the well-known model experiments on the silkworm and of some cases of natural triploidy. The assumption is borne out also by the features of the maturation process of unisexual females of *C. auratus gibelio*. The distribution pattern of univalents in haploid and diploid groups during the maturation process of the ovum may be regarded as a result of their cytological and genetic heterogeneity. Therefore, it is reasonable to presume that the genotype of gynogenetic females is of hybrid allopolyploid nature.

When studying the fecundation process of bisexual females of *C. auratus gibelio*, some instances of spontaneous diploidization of the female chromosome complex, caused by the return of the second polar body into the plasma of the ovum, were observed. This process resulted in spontaneous emergence of triploids in the progeny of bisexual females. Thus there is a second channel of triploidy of *C. auratus gibelio* which is not connected with hybridization, namely, autotriploidy.

It is comparatively easy to distinguish females of the unisexual and bisexual forms of *C. auratus gibelio*. This is of importance in investigations on mixed populations of the species. Distant crossings and crossings with the use of irradiated sperm were most suitable for distinguishing the two forms. In both cases the difference between the females is quite distinct during the process of embryogenesis; and, after hatching, by malformation in the progeny of the females of the bisexual form. Identification of females by the size of erythrocytes is also possible.

Natural Gynogenesis of *Mollienesia formosa*

Another case of natural gynogenesis among fishes is that of the live-bearing fish, *M. formosa*, belonging to Poeciliidae. In nature this species is represented only by a gynogenetic form. In natural conditions, females of *M. formosa* live together with two related species *M. sphenops* and *M. latipinna*, and propagate with the help of males of these species. In morphological characters, females of *M. formosa* occupy an intermediate position between *M. sphenops* and *M. latipinna*. This led to the belief, in spite of evidence of gynogenesis, that *M. formosa* was a hybrid form.

Like the gynogenetic females of *C. auratus gibelio*, females of *M. formosa* also cannot propagate by parthenogenesis. When crossed with males of other species they give only female progeny of *M. formosa*.

During the entire period of investigations on natural populations of *M. formosa*, only three males were found. Phenotypically they resembled F_1 males obtained by crossing *M. sphenops* and *M. latipinna* with females of *M. formosa* and masculinized by methyltestosteron. It was assumed that emergence of males may result from three types of development: phenotypical sex transformation of females; disturbances in their maturation; and introgression of paternal chromatin. No detailed investigation of these males was carried out.

Experimental crossings over a period of 14 years from 1932 to 1946 yielded 8 000 gynogenetic specimens of 20 generations. Males of 50 different species, subspecies and races of Mollienesia and males of five other species belonging to Cyprinodontidae were used for crossing.

Much information on the nature of gynogenesis of *M. formosa* was obtained as a result of studies in the laboratory, and observations of natural population of the fish, using the method of transplantation of tissue. As known, compatibility of tissue is possible when the donor and the recipient are of related genetic constitution. In experiments with *M. formosa* the transplants were fins, spleen and heart. Transplantation was considered successful if no detachment of the transplants was observed during the whole period of the experiment. The degree of incompatibility of tissue, i.e. the degree of genetic relationship of the donor and the recipient, was determined by the time interval between transplantation and detachment.

When investigating laboratory populations, two types of transplantation were carried out: (1) transplantation of tissue between parents and their offspring, i.e. between P and F_1; and (2) between the progeny of the first and following generations obtained from the initial parent pair. In crossings, males of *M. sphenops* were used. The results allowed the following conclusions:

1. Gynogenetic progeny of *M. formosa* do not inherit any antigens from the father.
2. Progeny obtained from a female is a clone genetically identical with the mother form and with one another.

The method of transplantation of tissue was used for investigating the clone structure of some natural populations. In these experiments the natural populations of *M. formosa* were found to include different clones. Researchers assumed that different clones in gynogenetic populations may be regarded as the result of the gradual effect of mutation.

Cytological observations of natural gynogenesis of *M. formosa* show that gynogenetic females of *M. formosa* are diploid, the number of chromosomes being 45-46 which corresponds with the diploid number determined for related *M. sphenops*, *M. latipinna* and M. selifera. The diploidy of female *M. formosa* is also confirmed by the size of cells and their nuclei in fin regenerates and the amount of DNA in somatic cells of *M. formosa*, *M. sphenops* and *M. latipinna*.

The cytological pattern of maturation of ova of female *M. formosa* has not been studied. However, the results of transplantation experiments, that allowed the clone structure of gynogenetic populations of the species to be determined, show that the maturation process of *M. formosa*, as well as that of *C. auratus gibelio*, is of the ameiotic type.

Some crossings of females of *M. formosa* with males of *M. sphenops* and M. vittata (in these cases genetic markers were used) gave hybrids. F_1 and F_2 were constituted by females and males having characters of both parents. When mature they produced no progeny.

It was observed that hybrids accepted transplants from donors belonging to the mother clone, whereas graftings from the father to hybrids were not successful. Kallman therefore observed that all tissue antigens of the hybrids were characteristic of the mother form. Regarding these hybrids as amphimictic diploids and trying to explain the mechanism of inheritance, Kallman came to the wrong conclusion, that females of *M. formosa* were entirely homozygous.

Also, as a logical proceeding from these calculations was another wrong conclusion about restitution of the diploidy of gynogenetic females of *M. formosa* in the first division of segmentation. His conclusions contradicted the data on clone multiplication of female *M. formosa*. By cytophotometric analysis, other investigators have shown that these hybrids are triploids, with the diploid set of chromosomes from the mother and the haploid set from the father. Study of spermatogenesis of hybrid males from F_1 has shown that their sterility is connected with the irregular distribution of chromosomes in meiosis. This explains the results of crossing.

Analysis of the Phenonemon of Natural Gynogenesis in Fish

It is necessary to emphasize that triploidy is characteristic of gynogenetic females of *C. auratus gibelio* and diploidy of *M. formosa*. This accounts for the difference in the nature of gynogenesis in the two species. Gynogenesis of *M. formosa* is evidently a result of mass hybridization between two related species (*M. sphenops* and *M. latipinna*) and the subsequent effect of natural selection. The origin of gynogenesis of *C. auratus gibelio* is not so clear. We assume that gynogenesis of *C. auratus gibelio* is of hybrid nature, and the emergence of triploidy may have been preceded by diploid gynogenetic forms.

In the case of autotriploidy, the possibility of spontaneous emergence of triploids in bisexual populations and their further destiny is of great interest. However, mention should be made of the well-known thesis that gynogenesis under any condition is determined by the presence in the genetic pool of populations of initial species of genes that control individual stages of gynogenetic propagation.

Studies of some general features of natural gynogenesis of *M. formosa* and *C. auratus gibelio* reveals the evolutionary importance of this method of reproduction of fish.

Change over from the usual bisexual multiplication to parthenogenesis has two main advantages: (i) Possibility of somatization during meiosis, resulting in preservation of the initial valuable genotype in a number of generations; (ii) rise in effectiveness of reproduction. Both these are true in natural gynogenesis of fish. The assumption that genetic advantages of unisexual lines of *C. auratus gibelio* may be connected with their hybrid constant heterozygous nature and heterosis (which are inherited in a number of generations due to gynogenesis) was made in the first study of natural gynogenesis of this species. Cytological studies have shown that there is no crossing-over and reduction during the process of maturation of ova of gynogenetic females of *C. auratus gibelio*; and this supports the above assumption. The above is true for females of *M. formosa*; clone reproduction of this species has been proved with the help of tissue transplantations.

Increasing effectiveness of reproduction under parthenogenesis is connected firstly with the doubling of the rate of reproduction due to wholly female structure of populations, and secondly with the removal of the loss of time necessary for meeting of two specimens (a male and a female) under usual reproduction. The latter is especially important for animals with low mobility (lizards, for instance) and when the imaginal stage is short (as in the case of some insects).

For gynogenesis fecundation is necessary, and therefore the main factor determining the numerical strength of gynogenetic populations is the number of males able to ensure the reproduction of gynogenetic forms.

Study of natural gynogenesis of different groups of animals has shown that usually reproduction of gynogenetic females depends greatly on males of the bisexual form of the same or related species. This factor limits their population and dispersion, thus depriving them of the main advantages of parthenogenesis. Natural gynogenesis of fish is peculiar in this respect.

The results of numerous crossings of unisexual females of *C. auratus gibelio* and *M. formosa* have shown their ability to multiply with the participation of males of various species of fish. In experimental conditions, *C. auratus gibelio* gave progeny when crossed with males of ten different species representing three families. A large variety of species gave gynogenetic progeny of *M. formosa* It is therefore possible to conclude that in experimental conditions (under artificial fertilization) fish have no limitations in producing gynogenetic progeny by males of different species.

Under natural reproduction it is necessary that the ecology of reproduction of gynogenetic females and of the bisexual species living with them should be adequate. It is also necessary that the numerical strength of the latter should be great. We have evidence of abrupt decrease in the numerical strength of gynogenetic populations of *C. auratus gibelio* when these conditions were not manifest.

In spite of the limiting effect of these factors, because of their ability to propagate with males of many species, natural gynogenesis of fish leads to increase in the numerical strength of the species, which is characteristic of natural parthenogenesis of other animals. The marked predominance of the unisexual form of *C. auratus gibelio* within the area of its natural distribution may be attributed to the above considerations.

Further investigations of the origin of triploidy of *C. auratus gibelio* and cytological analysis of natural populations connected with them, as well as research programmes of experimental triploidy would help clearer understanding of the phenomenon of natural gynogenesis of this fish. It is also necessary to study factors regulating the numerical strength of the unisexual and bisexual forms of *C. auratus gibelio* in mixed populations of this species and analyse the possibility of change over from one manner of reproduction to another. It is necessary to investigate the cytology of gynogenesis of female *M. formosa.*

No investigations have been carried out to determine the function of sperm under natural gynogenesis of fish or other animals. The assumption that the inactivated nucleus of sperm has no effect on the characteristics of the gynogenetic progeny is based only on morphological analysis.

Experiments in transplantation of tissue were a considerable step forward in this respect. For *C. auratus gibelio* this problem is of great practical importance, as males of different species are used in commercial raising of the unisexual form of this fish. There is no information available on the nature of the biological inactivator causing inactivation of the male nucleus under gynogenesis. These and many other problems need further investigation.

Spontaneous emergence of gynogenetic offspring in natural conditions was observed under distant hybridization. Hybrid gynogenesis is based on the incompatibility of plasma and chromosomes of the crossing species. Chromosomes brought in by sperm get eliminated to some extent in the primary phase of development of hybrid embryos.

Most of these hybrid offspring are haploids and mosaics with grave malformations. Their development is usually more regular when they resemble the mother type. Individual viable specimens emerging among hybrid offspring fully repeat the mother form. Their emergence may be accounted for by complete elimination of male chromosomes and diploidization of the female chromosome complex.

Hybrid gynogenesis of fish is described in the experiments in distant hybridization carried out by the researchers.

Hybrid gynogenesis proves that in natural conditions mass hybridization facilitates the detection of specimens with hereditary tendency to gynogenesis.

HYBRIDOGENESIS IN FISH

Hybridogenesis was discovered in the late 1950s by Robert R. Miller and R. Jack Schultz in small fish of the genus *Poeciliopsis* from north-western Mexican streams. These fish attracted attention because entire populations consisted only of females, with apparently no males. In the absence of conspecific males, the females mated with males from a co-occurring sexual species (*Poeciliopsis lucida*) to reproduce. A decade later, the puzzle of the reproductive strategy of these all-female fish was solved. Researchers discovered that the females were the result of an ancient hybridization event between *P. lucida* and a second, closely related species, *Poeciliopsis monacha.*

Even though they systematically mate with *P. lucida* males, which should lead to a largely *lucida*-like genome, the genome of the females looks like that of a first generation hybrid between the two species. The hybrid composition is maintained because the females only transmit their maternal *P. monacha* genome to their eggs, discarding all *P. lucida* chromosomes in the process. In the offspring, the F_1 hybrid configuration persists, because the eggs are fertilized via matings with *P. lucida* males (Fig. 1A).

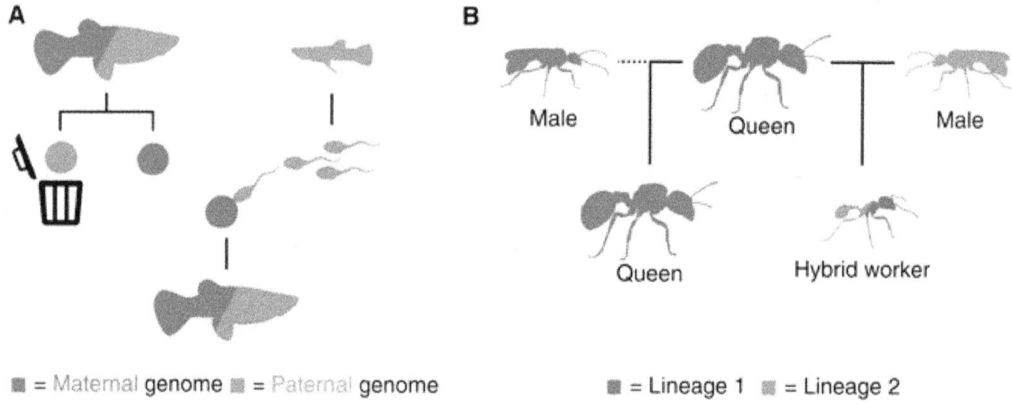

Fig. 1. Hybridogenesis.

(A) Classic hybridogenesis as in *Poeciliopsis*. The hybridogens are F1 hybrid females. They mate with males from their paternal ancestor and clonally transmit the genome of their maternal ancestor while discarding all paternally inherited chromosomes. (B) Social hybridogenesis. Queens are produced either via parthenogenesis (e.g. in the genus *Cataglyphis*) or from intra-lineage matings (dashed line; e.g. in the genus *Pogonomyrmex*). Workers are always produced from hybrid crosses between genetically diverged lineages.

P. lucida males gain no benefit from mating with hybridogenetic females, as their genome will be eliminated in their daughter's eggs and will not contribute to future generations. As a consequence, hybridogenetic females are often considered sexual parasites that exploit the reproductive effort of their sexual host. The term hybridogenesis was coined by Schultz in 1969 to describe this novel reproductive mode. It is also commonly referred to as 'hemiclonal reproduction', to reflect the fact that half of the genome - in our example, the maternal *P. monacha* one — is clonally transmitted between generations, while the other half - the paternal *P. lucida* one — is replaced in each generation. The term 'hybridogenesis' has sometimes been used to describe the generation of new species via hybridization. We concentrate here on the original meaning of hybridogenesis.

Distribution of Hybridogenesis

Since its first discovery in *Poeciliopsis*, hybridogenesis has been found in a diverse range of animals, including other fishes, frogs, stick insects and ants. Reproduction in many of these species is similar to the situation in *Poeciliopsis*: hybrid females clonally transmit their maternal genome and fertilize their eggs via matings with males from their paternal ancestor. However, in some species, notably *Hypseleotris* fish and *Pelophylax* pool frogs, hemiclonal genome transmission occurs in males as well as in females.

In ants, hybridogenesis, referred to as 'social hybridogenesis', varies among species and differs from the originally described form. There are three types of social hybridogenesis. In the first type, found in *Cataglyphis hispanica*, two genetically distinct lineages coexist. Both produce new queens by parthenogenesis, males from unfertilized eggs (as is generally the case in ants), while workers are produced through hybridization between the two lineages. Although there is no genome elimination in this case, the result is similar to classical hybridogenesis: the maternal genome is always passed on to the next generation, while the paternal one is only transmitted to the sterile workers (the "soma" of the colony) and therefore does not contribute to future generations (Fig. 1B).

In the second type of social hybridogenesis, found, for example, in the little fire ant (*Wasmannia auropunctata*), new queens are also produced by parthenogenesis and workers via hybridization with a different lineage. The difference with the first type is that one lineage contains only females, while the other contains only males. This is due to the latter being produced via androgenesis.

Finally, the third type of social hybridogenesis differs more extensively from classical hybridogenesis, and is found, for example, in the genus *Pogonomyrmex*. Here, workers are also produced through hybridization between two lineages, but new queens are produced sexually, instead of via parthenogenesis, through within-lineage matings. While there is still only one type of genome transmitted to future generations, this genome is not asexually transmitted, which is expected to have a positive effect on the long-term persistence of this form of hybridogenesis.

To date, classical and social hybridogenesis have been reported in fewer than 20 animal genera. It has not been found outside the animal kingdom yet. Several of these genera contain multiple, independently derived hybridogenetic lineages, created through repeated hybridization or host shifts. Although hybridogenesis is most likely rare overall, the currently known examples certainly represent only a small fraction of the extant hybridogenetic taxa.

Most, if not all hybridogenetic systems were discovered serendipitously when researchers unexpectedly observed strongly biased sex ratios and extreme patterns of heterozygosity. With the increasing availability of genotyping methods for non-model organisms, the number of taxa being studied is growing. This is likely to lead to future discoveries of hybridogenesis and other unusual reproductive strategies.

The rise and Fall of Hybridogenesis

At first glance, it might seem surprising that a reproductive mode as complicated as hybridogenesis can evolve. Selectively transmitting a specific parental genome that is then combined with the genome of a different species requires several modifications to classical sex and is certainly not the most parsimonious way to achieve reproduction. Whether hybridogenesis evolved from classical sex through multiple intermediate steps or whether it appeared suddenly via major-effect changes is unknown. Nevertheless, hybridization appears to play a key role, given that hybridogenesis is solely known in hybrids.

While the evolutionary origin of hybridogenesis remains elusive, it is easier to understand how it can spread in a population. By bypassing the standard meiosis roulette, which for each locus would generate a 50% chance of being transmitted, the hybridogenetic genome gains a transmission advantage, as it is being transmitted to 100% of the offspring. In addition to this transmission advantage, hybridogens are expected to enjoy benefits of their hybrid ancestry, including heterosis and the possibility of transgressive phenotypes (Box 1). Hybridogens could therefore quickly establish themselves among their sexual hosts.

The initial success of hybridogenesis is nonetheless double-edged. Hybridogens rely on their hosts for reproduction, and therefore have to co-occur with them. Inevitably, this leads to competition for space and resources. Because of the transmission advantage, hybridogens are expected to rise in frequency at the expense of the hosts, ultimately supplanting them. With no hosts left, hybridogens cannot reproduce and may go extinct. Accordingly, many newly formed hybridogenetic complexes might be short-lived and go extinct before being noticed.

Are current hybridogenetic systems transient, or could there be factors maintaining them over longer evolutionary times? The answer is likely to be system-specific, but it is thought that some mechanisms contribute to the maintenance of hybridogens and their hosts in general. For example, ecological differences between the host and the hybridogens provide a possible explanation for extended coexistence.

Thus, an incomplete niche overlap, whereby the hosts could thrive under conditions inhospitable for the hybridogens, could allow the hosts to persist. Assortative mating, whereby hosts tend to mate preferentially with conspecific females rather than with the hybridogens, can also contribute to maintenance of sexual species and hybridogens. Hybridogens could also switch to a different host when the primary one becomes rare. Finally, at least in *Bacillus* stick insects and some forms of social hybridogenesis, the combination of hybridogenesis with androgenesis is also likely to contribute to long-term persistence. In these insects, hybridogenetic females produce a low proportion of males through androgenesis. This can facilitate their maintenance, because it reduces the probability of sperm limitation.

The maintenance of social hybridogenesis is difficult to explain for most systems as well. In *Cataglyphis hispanica*, queens gain no benefit from producing males. As males only father sterile worker offspring, they do not contribute any genes to future generations. Not producing males should therefore be beneficial and spread in the population as the saved resources could be reallocated to producing more queens and workers. This would, however, cause the extinction of the co-dependent lineages since they rely on each other for worker production.

The *prima facie* success of hybridogenesis is further tarnished by the long-term detrimental effects of clonal transmission. Non-recombining genomic regions are expected to gradually accumulate deleterious mutations. Such a build-up of deleterious mutations should result in the fitness of the hybridogenetic genome decreasing over time, which may lead to its extinction.

How long it would take until a fitness decrease could become apparent is not clear, especially since the hybridogenetic genome is usually not in a homozygous state, but together with the genome of a host species. Observations in *Pelophylax* and *Bacillus* indeed suggest that hybridogenetic genomes harbour deleterious mutations, but that these are largely masked when in the heterozygous state.

The picture is different in *Pogonomyrmex* and *Messor*, which perform the type of social hybridogenesis that involves the production of new queens through intra-specific sexual reproduction. Unlike in *Cataglyphis*, ceasing to produce males is not beneficial in these species, as males transmit their genes to new queens. In addition, the two co-existing lineages each use sperm from the other one for worker production.

If one lineage increases in frequency, it will suffer from mate limitation for the production of new workers, which generates an advantage for the lineage that is rare. This negative frequency-dependent selection enables the long-term maintenance of both lineages.

Proximal Mechanisms of Hybridogenesis

While the evolutionary mechanisms allowing for the spread and maintenance of hybridogenesis start to be reasonably well understood, there is little knowledge of the proximal mechanisms by which specifically only one of two available genomes is transmitted. The two genomes have to be distinguished somehow, and one genome has to be eliminated at some point during gametogenesis. This elimination needs to be both specific, because always the same genome is targeted, and complete, because all non-transmitted chromosomes have to be eliminated at once. Incomplete elimination would result in aneuploidy, which is often lethal. Because hybridogenesis has evolved independently in the different species, the proximal mechanisms might vary among them. This further complicates their study, as conclusions from one system might not apply to another.

5

INHERITANCE OF QUANTITATIVE TRAITS IN FISH

After the rediscovery of Mendel's law, many scientists conducted similar experiments and found that there was fluctuation in Mendel's law of inheritance. However, they were unanimous that genes inherit characters. Gene is a stable entity. It controls the phenotype of the individual.

An organism carrying the normal gene is called wild type, the organism carrying an altered gene is called a mutant. For example, wild type Drosophila has bright red-eye colour while mutant has white-eye colour. The body colour of Cyprinis carpio is grey colour but the mutant has body colour golden. These are the instances of qualitative characters.

On the other hand, the body weight at age or the body weight of Fish/organisms and milk production of cow and length of fishes of similar age group are not discrete entity (colour never be equal) but is continuous variable and, therefore, the examples of quantitative characters. The quantitative trait phenotype is determined by alleles of many genes acting together. The phenomena are called as polygenic.

The quantitative character are not only under polygenic (multiple alleles) control but the environment influences the quantitative trait too. Two individuals with similar genotype if raised in different environmental have a different quantitative phenotype.

For instance, fishes having the same genotype but, if one group is raised in environment having rich fed while other group is reared in environment having less rich fed, the former group will naturally grow faster showing quantitative phenotype. It is clear that quantitative traits would depend upon genotype as well as environment.

CHARACTERISTICS FEATURES OF QUANTITATIVE INHERITANCE

There are three characteristic features of quantitative inheritance:

1. They are continuous variable. There is no distinct phenotype in the offspring. For example, if the cross is made between white and red, the offspring will be having continuous distribution of colour intermediate between the colours of the parents with different intensities.
2. It is polygenic in nature. The single character is controlled by alleles of many different genes. The population is having large numbers of different genotype and different genotype may have the same phenotype.
3. The environment influences the quantitative traits as described above. When we use the term environment, it covers all aspects in which organisms interacted with physical and biological environment around it during its lifetime.

Here, it is clear that the genotype environment also plays important role in the appearance of phenotype of quantitative character. The example is again cited that a well-fed fish, Cyprinus carpio will grow faster than the poorly fed carp irrespective of genotype.

Hence it is established that environment and genotype both play role in the determination of quantitative character. Based upon the researches done between 1903-1918, it is accepted that inheritance of quantitative traits follow Mendelian fashion.

MEASUREMENT OF QUANTITATIVE INHERITANCE:

R.A. Fisher, a British statistician, described the biometrical approach and suggested that Analysis of variance and Heritability are needed for measurement for quantitative inheritance.

Analysis of Variance

Before understanding the variance it is essential to know that there is relationship between the mean (X) and standard deviation (S) in normal distribution curve. It is also necessary to know that there is a relationship between the standard deviation and variance. This relationship is given as under.

Standard deviation square $(S)^2$ = Variance

Standard deviation $S = \sqrt{V}$

The quantitative traits are continuous variable and distributed normally.

The normal distribution graph is bell shaped.

The following are the characteristics of the normal distribution:

1. It is a symmetrical distribution.
2. In the normal distribution the mean of the population (X) occurs at the peak

of the curve or the ordinate at the mean is the highest ordinate. Similarly the height of the mean ordinate and the height of other ordinates at various standard deviations from the mean happen to be a fixed relationship with the height of the mean ordinate.

3. The curve is asymptotic to the base line, which indicates to continuity to approach but it never touches the horizontal axis.
4. The standard deviation or variance or $(S)^2$ defines the spread of the curve. So if we examine two graphs having the same mean (X) but showing the difference of variance we can conclude whether the variance is more or less.

Formula for the Calculation of Variance:

In quantitative character, the variance is known as phenotypic variance and is represented as V_p. It is well accepted tool used in quantitative genetics, which includes the analysis of the components of the phenotypic variance.

$V_p = V_g + V_e$.

The formula is based on the following concept. The phenotypic variance (V_p) has three additive components: the genetic variance (V_g), the environmental variance (Ve) and the interaction variance (V_i).

The formula actually is as follows:

$V_p = V_g + V_e + V_i$.

Vp = phenotypic variance

V_g = genetic variance/variance due to genes (different alleles and loci, QTL)

V_e = environmental variance; we have no way of measuring interaction variance Vi, it is usually assumed to be zero, So $V_p = V_g + V_e$.

How to Calculate Genetic Variance (V_g)?

The variations caused by genes may come from three different sources. Firstly, the variation may be related to straightforward presence or absence of particular allele at quantitative trait loci (QTL). This is additive genetic variance and denoted by (V_a). It is most important as the presence or absence of a particular allele which is passed unchanged to the next generation.

Secondly, in some cases there is presence or absence of particular genotype at QTL. For instance, a particular heterozygous combination of alleles at a locus may confer an advantage on an individual with respect to a particular trait. This is called as dominant genetic variance and denoted as V_d.

It is less manageable to simple artificial selection. Because during meiosis segregation and independent assortment of alleles will take place and in the next generation instead of the same heterozygous combination (which is advantageous), different combination may appear hence there is no guarantee of inheritance.

Thirdly, genetic variation is produced by interactions between loci called epistatic or non-allelic interaction genetic variance, denoted by Vi.

Hence the genetic variance V_g can be calculated by the following formula:

$V_g = V_a + V_d + V_i$

V_a = additive genetic variance

V_d = dominant genetic variance

V_i = interaction genetic variance

Explanation:

The genetic variance in population is caused mostly due to the:

i. Allelic difference in the genotype. They are additive genetic variance (V_a),
ii. The dominance genetic variance (V_d), genes are dominant or recessive
iii. And genetic, interaction variance (V_i).

The genetic variance (V_g) can be calculated by subtracting the environmental variance from the variance of the F_2 individuals ($V_g = VF_2 - V_e$).

V_a is calculated as Follows:

The phenotypic variation that is caused by individuals is due to different alleles of gene at QTL that effect the phenotype. It is most important element of genetic variation for the aquaculture breeder as the presence or absence of particular alleles is a character.

The additive variance (Va) can be calculated by subtracting the backcross variance (VB_1 and VB_2) from the F_2 variance using the following formula:

$V_a = 2\ (VF_2 - (VB_1 + VB_2)/2\ )$

The dominant variance (V_d) is that portion of the phenotypic variance caused by individual having different alleles of genes that effect the phenotype. The dominance variance (V_d) is much less amenable to simple artificial selection because genotype are broken down during meiosis and put together again in different combinations in the next generation.

Hence crossing between inbred, highly homozygous lines does not guarantee predictable heterozygosity in the offspring and such F_1 hybrid are commonly used in plants and animal breeding. The development of inbred lines in aquatic organisms is still in infancy but ploidy manipulation techniques are quite successful.

The third component of genetic variation is produced by interactions between loci and is called epistatic or non alleleic genetic interaction variance (V_i). It includes epistasis, enhancement, suppression, etc. It includes an individual may be highly ranked for a trait as it possesses particular combinations of genotypes across two or more QTL.

The additive genetic variance and dominance variance can be estimated by measuring the variances of groups of individuals that have known genetic relationship. The values of these variances can be then used to make deductions about the alleles present in the population. The two types of F_1 crosses (F_1 individuals crossed with both sets of parents) are especially valuable for this type of analysis.

(ii) The dominance variance (V_d) can be calculated by subtracting the additive variance from genetic variance. In these calculations, the interactive component is ignored.

The environment variance (Ve) is the measure of phenotypic differences that are produced by the different environment such as quality of water, food quality and quantity, temperature and stocking density in which the individuals in the population live. For example a well-fed fish will grow faster than a poorly fed fish.

The growth of the plants will be more in the soil where the nutrients in the soil are more in comparison to the plant where the nutrients are low in the same field. So, there are physical and biological environment interacted in a close fashion. It also includes the cellular environment, which is responsible for coding the protein.

So the environmental variance should be measured using a genetically uniform population. This could be obtained by inbreeding. Such population would be genetically same and therefore $V_g = 0$. And all phenotypic variation must be due to environment and therefore $V_p = V_e$.

If two crosses namely P_1 and P_1 are crossed having the same genotype then the F_1 individuals will be genetically uniform and the total phenotypic variance is an estimate of the environmental variance. Assuming that all the populations are raised in the same environment, environmental variance is the average of the variance of parents and the F_1.

$V_e = (V_p^{\ 1} + V_p^{\ 2} + V_F^{\ 1})/3.$

Heritability

The proportion of the variance of a trait which is under genetic control is termed as the heritability. Heritability is a measure of the genetic component of the variance, and through this technique it is used in predicting the phenotypes of offspring.

So first the mean and variance of a quantitative trait is obtained from the parental population then this information can be used to make prediction about the mean of the phenotypic distribution in the offspring (generation). And we can determine how strongly the phenotype traits of the offspring will resemble the phenotype of the parents.

There are two widely used numerical measures of heritability. One is the coefficient of heritability (h^2), also called heritability in the narrow sense. The other is the degree of genetic determination (H^2), also called heritability in the broader sense.

Both values depend upon the ratio of the genetic variance to phenotype variance, h^2 is the ratio of the additive variance to the total phenotypic variance, and H^2 is the ratio of total genetic variance to total phenotypic variance.

The following are the formulae of measuring the heritability:

$h^2 = V_a/V_p$

$H^2 = V_g/V_p$

Both these values indicate that what portion of the variation in a population is the result of genetic variation? Both these values theoretically vary from 1 to 0, if the value is high it shows that large portion of the phenotypic variation is the result of genetic variations, h^2 and H^2 have important limitations.

Their values are calculated for one population in one environment, so they cannot be used for other generations of the same population raised in different environments or for other populations.

Each population has a different set of genotypes and will have a different proportion of the phenotypic variance caused by genetic variance and different values of h^2 and H^2. An example is cited to explain how h^2 can be used for predicting the phenotype of the offspring.

QUANTITATIVE PHENOTYPES

- The branch of genetics dealing with the genetic model for quantitative traits and its applications is called quantitative genetics.
- Quantitative phenotypes are those that are measured, such as length, weight, and fecundity. The important production phenotypes are quantitative ones, although some qualitative phenotypes are quite important, and can greatly increase the value of a crop.
- Because quantitative phenotypes are measured each phenotype is a single category, such as length.
- Individuals do not get segregated into alternate phenotypic categories, such as long vs short instead, they are distributed along a continuum, and differences among individuals are determined by the unit of measure that is used to assess the phenotype; millimeters, grams, etc.
- Because each fish's phenotype is determined by a measurement, quantitative phenotypes form what are called "continuous distributions", which are described by the population's mean and the distribution about the mean (variance and standard division).
- In a population, these phenotypes form what are called "normal" or "bell-shaped" distributions.
- The reason why quantitative phenotypes exhibit continuous distributions and why individuals do not get segregated into descriptive categories is that quantitative phenotypes are far more complicated genetically than qualitative phenotypes.

- Each quantitative phenotype is controlled by dozens to hundreds of genes, each of which makes a small contribution to the production of the phenotype. The exact number of gene is usually never known.
- Additionally, each phenotype is strongly influenced by the environment (e.g., date of birth, access to food, age of mother, etc), and the influence varies both from family to family and from individual to individual.
- The simultaneous action of many genes and the environmental effects creates single phenotypic categories (e.g. length), in which the only way to describe an individual is to measure it.
- Because each quantitative phenotype is controlled by numerous genes, as well as environmental variables, the only way to work with these phenotypes is to analyze the phenotypic variance that exists in the population and to divide the phenotypic variance into its component part.

GENETICS OF QUALITATIVE AND QUANTITATIVE PHENOTYPES

The mechanism of inheritance in fishes is known primarily on intensive studies of very few members of a single order, the Cyprinodontiformes – particularly the guppy (*Lebistes reticulates*), platyfishes and sword tails (*Xiphophorus),* and the medaka or Asiatic rice fish (*Oryzias latipes*).

In recent years, knowledge of the field has been enlarged through researches on carp and goldfish (Cyprinidae), Anatolian cyprinodontids, experiments with trouts and chars (Salmoninae), and on scattered representatives from a few other groups.

Although much is known of the genetics of the goldfish (*Carassius auratus),* the documentation is chiefly in Japanese. There remains the vast majority of fishes including the primitive, jawless vertebrates (Cyclostomata), for which almost no genetic information is available.

6

BIOCHEMICAL AND MOLECULAR (DNA) GENETIC MARKERS

INTRODUCTION

All organisms are subject to mutations because of normal cellular operations or interactions with the environment, leading to genetic variation (polymorphism). Genetic variation in a species enhances the capability of organism to adapt to changing environment and is necessary for survival of the species. In conjunction with other evolutionary forces like selection and genetic drift, genetic variation arises between individuals leading to differentiation at the level of population, species and higher order taxonomic groups.

Molecular genetic markers are powerful tools to detect genetic uniqueness of individuals, populations or species. These markers have revolutionized the analytical power, necessary to explore the genetic diversity. The conclusion from genetic diversity data has varied application in research on evolution, conservation and management of natural resources and genetic improvement programmes, etc.

Molecular markers can be classified into type I and type II markers. Type I markers are associated with genes of known function, while type II markers are associated with anonymous genomic regions. Under this classification, allozyme markers are type I markers because the protein they encode has known function. RAPD markers are type II markers because RAPD bands are amplified from anonymous genomic regions via the polymerase chain reaction (PCR).

Microsatellite markers are also type II markers unless they are associated with genes of known function. The significance of type I markers is becoming extremely important for aquaculture genetics. Type I markers serve as a bridge for comparison and transfer of genomic information from a map-rich species into a relatively map-poor species. In general, type II markers such as RAPDs, microsatellites, and AFLPs are considered non-coding and therefore selectively neutral. Such markers have found widespread use in population genetic studies to characterize genetic divergence within and among the populations or species.

The taxonomy of *Salmo* is still a matter of controversy since more than 60 synonyms for varieties of brown trout and more than 20 for varieties of Atlantic salmon were described. The number of *Salmo* species recognized varies considerably not only because of highly phenotypic variation (body shape, colour, etc.), but also because the species inhabit and are adapted to very different habitats over large distribution areas. Due to taxonomic ambiguities, authors often refer to brown trout as *Salmo trutta* species complex.

Salmo trutta comprises several distinct ecological and geographical morphs and with respect to this is still controversy as far as their classification as species or subspecies is concerned. Based on morphological and ecological variations, the existing populations of *Salmo trutta* from distinct areas are grouped into different taxa:

i. Black Sea populations – *Salmo labrax*,
ii. Caspian Sea populations – *Salmo caspius*,
iii. Aral Sea populations – *Salmo oxianus* and
iv. Mediterranean Sea populations – *Salmo macrostigma*.

The brown (common) trout (*Salmo trutta* morpha *fario* and *Salmo trutta* morpha *lacustris*) and the sea trout (*Salmo trutta* morpha *trutta*) are fish of the same species, considered by some taxonomists different subspecies in order to distinguish the anadromous *Salmo trutta trutta*, living in the sea and migrating in freshwater only to spawn, from *Salmo trutta fario*, residing in freshwater and the lake dwelling form *Salmo trutta lacustris*. Instead, other authors consider that these do not necessarily represent monophyletic groups.

In Romania, *Salmo trutta fario* (Linnaeus, 1758) is widely spread in a large number of water streams from the mountain area, along the Carpathian Arch, whereas the Black Sea trout, *Salmo labrax* is endemic to the Black Sea area and migrates for reproduction in the Danube River and its tributaries. Nowadays only few individuals are captured annually along the Black Sea coast and sporadically in the Lower Danube River. Due to the present situation of the species several measures were adopted to protect it. Thus, the fishing is completely prohibited and the Black Sea trout is included on the Red List of Danube Delta Biosphere Reserve.

Salvelinus fontinalis is predominantly raised in fish farms for food consumption, but a low number of wild populations are still present in the Romanian mountain waters.

Thymallus thymallus (Linnaeus, 1758) is the only native salmonid species for which no imports of biological material and restocking programs were completed in Romania.

Until two or three decades ago, excellent habitat conditions for Danube huchen (*Hucho hucho*, Linnaeus, 1758) still existed in many rivers in Romania. The historical range of this species included the majority of the Carpathian river systems, but hydropower development, river pollution, and overfishing and poaching led to drastic declines in the area inhabited by huchen, and the species is now extinct in the Mures, Timis, Cerna, Olt, Arges, and Ialomita river systems. Data on the current area of huchen occurrence in Romania are fragmentary, and the species only occurs in a few rivers, including, among others, the Tisa River and its tributaries (Viseu, Ruscova, Crasna, Bistra, Vaser, Somes, and Cris) and in the Siret River.

In relation to the main rivers from Romania, there are 10 hydrographic basins in which significant salmonid populations are found. Nowadays is a well-known fact that the most prevalent salmonid species is the brown trout, while the most threatened are the Danube huchen and the Black Sea salmon.

Both sturgeons and salmonids are of crucial importance for Romanian fauna, having also a significant socio-economic value. In this context, well documented studies regarding the biology, taxonomic classification and ecology of the species were performed. The development of technologies based on DNA markers has had a tremendous impact on animal genetics in generally, and changed the way in which studies were conducted inclusive in population genetics, phylogeny, phylogeography and conservation. Thus, in a progressive way, studies aiming at analyzing molecular aspects in sturgeons and salmonids populations from Romania founded their place in the research field, although the number of such studies is still relatively low.

MOLECULAR MARKERS FOR POPULATION GENETICS AND CONSERVATION

The definition of "marker" in wide sense is "something that serves to identify, predict, or characterize". In biology, the markers refer to any stable variation, which is heritable and can be measured or detected by an appropriate method. Such variations are produced at different levels for example, morphological, gene, chromosomal, biochemical or genomic. The markers that represent variations that appear at the DNA level are so-called *molecular markers*. The molecular markers occupy specific places in the genome and possess the role to "mark" the position of a specific gene or the inheritance of a particular character.

All organisms are subject to mutations that appear when genetic material fails to copy accurately or as a consequence of interactions with the environment, leading to genetic variation or polymorphism.

Individuals of a species are distributed into more or less separate groups called populations, distributed over the species range. The genetic variation of a species is distributed both within and between populations. The diversity at the gene level is often referred to as intra-specific variability as it represents biological diversity within a single species. The researchers are very interested in assessing genetic variation within and between populations and detecting similarities as well as differences between individuals/ populations in order to establish optimum conservation strategies.

For this variation to be useful in different genetic studies it is necessary to be inherited and to be perceptible for the researchers, whether is recognizable as phenotypic variation or as a genetic mutation detectable by different molecular techniques. At the DNA level the genetic variation is represented by point mutations called also single nucleotide polymorphisms (SNPs), indels (insertions or deletions of nucleotide sequences), inversion of a segment of DNA within a locus, and rearrangement of DNA segments around a locus of interest. As the mutation rate is very low the evolution through mutation is extremely slow. The process of mutation is the only way in which genetic variation is created and in the lack of mutations there would be no biological diversity.

DNA marker technology can be applied to reveal these mutations. Large deletions and insertions determine shifts in the size of DNA fragments resulted consequently digestion by restriction enzymes, and are among the easiest type of mutations to detect by electrophoresis on agarose gel; smaller indels require DNA sequencing or more elaborate electrophoretic techniques, while the SNPs can be easily detected by DNA sequencing.

By using molecular markers is possible to observe and exploit genetic variation across the entire genome. Thus, the application of molecular markers in fish allowed recording rapid progress regarding the study of genetic variability and inbreeding, parentage determination, species identification, genetic linkage map construction for aquaculture species, identifying Quantitative Trait Loci (QTL) related to specific traits for marker assisted selection. One of the classifications of the molecular markers refers to markers of type I as markers associated with genes having a known function and markers of type II as markers associated with unknown genomic segments.

The type I of DNA markers were not initially consider to be appropriate for genetic studies in fishes, but over time it became clear that these markers are very important both for the study of wild populations and aquaculture. Thus, this of markers has become very important for studying the phenomenon of linkage and for QTL mapping, being of great use in comparative genomic studies and for identification of candidate genes for quantitative traits in different fish species raised in aquaculture.

An important number of studies were focused on the elucidation of the molecular basis of economically important traits in different species of salmonids. For example in the case of Atlantic salmon the mapping the QTLs was performed for the loci correlated with viral disease resistance, flesh color and growth traits, salinity tolerance, late sexual maturation, etc.

The type II of molecular markers (RAPD, AFLP, microsatellites) is considered to be non-coding. Such markers are used in population genetic studies aiming to characterize the genetic diversity. The type II of DNA markers proved useful in identifying species, populations and subpopulation, but also in identifying interspecific hybrids.

A second classification of markers is the one regarding their position in the cell. Thus we can distinguish between nuclear and mitochondrial DNA (mtDNA) **markers**, depending on their localization in the nuclear or mitochondrion genome.

The molecular markers might be highlighted by a variety of techniques that differ by difficulty, repeatability, cost and nature of polymorphism that is detected. The DNA markers detected by Polymerase Chain Reaction (PCR) are the most frequent used in assessing the genetic diversity and have a particular role in conservation. By PCR is possible to amplify DNA sequences up to several million times, so these markers present the advantage of nonlethal sampling. Small amount of tissue (fin clips or scales) are sufficient for analysis and is not necessary that the individuals to be sacrificed for sampling.

This can be an important feature when evaluating genetic change in protected or declining populations and for providing access to DNA of ancient or archived tissue samples. It can provide information about genetic diversity over extensive temporal and spatial scales, especially for populations that no longer exist. Taken together, these characteristics suggest capability to monitoring populations that are small, exploited or declining.

Depending on the primers used for PCR amplification these markers can be divided in two groups: (i) PCR markers for target sequences – in this case the fragment of interest is amplified with two specific primers and (ii) PCR markers for arbitrary sequences – one primer with a arbitrary nucleotide sequence is used; the primer binds to randomly in the genome resulting unknown DNA fragments. The primer used in this type of techniques is usually short (of 8-10 nucleotides) and so the probability for it to bind at multiple sites in genome increase.

In the first category are found molecular markers like PCR-RFLP (PCR Restriction Fragment Length Polymorphism), PASA (PCR Amplification of Specific Alleles), SNP (Single Nucleotide Polymorphism), repetitive DNA sequences (minisatellites, microsatellites, etc.). The second category includes markers like RAPD (Randomly Amplified Polymorphic DNA), AP-PCR (Arbitrary Primed-PCR) and AFLP (Amplified Fragment Length Polymorphism), but these are less preferred comparing with PCR markers for target sequences due to the difficulty of analysis and lack of results accuracy and reproducibility.

ALLOZYME MARKERS

Analysis of allozyme loci remained one of the most popular approaches in examining population genetics and stock structure questions in fishes. The technique is rapid, relatively inexpensive and provides an independent estimate of level of variation within a population without an extensive morphological and quantitative survey. Isohyets are structurally different molecular forms of an enzyme system with qualitatively the same catalytic function encoded by one or more *loci*. Isohyets, which are encoded by different alleles of the same gene locus, are designated as "allozymes" or "alloenzymes". Amino acid differences in the polypeptide chain of the different allelic forms of an enzyme reflect changes in the underlying DNA sequence. Depending on the nature of the amino acid changes, the resulting protein products may migrate at different rates (due to charge and size differences) when run through a gel subjected to an electrical field. Differences in the relative frequencies of alleles are used to quantify genetic variation and distinguish among genetic units at the levels of populations, species, and higher taxonomic designations.

Disadvantages associated with allozymes include occasional heterozygote deficiencies due to null (enzymatically inactive) alleles and sensitive to the amount as well as quality of tissue samples. In addition, some changes in DNA sequence are masked at the protein level, reducing the level of detectable variation.

Some changes in nucleotide sequence do not change the encoded polypeptide (silent substitutions), and some polypeptide changes do not alter the mobility of the protein in an electrophoretic gel (synonymous substitutions). At present 75 isozyme systems representing several hundred genetic loci are known. With the strength as codominant marker, ease of use, and low cost, the allozyme markers are popular in population structure and phylogenetic studies, though has limited role in aquaculture genetics.

MITOCHONDRIAL DNA MARKERS

Mitochondrial DNA (mtDNA) analysis is being increasingly used in recent population and phylogenetic surveys of organisms. Studies of vertebrate species generally have shown that sequence divergence accumulates more rapidly in mitochondrial than in nuclear DNA. This has been attributed to a faster mutation rate in mtDNA that may result from a lack of repair mechanisms during replication and smaller effective population size due to the strict maternal inheritance of the haploid mitochondrial genome. Due to its rapid rate of evolution, mtDNA analysis has proven useful in clarifying relationships among closely related species. Different parts of the mitochondrial genome are known to evolve at different rates.

Almost the entire mtDNA molecule is transcribed except for the approximately 1-kb control region (D-loop), where replication and transcription of the molecule is initiated. In general, non-coding segments like the D-loop exhibit elevated levels of variation relative to coding sequences such as the cytochrome b gene, presumably due to reduced functional constraints and relaxed selection pressure. The 16S rRNA gene in the mitochondrial genome is one of the slowest evolving genes whereas rapidly evolving regions are control regions. Due to non-Mendelian mode of inheritance, the mtDNA molecule is considered as a single locus. In addition, because mtDNA is maternally inherited, the phylogenies and population structures derived from mtDNA data may not reflect complete picture of the nuclear genome if gender-biased migration or selection or introgression exists. Analyses of mtDNA markers have been used extensively to investigate stock structure in a variety of vertebrates including fishes, birds], mammals and reptiles.

PCR-RFLP

The analysis of these markers involves the amplification by PCR of a specific DNA region comprising one or more polymorphic sites for restriction enzymes (RE). With the increasing number of so-called "universal" primers available, can be targeted DNA regions that are relatively conserved among species. In addition, PCR products can be digested with restriction enzymes and visualized by ethidium bromide staining due to the increased amount of the resulting DNA amplification reaction. The ability of RFLP markers to highlight the genetic variation is relatively low compared with the one of other markers.

Substitutions, insertions, deletions and rearrangements of the regions containing restriction sites are probably quite widespread in the genomes of many species, but the probability that they exist at the locus that we want to study is quite low. Because the difference in size between the restriction fragments is usually large these can be easily separated by agarose electrophoresis. The disadvantage of PCR-RFLP is that presents a relatively low level of polymorphism and requires knowing the sequence of the fragment amplified by PCR, fact that makes difficult to establish new markers.

In genetic studies regarding sturgeons the PCR-RFLP markers were mainly used in two directions: to assess the genetic diversity of sturgeon populations and for species identification and implicit for caviar traceability. Researchers has identified species specific restriction profiles in a fragment from the mitochondrial gene *cytochrome b* (*cyt b*) in 10 species from *Acipenser* and *Huso* genera.

A similar study was performed by researchers for 22 sturgeon species, an accurate identification being possible by analyzing the restriction profiles resulted consequently the action of five REs on a fragment of 1121 bp from *cyt b* gene. Researchers proposed a molecular method based on PCR-RFLP to distinguish between Atlantic (*Acipenser oxyrinchus*) and European (*A. sturio*) sturgeon. The discrimination between the two species is difficult to be done exclusively by morphological traits, especially in their early life stages, while the application of two REs allowed the clear and unambiguous discrimination of 132 specimens of Atlantic and European sturgeon.

The identification of acipenserid species by PCR-RFLP presents the difficulty of correct diagnostic for species that are closely related from genetic and evolutionary point of view, like is the case of *A. gueldenstaedtii*/*A. persicus* or *Scaphirhynchus* genus species. Moreover, the analysis of length polymorphisms for mitochondrial DNA fragments put in evidence only the genetic variability originating from the maternal genitor, as the mtDNA is almost exclusively maternally inherited. This fact might lead to misinterpretations in case of inter-specific hybrids diagnostic and/or an ancient introgression, both phenomena occurring in sturgeons.

PCR-RFLP markers were used also in salmonid species for the genetic diversity and phylogeography analyses in different populations of brown trout, charrs, grayling, etc.

Microsatellites also referred to as "simple sequence repeat" (SSR) represent short repetitive sequences of 2-9 bp, wide spread in the genome and with a significant level of polymorphism. These markers are numerous in vertebrates, in fish appearing in every 10kb. The majority of microsatellite loci are relatively small sizes, being amplified easily by PCR. Up to 70% of loci present dinucleotide repeats, the (AC) motif being the most common in the vertebrate genome.

The main features of microsatellites are co-dominant inheritance, high degree of polymorphism, hypervariability, higher mutation rate than standard. Through microsatellites analysis is possible to infer the genetic profile of an individual (genetic fingerprint) and to establish the relationships between individuals. Some microsatellites have a high number of alleles for a locus within population and are very suitable to identify the genitors and their progeny in hybrid populations. The higher level of allelic variation at microsatellite markers make them useful for addressing questions related to genetic structure, particularly where genetic differentiation may be limited.

On the contrary other microsatellites present a low number of alleles or even a fixed allele and are more appropriate for phylogeny or species identification. The utility of microsatellite markers can be determined depending on the polymorphic information content (PIC), defined as the capacity of the marker to detect a polymorphism in the population. PIC is direct correlated with the number of detected alleles and their frequency, a higher number of alleles determining a higher value for PIC. By comparing the values for PIC in different markers is possible to obtain valuable information about their power and efficiency in population genetic and conservation studies.

The primers designed for microsatellites amplification in a species give also cross-amplification for similar loci in related species, this being an important benefit in analyzing populations that are small or at the brink of extinction. Microsatellite markers are widely used for population genetic and conservation studies in fishes.

The first studies based on microsatellite loci analysis were initiated at the North-American species, where it aimed at isolation and description of disomic *loci*.

The identification and characterization of new loci is complicated by the polyploidy of the sturgeon species. Lots of potential useful microsatellites were eliminated from analysis since these were polysomic and thus, they complicated the interpretation of the inheritance mode and of genetic variation within and between populations. Once a disomic set of microsatellites is established, this fact permits to analyze the genetic diversity and structure of wild populations and aquaculture stocks. Such disomic loci were isolated and characterized also for *A. naccarii* and *A. persicus*. In the case of Ponto-Caspian sturgeons, a few studies based on microsatellites analysis in *A. stellatus* from the Caspian Sea and, respectively, in *A. gueldenstaedtii* from the North-West of the Black Sea, North of the Caspian Sea and Azov Sea were performed.

Beside their high applicability in for inferring the genetic diversity and structure of the wild populations, the microsatellites are appropriate for aquaculture stocks evaluation, selection of breeders and proper conservation. Different studies were conducted in several sturgeon species (*A. transmontanus, A. fulvescens, A. naccarii, A. sturio*, etc.) for *ex-situ* conservation purposes. The obtained data were used in the management program to adopt appropriate conservation methods.

In salmonids, the microsatellites were successfully applied in phylogeography studies, in determination of genetic variation in wild and farmed fish populations, in inferring the genetic diversity within population, fine-scale genetic differentiation and relationship of populations, in assessment of stocking impact on wild populations.

The mitochondrial genome possesses certain characteristics (compact organization, maternal inheritance, hundreds to thousands of copies per cell, rapidly evolving, reduced recombination rate, and higher mutation rates compared to those of nuclear genes) that make it useful in population genetics and phylogeny studies. mtDNA analysis in sturgeons is a suitable method for the characterization of species and populations, providing useful information for the management of conservation activities.

For intraspecific studies, the most commonly used marker is the mitochondrial control region (D-loop) due to the relatively high degree of nucleotide variation. Studies to discriminate species/populations/aquaculture strains based on the analysis of D-loop region polymorphisms were performed for several species of sturgeon, such as *A. gueldenstaedtii, A. sinensis, A. stellatus* and *A. baerii.*

The gene coding for *cyt b* is a useful marker for identifying different species of sturgeon based on specific polymorphisms found at this level. A 648 bp fragment of the gene coding for *cytochrome oxidase subunit I (CO I)* is considered a real "barcode" for vertebrates and is useful for species identification by *DNA barcoding* technique.

Genes with highly conserved sequence and slow evolution like mitochondrial ribosomal genes (16SrRNA and 12SrRNA genes) are preferred when is about inferring the phylogenetic relationships of fishes at different taxonomic levels. 12SrRNA gene is considered a promising tool for tracing the history of more recent evolutionary events and it has been widely used to study the phylogenetic relationships among different levels of taxa such as families, genera and species.

In salmonid species mtDNA has proven to be useful for phylogeographical studies and for analysis of spatial and temporal population structure. Based on the analysis of the mitochondrial D-loop marker researchers showed that there are five different main lineages of *Salmo trutta* in Europe: Atlantic, Danubian, Marmoratus, Mediterranean and Adriatic. Analytical techniques for mtDNA include indirect methods such as the analysis of RFLP markers or the direct analysis of mtDNA sequences.

RANDOM AMPLIFIED POLYMORPHIC DNA (RAPD) MARKERS

RAPD markers are the amplified products of less functional part of the genome that do not strongly respond to selection on the phenotypic level. Such DNA regions may accumulate more nucleotide mutations with potential to assess inter-population genetic differentiation. The amplification of genomic DNA by PCR with arbitrary nucleotide sequence primers, RAPD can detect high levels of DNA polymorphisms. The technique detects coding as well as non-coding DNA sequences, and many of the most informative polymorphic sequences are those derived from repetitive (non-coding) DNA sequences in the genome.

Because 90% of the vertebrate nuclear genome is non-coding, it is presumed that most of the amplified loci will be selectively neutral. RAPD loci are inherited as Mendelian markers in a dominant fashion and scored as present/absent. RAPDs have all the advantages of a PCR-based marker, with the added benefit that primers are commercially available and do not require prior knowledge of the target DNA sequence or genome organization. Other advantages of RAPDs include the ease with which a large number of loci and individuals can be screened simultaneously.

Shortcomings of this type of marker include the difficulty of demonstrating Mendelian inheritance of the loci and the inability to distinguish between homozygotes and heterozygotes. Analysis follows the assumption that populations under study follow Hardy-Weinberg expectations. In addition, the presence of paralogous PCR product (different DNA regions which have the same lengths and thus appear to be a single locus), low reproducibility due to the low annealing temperature used in the PCR amplification, have limited the application of this marker in fisheries science.

EXPRESSED SEQUENCE TAGS (ESTS)

Expressed sequence tags (ESTs) are single-pass sequences generated from random sequencing of cDNA clones. The EST is use to identify genes and analyze their expression by means of expression profiling. It helps for rapid and valuable analysis of genes expressed in specific tissue types, under specific physiological conditions, or during specific developmental stages. ESTs offer the development of cDNA microarrays that allow analysis of differentially expressed genes to be determined in a systematic way, in addition to their great value in genome mapping.

For genome mapping, ESTs are most useful for linkage mapping and physical mapping in animal genomics such as those of cattle and swine, where radiation hybrid panels are available for mapping non-polymorphic DNA markers. A radiation panel is composed of lines of hybrid cells, with each hybrid cell containing small fragments of irradiated chromosomes of the species of interest. Typically, the cells from species of interest are radiated to break chromosomes into small fragments. The radiated cells are unable to survive by themselves. However, the radiated cells can be fused with recipient cells to form hybrid cells retaining a short segment of the radiated chromosome.

Characterization of the chromosomal break points within many hybrid cell lines would allow linkage and physical mapping of markers and genes. In spite of its popularity in mammalian genome mapping, radiation hybrid panels are not yet available for any aquaculture species. Development of radiation hybrid panels from aquaculture species is not expected in the near future, given the fact that physical mapping using BAC libraries can provide even higher resolution and the fact that BAC libraries are already available from several aquaculture species. Therefore, ESTs are useful for mapping in aquaculture species only if polymorphic ESTs are identified. The value of EST resources and applications of bioinformatics in aquaculture genetics/genomics is inevitable, and it is expected that various EST databases will serve as rich sources of genomic information not only for aquaculture geneticists, but also for aquaculture physiologists, immunologists and biotechnologists.

SINGLE NUCLEOTIDE POLYMORPHISM (SNP)

Single nucleotide polymorphism (SNP) describes polymorphisms caused by point mutations that give rise to different alleles containing alternative bases at a given nucleotide position within a locus. SNPs are becoming a focal point in molecular marker development since they represent the most abundant polymorphism in any organism's genome (coding and non-coding regions), adaptable to automation, and reveal hidden polymorphism not detected with other markers and methods. Theoretically, a SNP within a locus can produce as many as two alleles, each containing one of two possible base pairs at the SNP site. Therefore, SNPs have been regarded as bi-allelic. SNP markers are inherited as co-dominant markers.

Several approaches have been used for SNP discovery including SSCP analysis, heteroduplex analysis, and direct DNA sequencing. DNA sequencing has been the most accurate and most used approach for SNP discovery. SNPs are not without their limitations, however, might provide marginal additional, or even less, utility in some applications (e.g. relatedness).

MICROSATELLITE MARKERS

Microsatellites consist of multiple copies of tandemly arranged simple sequence repeats (SSRs) that range in size from 1 to 6 base pairs [e.g., ACA or GATA]. Abundant in all species studied to date, microsatellite motifs have been estimated to occur as often as once every 10 kb in fishes. Microsatellites tend to be evenly distributed in the genome on all chromosomes and all regions of the chromosome.

However, data from whole genome sequencing has somewhat contradicted this statement. They have been found inside gene coding regions, introns, and in the non-gene sequences.

Most microsatellite loci are relatively small, ranging from a few to a few hundred repeats. Regardless of specific mechanisms, changes in numbers of repeat units can result in a large number of alleles at each microsatellite locus in a population. Microsatellites have been inherited in a Mendelian fashion as codominant markers. Microsatellites were found to be informative in several species, which showed almost no variation at other markers.

However, use of microsatellite markers involves a large amount of up-front investment and effort. Each microsatellite locus has to be identified and its flanking region sequenced to design of PCR primers. Due to polymerase slippage during replication, small size differences between alleles of a given microsatellite locus (as little as 2 bp in a locus comprised of di-nucleotide repeats) are possible. Microsatellites recently have become an extremely popular marker type in a wide variety of genetic investigations.

DNA BARCODING

The principle of conservation biology is the preservation and management of biodiversity. The two major problems to such an endeavor are the difficulty of developing an assessment of this diversity for prioritization of hotspots of species richness and the identification of lineages particularly worthy, or in need, of preservation. Understudied taxa are greatly susceptible to extinction, suggesting there is a conservation penalty for our ignorance. Even there are millions of unidentified and unknown species.

DNA barcodes, segments of approximately 600 base pairs of the mitochondrial gene cytochrome oxidase I (COI), have been proposed as a fast, efficient, and inexpensive technique to catalogue all biodiversity. Barcoding is the use of universal polymerase chain reaction (PCR) primers to amplify and sequence an approximately 600-basepair fragment of the COI gene. That portion of sequence is then compared using distance-based algorithms with an existing database of "known" sequences from specimens previously identified by taxonomists.

DNA barcodes from a small portion of the mitochondrial genome might seem like an effective and rapid way to assess at least some, perhaps minimal, level of biodiversity. And for groups that are already relatively well known, especially birds and mammals, molecular studies based on barcode sized sequences have revealed cryptic DNA lineages and may be helpful.

MOLECULAR STUDIES FOR SPECIES IDENTIFICATIONS AND GENETIC DIVERSITY ASSESSING IN STURGEONS FROM ROMANIA

The markers analyzed in sturgeon species and populations from Romania were nuclear markers (microsatellites) and mtDNA. The methodologies applied in the studies included microsatellites genotyping, PCR-RFLP and sequencing of mitochondrial markers. Thus, the DNA was isolated from biological samples consisting in small pieces of fin sampled without harming the animals (aspect very important in vulnerable populations) by a classic protocol with phenol-chloroform-isoamylic alcohol.

For mtDNA analysis the primers were designed based on DNA sequences for the interest markers retrieved from GenBank data base. The primers for microsatellites amplification were described in literature in different species and the cross-amplification for the similar species from Romania was tested.

The molecular markers were amplified by PCR or multiplex PCR in specific conditions established consequently several steps of reaction optimization. The microsatellites were analyzed by capillary electrophoresis in ABI Prism 310 Genetic Analyzer (Applied Biosystems) and the mitochondrial markers were sequenced by Sanger method, *dye terminator* variant, in ABI Prism 3130 Genetic Analyzer (Applied Biosystems). The raw data were processed and edited with dedicated computer programs. The interpretation of genotypic data was done by using specialized computer software for population genetics and phylogeny.

In the context of severe decline of sturgeon population from the Lower Danube several studies based on molecular marker analysis were performed in the recent years in Romania. In consequence, the studies were directed to assessing the genetic diversity in Lower Danube sturgeon populations based on microsatellites and to evaluate genetic variability in aquaculture strains in order to sustain the efforts of conservation.

The analyses regarding the genetic diversity were preceded by the correct species identification based on molecular markers for each individual. As mentioned before, the accurate detection of sturgeon species encountered in the Danube River can be very difficult, due to the plasticity of various external morphological features. The number of hybrid individuals occurring in natural conditions is unknown. However, the number of hybrids in natural waters could increase due to escapes from commercial farms where exotic species or genotypes and inter-specific hybrids are regularly used for production purposes. So, decisions to initiate restocking programs for sturgeon conservation need to take into account, besides the socio-economical aspects and the assessment of genetic diversity, the correct diagnostic of fishes included in this type of programs.

For sturgeon species from the Lower Danube diagnostic a molecular methods was set-up by researchers based on the analysis of nuclear markers. In this study 84 individuals belonging to all four species from the Lower Danube (A. *stellatus, A. gueldenstaedtii, A. ruthenus* and *H. huso*) were analyzed. Initially 25 microsatellites were tested from which eight loci (LS19, LS34, LS39, LS54, Aox27, AoxD234, AnacE4, and AnacC11) that have demonstrated a good amplification, results repeatability and interspecific polymorphism in all of the four investigated species were selected. The genotypic data processing involves three successive statistical analyses including Factorial Correspondence Analysis (FCA), STRUCTURE assignation and NewHybrids status determination.

In a first step, the genotypes data were run in a FCA test, using GENETIX software only for the individuals considered being pure species based on morphology analysis. The FCA highlighted the differences between the four analyzed species. Four main clusters, each corresponding to one of the sturgeon species analyzed, were identified.

In the second step, the genotypes data of putative hybrids, diagnosed such as based on morphometric indices were included in FCA analysis as supplementary individuals. A fifth category was highlighted including the putative hybrids, but also individuals that were considered as pure species, based on their morphology. The analysis had showed that only three of the four species hybridize each other, the putative hybrids displaying an intermediate position between *A. stellatus*, *H. huso* and *A. gueldenstaedtii*. Consequently, we deduced that *A. ruthenus* individuals should be eliminated from further tests.

In consequence in the third step, FCA was performed only on three pure species and hybrid individuals included as supplementary elements in analysis. The analysis grouped the pure species and the hybrids in distinct clusters. The hybrids and some of the individuals labeled as pure were occupying an intermediate position between the pure species groups, some of these being placed approximately in the middle of the triangle delimited by the three pure species, while others appear to be closer to *H. huso* and *A. gueldenstaedtii.*

APPLICATION OF MOLECULAR MARKERS SPECIES IDENTIFICATION

The inter-specific genetic divergence established through species specific diagnostic molecular markers provides precise knowledge on phylogenetic relationships and also resolve taxonomic ambiguities. These markers can be used to detect hybrid and introgressed or backcrossed individuals, distinguish early life history stage of morphologically close species both in hatchery and in natural populations.

Species-specific allozyme markers have been identified in many fishes [Tilapia: 72,77,78; Sciaenid: 73; Anguilla sp: 79; Mugilidae: 80] Specific diagnostic allozyme loci were used for different species: apache trout (*Oncorhynchus apache*), cutthroat (*Oncorhynchus clarki*) and rainbow trout (*Oncorhynchus mykiss*) and Gambusia affinis and *G. holbrooki.* Allozyme markers have also been used for individual classification in cyprinid species *Zacco pachycephalus* and *Z. platypus*, in cyprinodontid species *V. letourneuxi* and *V. hispanica*, in mullets *Mullus barbatus* and *M. surmuletus* and hake species *Merluccius australis* and *M. hubbsi.*

Species-specific diagnostic RAPD fingerprints were generated in several fish species and their taxonomic relationship has been analyzed. The RAPD-PCR technique was employed to identify three endemic morphologically similar Spanish species of *Barbus*: *Barbus bocagei, B. graellsii and B. sclateri* that have similar morphologies. RAPD markers were characterized to identify five species of family Cyprinidae: *Chondrostoma lemmingii, Leuciscus pyrenaicus, Barbus bocagei*, Barbus comizo, all endemic in the *Iberian Peninsula*, and introduced *Alburnus alburnus*, for studying genetic relationship and diversities in four species of Indian Major carps (family Cyprinidae): rohu (*Labeo rohita*), kalbasu (*L. calbasu*), catla (*Catla catla*) and mrigal (*Cirrhinus mrigala*), for identification of three eel species, *A. japonica, A. australis and A. bicolor* and to estimate the population structure and phylogenetic relationships among the eight species of the genus *Barbus*.

Large variation in mtDNA sequences among species can be utilized to produce species-specific markers. Since the structures of mitochondrial RNA genes (tRNA and rRNA) and the functional molecule of the 16S rRNA are highly conserved among the animal taxa that are related even distantly, change of even few nucleotides in such a gene between closely related taxa might indicate a substantial degree of genetic divergence. Mt-DNA sequences have been used as useful marker for species-specific identification in many fishes [Tuna: 91; Billfish: 92 Snappers: 29, 93; Myctophidae: 94; Grey mullets: 95]. Comparable levels of divergence based on 12S rRNA and 16S rRNA sequences have been reported for several recently diverged fish species [genus Sternoptyx: 96; Cyclothone sp: 97]. Sequence variation in the control region (D-loop) of the mitochondrial DNA (mtDNA) was examined to assess the genetic distinctiveness of the short-jaw cisco, *Coregonus zenithicus* and revealed high similarity of *C. zenithicus* and the related species *C. artedi, C. hoyi, C. kiyi, and C. clupeaformis.*

Identification of *Astyanax altiparanae* (Teleostei, Characidae) in the Iguacu River, Brazil, was done on the basis of mitochondrial DNA and RAPD markers. Two species, Acipenser baeri, and A. stellatus, was studied using mitochondrial DNA (D-loop, cytochrome b (cyt-b) and ND5/6 genes) sequencing to determine whether traditionally defined subspecies correspond to taxonomic entities and conservation management units.

COMPARISON OF GENETIC VARIATION BETWEEN WILD AND HATCHERY POPULATIONS

Molecular markers also find application in aquaculture to assess loss of genetic variation in hatcheries through, comparison of variation estimates between hatchery stocks and wild counterparts. The information is useful obtained in monitoring farmed stocks against inbreeding loss and to plan genetic up gradation programmes. A major aspect such studies address is concerned with the assessment of farm escapes into the natural population and introgression of wild genome.

Brook trout *Salvelinus fontinalis* from ustocked waters, naturalized lakes, and hatcheries in New York and Pennsylvania were analyzed electrophoretically for allozyme expression. All wild-unstocked samples were highly differentiated populations and significantly different from each other and from hatchery samples.

Genetic diversity was investigated using microsatellites between farmed and wild populations of Atlantic salmo. Farmed salmon showed less genetic variability than natural source population in terms of allelic diversity.

Variation in allozymes and three microsatellite loci was assessed in populations of wild and cultured stocks of *Sparus aurata* and *Sparius auratus*. The microsatellite heterozygosity values were high in wild, but lower in the cultured samples.

ASSESSMENT OF DEMOGRAPHIC BOTTLENECK IN NATURAL POPULATION

Demographic bottlenecks occur when populations experience severe, temporary reduction in size. Because bottlenecks may influence the distribution of genetic variation within and among populations, the genetic effects of reductions in population size have been studied extensively by evolutionary biologist.

It may often be necessary to perform genetic analyses of temporal replicates to estimate the significance of spatial variation independently from that of temporal variation in order to ensure the reliability of estimates of a defined population structure. Such estimates provide understanding about changes in genetic variation, effective population size and other historical bottlenecks and can be extrapolated to define evolutionary trends of species.

Today various models are available that can resolve bottlenecks or effective population size changes through use of heterozygosity excess, linkage disequilibrium etc. However, estimates through temporal changes are considered more accurate.

Analysis of temporal changes is limited due to lack of historical data as well as samples. Therefore, such studies are limited and mostly use archived samples, wherever available. In vertebrates, a limited number of studies have specifically assessed the temporal changes in genetic variation for more than one generation.

Microsatellite DNA markers have been used to assess bottlenecks in many fish species. A microsatellite analysis of DNA was performed, from archived scales to compare the population structure among four sympatric landlocked populations of Atlantic salmon, Atlantic salmon , European hake and steelhead.

Researchers recommended close monitoring of negative effects on sea otter population based on the conclusion from mtDNA, D-loop, microsatellite variability comparison between prefur trade and present population. Prefur trade DNA samples were obtained from excavated bones.

PROPOGATIONASSISTEDREHABILITATIONPROGRAMMES

Habitat alterations and over harvesting have contributed to the decline or disappearance of numerous natural populations. In addition, reinforcement programs of wild populations based on releases of hatchery reared fish of non-native origin compromise the conservation of remnant native trout resources. Effect of these programmes through releases in natural populations has been studies in many fishes through molecular markers.

Researchers found through allozyme polymorphism that brown trout (*Salmo trutta* L.) in the Abatesco river basin on the eastern coast of Corsica restoration was mainly due to the populations of the tributaries, which had been less disturbed by the spate. This study has shown that the wild population was primarily restored by the surviving individuals, particularly those from the tributaries that escaped the spate.

To assess the levels of gene introgression from cultured to wild brown trout populations, four officially stocked locations and four non-stocked locations were sampled for one to three consecutive years and compared to the hatchery strain used for stocking. Allozyme analysis for 25 loci included providing allelic markers distinguishing hatchery stocks and native populations. Different levels of hybridization and introgression with hatchery individuals were detected in stocked drainages as well as in protected locations.

POLYMORPHISMS

The technology for such gene hunting has been revolutionized several times over the past thirty years. In fact, some techniques considered "hot" only a decade are now obsolete. The remarkable progress has been due to one major phenomenon– the ability to detect and cheaply genotype "spelling variations" in the human genome.

Suppose that we gathered DNA from a large number of individuals and examined the first 1,000 nucleotides on the first chromosome. There would be some sections in which the base pair sequence is the same for all of the DNA strands. These are called monomorphic sections. There will be other sections in which the sequence of nucleotides differs. These are called polymorphic sections or, simply, polymorphisms. Using the analogy of a four-letter DNA alphabet, polymorphisms are really "spelling variations." The term allele refers to a specific spelling variation.

Types of Polymorphisms

Protein/Enzyme Polymorphisms

In the early days of human genetics, the majority of polymorphisms were those associated with proteins and enzymes. To detect the polymorphism and a person's genotype, one performed assays for the gene product, i.e., the protein or enzyme produced by the genetic blueprint.

Most of these polymorphisms were detected in blood. When your blood is typed, you are informed that you are blood group O+ or AB- or A+, etc. The letter in this blood group gives your phenotype at the ABO locus, and the plus (+) or minus (-) sign denotes your phenotype at the Rhesus gene.

A number of other loci such as Kell, Duffy, MN, and Kidd can also be phenotyped from blood. These polymorphisms are still used today to assess suitability of donors and recipients for blood transfusions (ABO locus) and to assess Rhesus incompatibility between a mother and her fetus. However, blood group polymorphisms have given way to other, more sophisticated techniques in modern human genetic research.

DNA Polymorphisms

The other large class of polymorphisms are those that detect spelling variations at the level of DNA nucleotides. For our purposes, we can classify them into three types:

Single Nucleotide Polymorphisms

A single nucleotide polymorphism or SNP is a sequence of DNA on which humans vary by one and only one nucleotide (see Fig. 1). Because humans differ by one nucleotide per every thousand or so nucleotides, there are millions of SNPs scattered throughout the human genome. The major advantage of SNPs, however, lies in the fact that they can be detected in a highly automated way using specialized DNA "chips" usually called DNA arrays.

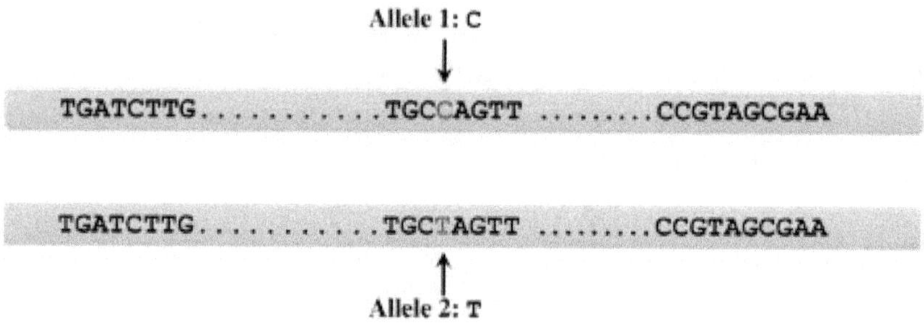

Fig. 1: Example of a single nucleotide polymorphism.

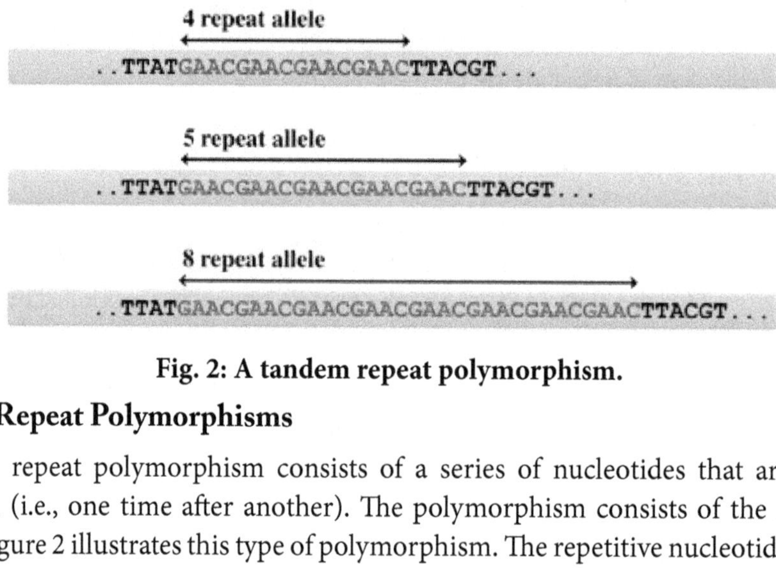

Fig. 2: A tandem repeat polymorphism.

Tandem Repeat Polymorphisms

A tandem repeat polymorphism consists of a series of nucleotides that are repeated in tandem (i.e., one time after another). The polymorphism consists of the number of repeats. Figure 2 illustrates this type of polymorphism. The repetitive nucleotide sequence is GAAC and the figure depicts three alleles–a four-repeat allele, a five-repeat allele, and an eight-repeat allele.

Tandem Repeat Terminology (graduate)

Unfortunately, even though the concept of the tandem repeat is quite simple, the terminology for referring to these polymorphisms can be confusing. When the number of repeats is small, usually five to six or fewer, then the polymorphism may be called a microsatellite, simple sequence repeat (SSR), or short tandem repeat (STR). When the number of repeated nucleotides is larger, then the polymorphism may be called a minisatellite, particularly when it is located in a telomere. Finally, the term variable number of tandem repeats (VNTR) polymorphism has been used equivocally. Sometimes it is generic and refers to any tandem repeat polymorphism. At other times, it refers to repeat with a large involving a large number of nucleotides.

Structural Variants

Here, we use the term structural variants to refer to spelling variations that involve deletions or insertions of a nucleotide sequence, inversions, and translocations. When the structural variant is somewhat large (some geneticists define "large" as 1 kilobase or more, others 10 kb), the polymorphism is called a copy number variant or CNV. There is considerable research being done on CNVs and medical disorders, including psychopathology.

Insertion-deletion polymorphisms or indels, an example of which is given in Fig. 3, are intuitive. Whether an allele is called an insertion or deletion, however, depends on the consensus nucleotide sequence of the human genome. If an allele is missing a nucleotide sequence that is present in the consensus sequence, then the allele is a called a deletion. If the allele contains a nucleotide sequence that is not in the consensus sequence, then it is an insertion.

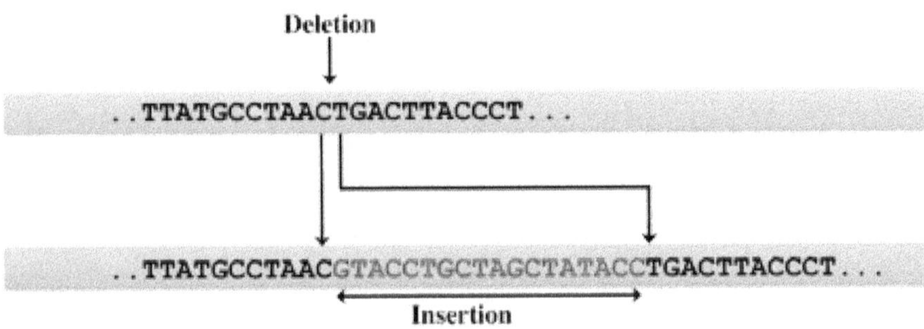

Fig. 3: An insertion-deletion polymorphism

A particularly important type of insertion occurs when a section of DNA is duplicated and inserted into the same region. These are sections of DNA with a nucleotide sequence very similar to a known gene but the DNA does not produce a functional polypeptide. Most pseudogenes resulted from duplications.

An inversion polymorphism occurs when one allele has a nucleotide sequence that is reversed in another allele. Fig. 4 presents an example. Assume that the spelling variation in the consensus sequence is the one on the top. The inverted allele has a section that has the same spelling but is "read in reverse" from right to left instead of the ordinary left to right order.

Inversions usually occur when a chromosome breaks in two places and DNA repair mechanisms mistakenly splice the middle fragment back but in reverse order. Typically they involve many thousands of base pairs. A translocation occurs when a section of DNA is deleted from one chromosome and then inserted into another chromosome.

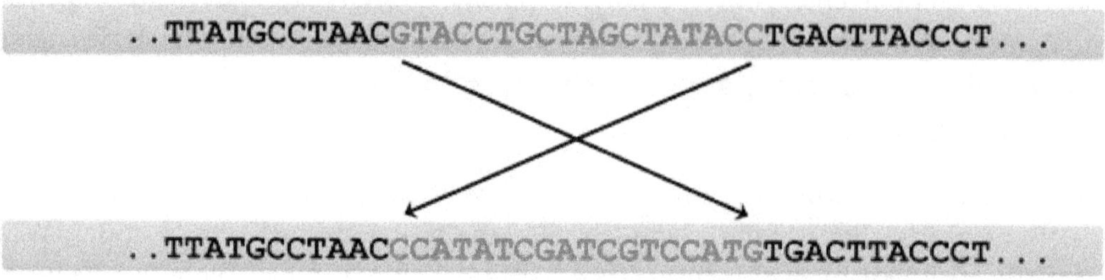

Fig. 4: An inversion structural variant

Sequence Polymorphisms

The ultimate polymorphism is to actually have the whole sequence of nucleotides for a region for a large number of DNA strands and then examine all of the differences among the strands. Here, the DNA differences could be a SNP, a tandem repeat or a structural change. There is no accepted term for this phenomenon, so we call them sequence polymorphisms. In effect, sequence polymorphisms subsume all known DNA polymorphisms.

7

DNA GENETIC MARKERS AND THEIR APPLICATION

INTRODUCTION

All organisms owing to their cellular and environmental properties are subjected to mutations, which ultimately lead to genetic variations (polymorphism). These mutations if occur within a particular nucleotide sequence might alter the whole amino acid string of a protein, result in a functionally new variant. In comparison to their original (wild type) counterparts, these variants might hold an increased or decreased metabolic efficiency, might perform poor or even get a new function. Similarly, gene expression may be pretentious by certain mutations in the regulatory regions, which alter the magnitude, and even pattern of expression e.g. turn on or off, over or under expression of proteins in certain tissues at various developmental or physiological stages of an organism.

Consequently, DNA markers have been applied widely to validate pedigrees, monitor populations in wild, to exploit multiplicity in founder animals and to display inbreeding level in reproductively effective populations; besides describe quantitative trait loci. Genetic diversity among the population ultimately provides evidence to their life histories and scale of evolutionary segregation.

Heritable variations are expressed as alterations in both the quality and quantity of alleles and gene arrangement on the chromosomes that are present within and among the basic populations. Molecular markers are in fact the powerful informational tools in hand which help us to reveal genetic rareness among individuals, species and populations. Even during practicing diverse aspects of aquaculture, molecular markers have provided valuable evidences like identification and discrimination of aquaculture stocks on genetic basis, examining the importance of onset and propagation of fish stocks in aquaculture, besides facilitating selection of breeding programs and evaluation of chromosomal and gene manipulations.

The aquaculture based fisheries management in most cases has been for the most part fruitful with the instant source of concern, for instance the large quantity and size of fish on hand during harvesting. Economically, this approach seems to be advantageous but in the end may result in the extermination of the respective populace. As part of a larger global concern, reduction of genetic resources in fish can in turn limit the genetic resources of the biosphere. It is therefore imperative to support molecular genetics for its indispensable role in sustainable management of fisheries genetic resources.

This approach will address conservation of both gene pools and genetic diversity; two slightly different aspects of genetic resources. Therefore, to develop an optimal strategy for effective management of genetic resources, genetic characterization of fish species in wild is of vital significance. Such a strategy once adopted will ensure the greatest economic benefits on one hand and will help to sustain natural populations on the other hand.

Fish biologists and aquaculturists should elucidate the significance of the factors and conditions; permitting maintenance of species and populations as such. Molecular tools, which are currently available, hold the promise of allowing identification of genes for traits like, adaptation, functional traits, growth, and disease resistance. This review discusses the development and use of these different types of molecular markers and examines the present status and future prospects in fisheries and aqua-farming.

Molecular markers can be characterized as Type I and Type II markers; Type I markers are associated with genes of known function and type II markers are associated with genes of unknown function. Allozyme markers are type I markers as the proteins they encode are associated with some functions. Microsatellites and other neutral markers are type II markers unless they are associated with genes of some known function.

APPLICATION OF DNA MARKERS IN AQUACULTURE AND FISHERIES

Allozyme

Allozyme electrophoresis is a method which can identify genetic variation at the level of enzymes that are directly encoded by DNA. Protein variants called allozymes originates from allelic variants and they will differ slightly in electric charge. Allozymes are codominant markers having been expressed in a heterozygous individual in a Mendelian way. Thus allozyme analysis provides us with data on single locus genetic variation which can answer many questions about fish and fish populations. To detect allozyme variation, the first step is to extract allozymes from tissues using specific protocols. Then the variation is detected through electrophoresis in an acrylamide or cellulose acetate gel. Individuals that are homozygous show a single band whereas heterozygous individuals show two bands. Allozymes are one of the most studied form of molecular variation due to their simplicity, low cost and the requirement of little specialized equipment. Any kind of soluble protein is suitable for allozyme analysis. A large number of loci can be screened at a time. The limitations of this technique include requirement of a large amount of tissue and consequently this method could not be applied when the organisms are small (for e.g.; larval forms). The tissue sampling method is invasive and so the fish needs to be sacrificed and the tissue needs to be stored cryogenically.

A point mutation in a nucleotide sequence may not result in a change in amino acid at all and thus could not be detected by protein electrophoresis. In addition to that, a change in DNA that results in a change in amino acid will not result in the overall charge of the protein and therefore is not detected. In spite of their limitations, the use of allozyme analysis has been widespread in fisheries mainly in fish systematics, population structure, conservation genetics, mixed stock fishery analysis and forensic analysis.

The general method for detecting allozyme variation includes extraction, electrophoresis and detection. Electrophoresis is a separation technique based on the motion of charged particles in an electric field. Tissue extracts are introduced into a solid support medium (a gel) at the Origin. As matrix (gel) for electrophoresis various media can be used such as acrylamide, cellulose acetate and hydrolyzed potato starch. Acrylamide typically has the best resolving power, however it is the most difficult to handle and also toxic, cellulose acetate electrophoresis is not only simpler and more rapid, but that it is also quite sensitive and provides good resolution.

Potato starch has been the predominant medium of choice particularly in applications requiring large amounts of data. Starch gels are easy to prepare and use can be sliced for the staining of 4-6 different enzymes, and 40-50 individuals can be analyzed per gel. When an electrical field is applied, most proteins have a net negative charge and will migrate from the Origin at the cathode ("negative") end towards the anode ("positive") end of the field. The positions of the protein products are detected either directly, or by coupled enzymatic reactions. For a monomeric enzyme, individuals that are homozygous each show a single band; those that are heterozygous show two bands.

The simplicity, speed, relatively low cost, little specialized equipment requirement and general applicability of the technique have made this the most widely studied form of molecular variation. Any source of soluble proteins, from bacterial cultures to animal fluids, is in principle suitable for allozyme analysis. The protocols of electrophoretic separation and staining are easily adjustable from species to species. No marker development phase is necessary and the analysis can start for any species immediately after samples have been collected from the field.

The genetic interpretation of allozyme profiles is also straightforward. It allows for screening a large number of loci, often more than 30-40. However, the technique has also some certain limitations. One major drawback has been the inability to read genotypes from small quantities of tissue, which makes allozymes inapplicable for small organisms (e.g. larvae). One disadvantage that appears to be difficult to overcome is that only a small fraction of enzyme loci appear to be allozymically polymorphic in many species. There are important demands concerning the freshness of tissue samples and many loci exhibit tissue-specific expression (e.g. some loci are only expressed in heart tissue).

Thus the invasive tissue sampling method, which requires sacrificing the fish and the need for cryogenic storage to preserve enzyme activity are serious constraints. In addition a given change in nucleotide sequence may not result in a change in amino acid at all, and thus would not be detected by protein electrophoresis. Furthermore, a change in the DNA that results in a change in an amino acid may not result in a change in the overall charge of the protein and, therefore, would also not be detected.

In spite of these limitations allozyme analysis has had a more profound effect on fisheries research and management than all most all other genetic tools. It has demonstrated that genetic markers can be useful in stock identification as evidenced by scores of studies that document differences in protein allele frequencies between stocks. As can be seen in the Practical Applications section below, there are wide spread applications of allozyme analysis in fisheries. These are; systematic, population structure, conservation genetics, mixedstock fishery analysis (e.g. salmonids), forensic (species, population and individual – specific identification for fisheries law enforcement), hybridization and inbreeding, individual identification and inheritance and genome mapping.

Mitochondrial DNA Markers

Mitochondrial DNA is non- nuclear DNA in the cell having located in within organelles in the cytoplasm called mitochondria. Mitochondrial DNA is maternally inherited with a haploid genome. The entire genome undergoes transcription as one single unit. They are not subjected to recombination and thus they are homologous markers. They are selectively neutral occurring in multiple copies in each cell. Mitochondrial DNA is physically separate from the rest of cell's DNA and so it is relatively easy to isolate from any tissue or blood sample. Due to the maternal inheritance of mitochondrial DNA, the effective population size is smaller than nuclear DNA and so mitochondrial DNA variation is more sensitive to population bottlenecks and hybridizations.

The differences in the nucleotide sequence of DNA molecule in the mitochondria can be determined directly or indirectly by several methods. Many population genetic studies have employed RFLP (Restriction Fragment Length Polymorphism) analysis of mitochondrial DNA for understanding population genetic variation either by digesting the whole purified mtDNA with restriction endonucleases or by DNA sequencing of small segments of mtDNA molecule obtained by PCR amplification. These techniques with increased resolution and maximum information have made mtDNA analysis very popular.

The newly emerged sequencing technologies have enabled direct sequencing of mitochondrial genes and several sets of universal primers have been developed from conserved sequence regions. Slow evolving gene regions are being used for inter species comparisons and fast evolving gene regions for population comparisons. The only non-coding region of mtDNA is D-loop region and this region is fast evolving and mostly used for population comparisons. The cytochrome b and ND-1 and ND-5/6 gene regions are also being used widely.

Mitochondrial Cytochrome C Oxidase I gene has been identified as the universal barcode for species level identification due to its conserved nature across a wide range of taxa. DNA barcodes are segments of approximately 600 base pairs of the mitochondrial COI gene which is a fast, efficient and inexpensive technique helpful in cataloguing the biodiversity. During the last two decades, mitochondrial DNA genes have found widespread application fish taxonomy, biology and population genetics.

mtDNA is so intensively studied and sequences in some parts of the molecule are highly conserved across species. Thus several sets of "universal primers" have been developed that allow for analysing the same mtDNA segments in a variety of species. In initial studies with fish (e.g. salmonids) universal primers of Cronin et al. for analysing the mtDNA ND-1 and ND-5/6 segments were used. Smaller fragments of the mitochondrial genome (Dloop region) have also been targeted by probing or PCR and findings have indicated that it may be best to concentrate on the 'slow evolving' coding sequences using 6 base cutters for species comparisons, and to use the 'fast evolving' non-coding regions with 4 base cutters for population investigations.

The D-loop region of the mtDNA is practically the only noncoding region in the entire mtDNA of vertebrates. Some markers have targeted the D-loop region is highly variable in mammals but for some fish species little polymorphism is shown in the D-loop. Studies with a number of fish species have actually shown less variability in the non-coding D-loop region than elsewhere in the mitochondrial genome. Thus, the cytochrome b and dehydrogenase genes may be examined more profitably.

Application of mtDNA in animals, including fishes has some major problems as well. The major disadvantages are the strict requirement of fresh or frozen tissue, the need for more sample than most DNA methods, the necessity to sacrifice the animals, and the low level polymorphism in some species and populations. The recent demonstration of the presence of mitochondrial pseudogenes in the nuclear genome of a wide range of organisms is, for population studies, an unwanted reality. The effectiveness of using mtDNA in population-genetic studies has been greatly weakened by this fact. In addition mtDNA data on their own have some important limitations.

Firstly, mtDNA represents only a single locus. So we can look only through a single window of evolution. This window reflects at best only the maternal lineal history, which could well differ from that overall of populations or species. Therefore, the inference we make on species/population history is likely to be highly biased and the need for independent, genomic molecular markers to support mtDNA analysis is clear. Second, the effective population size of mtDNA is only a fourth of that of nuclear autosomal sequences; that means mtDNA lineages have a much faster lineage sorting rate and higher allele extinction rate.

mtDNA have a number of applications in fisheries biology, management and aquaculture. In the past 15 years mtDNA has attracted a lot of attention in many species, especially for population and evolutionary studies. It has become a very popular marker and dominated genetic studies designed to answer questions of phylogeny and population structure in fish for more than a decade. mtDNA studies particularly can contribute to identification of stocks and analysis of mixed fishery, provide information on hybridization and introgression between fishes, serve as a genetic marker in forensics analysis and provide critical information for use in the conservation and rehabilitation programmes.

Arbitrary Nuclear DNA markers

Arbitrary markers are used when we target a segment of DNA of unknown function. The widely used methods of amplifying unknown regions are RAPD (Random Amplified Polymorphic DNA) and AFLP (Amplified Fragment Length Polymorphism) DNA. RAPD uses an arbitrary primer which can amplify anonymous loci. It is fast, cheap and shows very high amount of polymorphism and this marker does not require knowledge of the genetic makeup of the organism. The major drawback with RAPD markers is the lack of reproducibility and repeatability and the large number of products generated.

RAPD is a dominant marker and so homozygous and heterozygous states cannot be differentiated and these patterns are sensitive to slight changes in amplification conditions. Amplified Fragment Length Polymorphism (AFLP) markers combine the benefits of both RFLP and RAPD. The total genomic DNA is digested using two restriction enzymes. Double-stranded nucleotide adapters are ligated to the ends of DNA fragments to serve as primer binding sites for PCR amplification. Primers complementary to the adapter and restriction site sequence, with additional nucleotides at the 3'-end, are used as selective agents to amplify a subset of ligated fragments. The presence of absence of DNA fragments are detected on polyacrylamide gels and thus polymorphisms are studied.

Randomly Amplified Polymorphic DNA (RAPD)

RAPD is a random amplification of anonymous loci by PCR. It has several advantages and has been quite widely employed in fisheries studies. The method is simple, rapid and cheap, it has high polymorphism, only a small amount of DNA is required no need for molecular hybridization and most importantly, no prior knowledge of the genetic make-up of the organism in question is required. RAPD markers allow creation of genomic markers from species of which little is known about target sequences to be amplified. This methodology has some disadvantages which include difficulty in reproducing results, subjective determination of whether a given band is present or not, and difficulty in analysis due to the large number of products. This is because RAPDs are not sensitive to any but large-scale length mutations. Therefore, variation might be underestimated.

The technique is a based on the PCR amplification of discrete regions of genome with short oligonucleotide primers of arbitrary sequence. First of all RAPD is generated by PCR. A small amount of genomic DNA, one or more oligonucleotide primer (usually about 10 base pair in length), free nucleotides and polymerase with a suitable reaction buffer are major requirements. The main drawback with RAPDs is that the resulting pattern of bands is very sensitive to variations in reaction conditions, DNA quality, and the PCR temperature profile. Even if the researcher is able to control the major parameters, other drawbacks of RAPD will remain: homozygous and heterozygous states cannot be differentiated and the patterns are very sensitive to slight changes in amplification conditions, giving problems of reproducibility.

These problems have limited the application of RAPDs in fish studies. RAPDs have gained considerable attention particularly in population genetics, species and subspecies identification, phylogenetics, linkage group identification, chromosome and genome mapping, analysis of interspecific gene flow and hybrid speciation, and analysis of mixed genome samples, breeding analysis and as a potential source for single-locus genetic fingerprints.

RAPD analysis has been used to evaluate genetic diversity for species, subspecies and population/stock identification in guppy, tilapia, brown trout and Atlantic salmon, largemouth bass, Ictalurid catfishes, common carp and Indian major carps. Researchers found the technique useful in detecting diversity within and between strains of Oreochromis niloticus.

Amplified Fragment Length Polymorphism (AFLP)

AFLPs are dominant and, several markers and alleles are confounded in the same polyacrylamide gel. It combines the strengths of RFLP and RAPD markers and overcome their problems. The approach is PCR-based and requires no probe or previous sequence information as needed by RFLP. It is reliable because of high stringent PCR in contrast to RAPD's problem of low reproducibility. The major advantage of AFLPs is that a large number of polymorphisms can be scored in a single polyacrylamide gel without the necessity for any prior research and development.

AFLP seems to be much more efficient than the microsatellite loci in discriminating the source of an individual among putative populations. Similar to RAPD, AFLP analysis allows the screening of many more loci within the genome in a relatively short time and in an inexpensive way. Although the dominant nature of inheritance of AFLP markers currently limits their utility for applied population genetics, the method appears to be ideal for obtaining genomic support for intraspecific mtDNA analyses. The weakness is that they are dominant markers, thus on average half of which are useful for a given backcross reference family. The methodology is also difficult to analyze due to the large number of unrelated fragments that are visible (on the gel) along with the polymorphic fragments (as with RAPD's).

Specific Nuclear DNA markers

Variable Number of Tandem Repeat is a segment of DNA that is repeated tens or even hundreds to thousands of times in nuclear genome of eukaryotes. They repeat in tandem; vary in number in different loci and differently in individuals. There are two main classes of repetitive and highly polymorphic DNA; minisatellite DNA referring to genetic loci with repeats of length 9-65 bp and microsatellite DNA with repeats of 2-8 bp (1-6) long.

Microsatellites are much more numerous in the genome of vertebrates than mini satellites. They are widely used in population genetics of fishes and aquatic invertebrates. Minisatellites can be classified into multilocus and single-locus minisatellites. Multilocus minisatellites are composed of tandem repeats of 9-65 base pair and have a total length ranging from 0.1 to 7kb.

Minisatellite loci are used mainly in parentage analysis. They are less useful for population genetic analysis unless we use large sample sizes. The complexity of mutation processes undergone by minisatellite loci is also a limitation. Due to the difficulties in the interpretation of multilocus fingerprints, the research work were concentrated on single locus minisatellite probes and this procedure required reasonable quantities of high-quality DNA.

These single locus minisatellite probes have been very useful and successful in detecting genetic variations within and between populations. It has also been used in fisheries for forensics, parentage, genetic identity, estimating mating success and confirming gynogenesis.

Single Nucleotide Polymorphisms

Single nucleotide polymorphisms arise due to single nucleotide substitutions (transitions/ transversions) or single nucleotide insertions/deletions. These point mutations give rise to different alleles with alternative bases at a particular nucleotide position. SNP,s are the most abundant polymorphisms in the genome (coding and non-coding) of any organism.

These single nucleotide variants can be detected using PCR, microchip arrays or fluorescence technology. They are considered as next generation markers in fisheries and can be employed for population genetics studies, genomics studies and for detection of diseases.

DNA Microarrays or DNA Chips

DNA microarray consists of small glass microscope slides, silicon chip or nylon membranes with many immobilized DNA fragments arranged in a standard pattern. A DNA microarray can be utilized as a medium for matching a reporter probe of known sequence against the DNA isolated from the target sample which is of unknown origin. Species-specific DNA sequences could be incorporated to a DNA microarray and this could be used for identification purposes. DNA extracted from a target sample should be labeled with a specific fluorescent molecule and hybridized to the microarray DNA. When the hybridization is positive a fluorescent signal is detected with appropriate fluorescence scanning/imaging equipment.

Expressed Sequence Tags (ESTs)

ESTs are single-pass sequences which were generated from random sequencing of cDNA clones. ESTs can be used to identify genes and analyze their expression by means of expression analysis. Fast and reliable analysis can be made for the genes expressed in particular tissue types under specific physiological conditions or developmental stages. Differentially expressed genes could be identified using cDNA microarrays in a systematic way. ESTs are most valuable for linkage mapping.

APPLICATIONS OF MOLECULAR MARKERS IN FISHERIES

Inter Specific and Intra Specific Variations

Molecular genetic markers can be used as a supplementary marker system which will increase resolution in taxonomic research. The molecular evolution among taxa is highly variable and the extent of divergence in DNA or genes can be taken as the basis for differentiation among species. The morphological and ecological characters may diverge at a faster rate compared to genetic differentiation at neutral loci and so sometimes we may observe poor correlation between morphological traits and gene divergence. Thus molecular markers have to be interpreted critically among different sets of traits. Molecular markers are most useful when some species which occur in mixed catches is to be identified and where morphological identification is very difficult.

Processed fish products (filleted, smoked or salted) and early life stages of fishes like planktonic eggs and larval forms cannot be identified using morphological characters and molecular markers can be employed in such instances. Endangered and threatened whales, sharks and dolphins which are dead and stranded can also be identified using these methods as in most cases morphological identification is not possible. Within a species, genetic differences are more than between populations and so identification is possible even when sample size is small. Molecular markers have also been used in sub-species identification. Some of the high evolving loci will show more divergence within species and so these loci can be used for finding out intra-specific variations.

Phylogenetic and Phylogeograhical Studies

Phylogenetic studies assess the historical processes which affect relationships and phylogeographic studies assess the geographical distributions. Phylogenetic and phylogeographic studies started with the introduction of mtDNA markers in population genetic analyses. The evolutionary history of groups of fishes could be reconstructed which will give vital information regarding historical demography. Information regarding conservation units and ecological patterns could also be derived using phylogenetic studies. Mitochondrial DNA analysis has been used widely as a powerful tool for intraspecific phylogenetic patterns inference in many animal species. The high levels of mutation rate, smaller levels of effective population size and predominantly maternal inheritance of mtDNA will provide greater power to identify population structure. The lack of recombination and low efficiency of repair mechanisms induces high rate of evolution in mtDNA which makes this molecule highly useful in phylogenetic analyses. MtDNA has been used to resolve relationships among species that had diverged as much as 8-10 million years.

Genetic Tagging/Marking

There are instances when we need to mark individual fish for various purposes like tracking movement or migration, estimating population size or evaluating contributions of individual stock to mixed fishery. Physical tags are not heritable and so cannot be employed for generations. Genetic marking by finding a rare allele in an individual or populations and following them over generations will provide information regarding the contribution of hatchery programme on harvest, proportion of stocked individuals on growth of targeted population and identifying migrants from source populations.

Forensic Investigation

In samples where morphological identification is not possible like dead or stranded fishes, preserved or canned fish flesh and fish fillets, molecular markers could be employed for identification and certification. Forensics uses scientific methods to draw inferences about past events and it is being increasingly used in certification of fishery products and detection of illegal trading in fish and fisheries products. Deliberate or accidental release into natural waters also could be monitored using molecular tools.

Structure of Populations: Between and within Populations

Identification of stock structure is a very important issue in fisheries management and conservation programmes. Stocks are groups within species which are reproductively isolated with different physiological and behavioural patterns. Morphological and meristic features have genetic and environmentally influenced components and so morphological and meristic data should be used in conjunction with genetic data. Thus when a combination of these methods is used, we get information regarding actual genetic differences and important environmental effects on phenotype. Microsatellite DNA is the most favored DNA marker for stock structure studies due to their high rate of evolution. The non-coding region of mtDNA (D-loop) has also been used for stock structure studies due to their high rate of evolution. Mixed fisheries are comprised of subunits of different populations, different life stages or individuals from different stocks. Data from microsatellite markers, allozymes or mitochondrial DNA markers is useful for identifying the origin of stock components from mixed stock fisheries.

Applications in Aquaculture

Molecular markers have wide range of applications in aquaculture mainly; in genetic identification and discrimination of hatchery stocks, finding out inbreeding events, assignment of progeny to parents using genetic tags, finding out quantitative trait loci, marker assisted selection for selective breeding trials and assessment of the effect of polyploidy induction and gynogenesis. Genetic variability can also be assessed between and within stocks using molecular markers. They are also useful in determining the contribution of possible parents in mass spawning events. Genome mapping and identification of quantitative trait loci (QTL), (the locations of commercially important quantitative traits) is also possible. QTLs are important in breeding programmes and they are detected by analyzing phenotypes with linked marker maps and identification of markers related to QTLs can provide information on relatedness between strains and families and this knowledge is useful in marker-assisted programmes to improve production related traits. Disease diagnosis is another major area where the benefits of molecular markers can be successfully harnessed.

PCR assays for different pathogens have become inexpensive, safe and user friendly in many diagnostic laboratories. Several PCR based disease diagnosis methods have been developed for pathogens like white spot syndrome virus (WSSV), channel catfish virus (CCV), infectious hematopoietic necrosis virus (IHNV), infectious pancreatic necrosis virus (IPNV), viral hemorrhagic septicemia virus (VHSV), viral nervous necrosis virus (VNNV) and several other diseases. Genes that are resistant to pathogens like Major Histo Compatibility genes (MHC) can be identified and used for selection programmes to produce disease resistant strains of fin or shell fishes.

TYPES OF DNA MARKERS

"The genetic markers are often known as DNA marker is a known DNA sequence located in the genome, used in the molecular genetics for identification of alteration or DNA sequence from the unknown DNA sample."

The genetic marker is a known DNA sequence or gene located on the chromosome which can be applied in the identification of individual species or organisms or we can use it in the identification of other genes or DNA sequences.

Before that let me give you a brief idea about how different genetic markers are developed.

Restriction Fragment Length Polymorphisms (RFLPs)

RFLPs refer to variations found within a species in the length of DNA fragments generated by specific endonuclease. RFLPs are first type of DNA markers developed to distinguish individuals at the DNA level. RFLP technique was developed before the discovery of Polymerase Chain Reaction (PCR).

In the PCR-based RFLP method, the gene of our interest is amplified using the polymerase chain reaction and proceed for the restriction digestion using the specific REase.

- The results are analyzed using agarose gel electrophoresis.
- No tedious labeling and hybridization techniques are required in this method.
- The PCR-RFLP method is easy, simple, time-efficient and reliable.
- Polymorphism in the alleles of a gene such as MTHFR and HBS can be detected using PCR- RFLP.

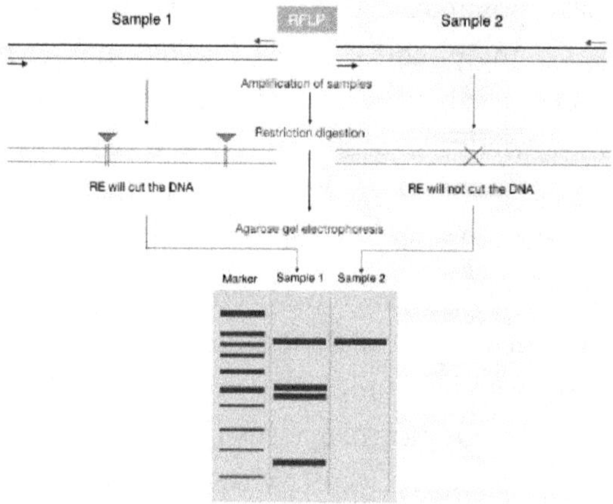

Fig. 1 Graphical illustration of Restriction Fragment Length Polymorphisms

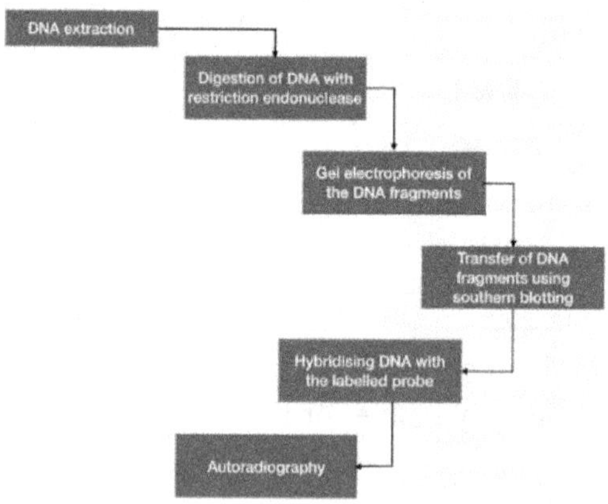

Fig. 2. Graphical illustration of different steps in the restriction fragment length polymorphism: RFLP

The advantages, disadvantages and uses of this technique are presented below:

Advantages:

RFLP technique has several advantages. It is a cheaper and simple technique of DNA sequencing. It does not require special instrumentation. The majority of RFLP markers are co-dominant and highly locus specific. These are powerful tools for comparative and synteny mapping.

It is useful in developing other markers such as CAPS and INDEL. Several samples can be screened simultaneously by this technique using different probes. RFLP genotypes for single copy or low copy number genes can be easily scored and interpreted.

Disadvantages

Developing sets of RFLP probes and markers is labour intensive. This technique requires large amount of high quality DNA. The multiplex ratio is low, typically one per gel. The genotyping throughput is low. It involves use of radioactive chemicals. RFLP finger prints for multi-gene families are often complex and difficult to score. RFLP probes cannot be shared between laboratories.

Uses

They can be used in determining paternity cases. In criminal cases, they can be used in determining source of DNA sample. They can be used to determine the disease status of an individual. They are useful in gene mapping, germplasm characterization and marker assisted selection. They are useful in detection of pathogen in plants even if it is in latent stage.

Amplified Fragment Length Polymorphism (AFLP)

AFLPs are differences in restriction fragment lengths caused by SNPs or INDELs that create or abolish restriction endonuclease recognition sites. AFLP assays are performed by selectively amplifying a pool of restriction fragments using PCR. RFLP technique was originally known as selective restriction fragment amplification.

The process of the AFLP is restriction digestion of the genomic DNA, adapter ligation of the digested DNA fragments and amplification of DNA fragments in PCR using the adapter specific primers.

The method was originally developed by the keygen in 1990.

The genomic DNA is extracted using one of the DNA extraction methods.

The DNA is digested with the help of the restriction endonuclease

The digested DNA is ligated with the short adapter sequences. The sequence of the adapters is known.

The fragments are amplified using a set of primers.

The results are analyzed on agarose gel electrophoresis followed by autoradiography or we can use the PAGE technique to differentiate multiple fragments of DNA.

The AFLP technique is highly applicable in Crop improvement programs and genomic analysis of different types of crop varieties.

As the primers are based on the adapter data, no previous data on the DNA sequences are required to construct a marker.

Due to the high priming capacity of the primers, it is also used in phylogenetic studies for different plant varieties.

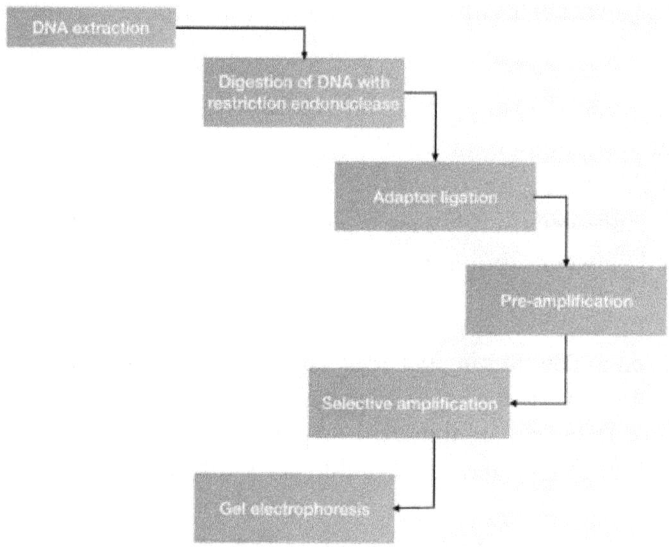

Fig. 3. Graphical illustration of different steps in amplified fragment length polymorphism

Advantages

It provides very high multiplex ratio and genotyping throughput. These are highly reproducible across laboratories. No marker development work is needed; however, AFLP primer screening is often necessary to identify optimal primer specificities and combinations.

No special instrumentation is needed for performing AFLP assays; however, special instrumentation is needed for co-dominant scoring.

Start-up costs are moderately low. AFLP assays can be performed using very small DNA samples (typically 0.2 to 2.5 pg per individual). The technology can be applied to virtually any organism with minimal initial development.

Disadvantages

The maximum polymorphic information content for any bi-allelic marker is 0.5. High quality DNA is needed to ensure complete restriction enzyme digestion. DNA quality may or may not be a weakness depending on the species. Rapid methods for isolating DNA may not produce sufficiently clean template DNA for AFLP analysis.

Proprietary technology is needed to score heterozygotes and ++ homozygotes. Otherwise, AFLPs must be dominantly scored. Dominance may or may not be a weakness depending on the application.

The homology of a restriction fragment cannot be unequivocally ascertained across genotypes or mapping populations. Developing locus specific markers from individual fragments can be difficult and does not seem to be widely done.

The switch to non-radioactive assays has not been rapid. Chemiluminescent AFLP fingerprinting methods have been developed and seem to work well.

The fingerprints produced by fluorescent AFLP assay methods are often difficult to interpret and score and thus do not seem to be widely used. AFLP markers often densely cluster in centromeric regions in species with large genomes, e.g., barley (Hordeum vulgare L.) and sunflower (Helianthus annuus L.).

Uses

This technique has been widely used in the construction of genetic maps containing high densities of DNA marker. In plant breeding and genetics, AFLP markers are used in varietal identification, germplasm characterization, gene tagging and marker assisted selection.

Random Amplified Polymorphic DNA (RAPDs)

RAPD refers to polymorphism found within a species in the randomly amplified DNA generated by restriction endonuclease enzyme. RAPDs are PCR based DNA markers. RAPD marker assays are performed using single DNA primer of arbitrary sequence.

By using the short, arbitrary DNA primers, the part of the genomic DNAs is randomly amplified.

Pronounced as a "rapid", it does not require any previous sequence information, the primer of 8 to 12mer randomly amplifies the region of the genomic DNA.

If any mutation or alteration occurred in the region of the primer binding site, the primer can not bind to that location or the amplified fragment length decreases or increases.

A different pattern of amplification is observed in different sample types. Hence the RAPD can help in the polymorphism studies.

Detection of RAPD

The RAPD fragments are between 0.2 to 5.0kb. It can be observed using the ethidium bromide-stained agarose gel or it can also be analyzed on polyacrylamide gel electrophoresis.

Advantages

RAPD primers are readily available being universal. They provide moderately high genotyping throughput. This technique is simple PCR assay (no blotting and no radioactivity). It does not require special equipment. Only PCR is needed. The start-up cost is low.

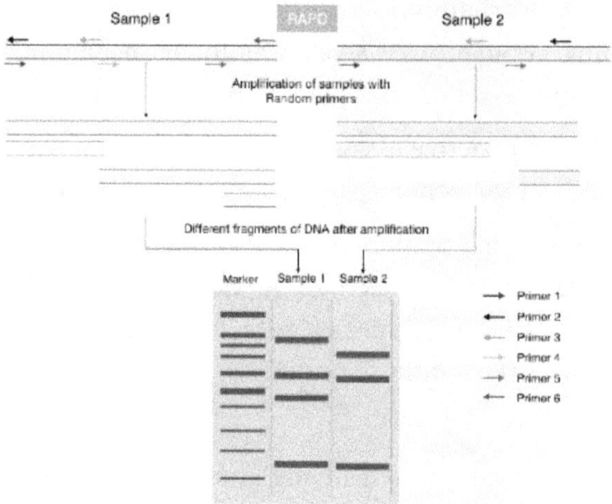

Fig. 4. The graphical illustration of different steps in the random amplified polymorphic DNA: RAPD.

RAPD marker assays can be performed using very small DNA samples (5 to 25 ng per sample). RAPD primers are universal and can be commercially purchased. RAPD markers can be easily shared between laboratories. Locus-specific, co-dominant PCR-based markers can be developed from RAPD markers. It provides more polymorphism than RFLPs.

Disadvantages

The detection of polymorphism is limited. The maximum polymorphic information content for any bi-allelic marker is 0.5. This technique only detects dominant markers. The reproducibility of RAPD assays across laboratories is often low. The homology of fragments across genotypes cannot be ascertained without mapping. It is not applicable in marker assisted breeding programme.

Uses

This technique can be used in various ways such as for varietal identification, DNA fingerprinting, gene tagging and construction of linkage maps. It can also be used to study phylogenetic relationship among species and sub-species and assessment of variability in breeding populations.

Cleaved Amplified Polymorphic Sequences (CAPS)

CAPS polymorphisms are differences in restriction fragment lengths caused by SNPs or INDELs that create or abolish restriction endonuclease recognition sites in PCR amplicons produced by locus-specific oligonucleotide primers.

CAPS assays are performed by digesting locus-specific PCR amplicons with one or more restriction enzymes and separating the digested DNA on agarose or polyacrylamide gels.

CAPS analysis is versatile and can be combined with single strand conformational polymorphim (SSCP), sequence-characterized amplified region (SCAR), or random amplified polymorphic DNA (RAPD) analysis to increase the chance of finding a DNA polymorphism.

Researchers proposed a variant of the CAPS method called dCAPS based on SNPs.

Advantages

The genotyping throughput is moderately high. It is a simple PCR assay. Markers are developed from the DNA sequences of previously mapped RFLP markers. Most CAPS markers are co- dominant and locus specific. No special equipment is needed to perform manual CAPS marker assays.

CAPS marker assays can be performed using semi-automated methods, e.g., fluorescent assays on a DNA sequencer (e.g., ABI377). Start-up costs are low for manual assay methods. CAPS assays can be performed using very small DNA samples (typically 50 to 100 ng per individual). Most CAPS genotypes are easily scored and interpreted. CAPS markers are easily shared between laboratories.

Disadvantages

Typically, a battery of restriction enzymes must be tested to find polymorphisms. Although CAPS markers still nave great utility and should not be over looked, other methods have emerged as tools for screening locus-specific DNA fragments for polymorphisms, e.g., SNP assays. The development of easily scored and interpreted assays may be difficult for some genes, especially those belonging to multi-gene families.

Uses

This is straightforward way to develop PCR-based markers from the DNA sequences of previously mapped RFLP markers. It is a simple method that builds on the investment of an RFLP map and eliminates the need for DNA blotting.

Simple Sequence Repeats (SSRs)

Simple sequence repeats (SSRs) or microsatellites are tandemly repeated mono-, di-, tri-, tetra-, penta-, and hexanucleotide motifs. SSR length polymorphisms are caused by differences in the number of repeats. SSR loci are individually amplified by PCR using pairs of oligonucleotide primers specific to unique DNA sequences flanking the SSR sequence.

Jeffreys (1985) showed that some restriction fragment length polymorphisms are caused by VNTRs. The name ”mini satellite” was coined because of the similarity of VNTRs to larger satellite DNA repeats.

Advantages

SSR markers tend to be highly polymorphic. The genotyping throughput is high. This is a simple PCR assay. Many SSR markers are multi-allelic and highly polymorphic. SSR markers can be multiplexed, either functionally by pooling independent PCR products or by true multiplex- PCR. Semi-automated SSR genotyping methods have been developed. Most SSRs are co-dominant and locus specific.

No special equipment is needed for performing SSRs assays; however, special equipment is needed for some assay methods, e.g., semi-automated fluorescent assays performed on a DNA sequences. Start-up costs are low for manual assay methods (once the markers are developed). SSR assays can be performed using very small DNA samples (~100 ng per individual). SSR markers are easily shared between laboratories.

Disadvantages

The development of SSRs is labor intensive. SSR marker development costs are very high. SSR markers are taxa specific. Start-up costs are high for automated SSR assay methods. Developing PCR multiplexes is difficult and expensive. Some markers may not multiplex.

Uses

SSR markers are used for mapping of genes in eukaryotes.

Single Strand Conformational Polymorphisms (SSCPs)

SSCPs refer to DNA polymorphisms produced by differential folding of single-stranded DNA harboring mutations. The conformation of the folded DNA molecule is produced by intra-molecular interactions and is thus a function of the DNA sequence.

SSCP marker assays are performed using heat-denatured DNA on non-denaturing DNA sequencing gels. Special gels (e.g., mutation detection enhancement gels) have been developed to enhance the discovery of single-strand conformational polymorphisms caused by INDELs, SNPs, or SSRs.

Advantages

It is a simple PCR assay. Many SSCP markers are multi-allelic and highly polymorphic. Most SSCPs are co-dominant and locus specific. No special equipment is needed. Start-up costs are low. SSCP marker assays can be performed using very small DNA samples (typically 10 to 50 ng per individual).

SSCP markers are easily shared between laboratories. SSCP gels can be silver stained (no radioactivity). The complexity of PCR products can be assessed and individual fragments can be isolated and sequenced.

Disadvantages

The development of SSCP markers is labor intensive. SSCP marker development costs can be high. SSCP marker analysis cannot be automated.

Uses

SSCPs have been widely used in human genetics to screen disease genes for DNA polymorphisms. Although SSCP analysis does not uncover every DNA sequence polymorphism, the methodology is straight forward and a significant number of polymorphisms can be discovered. SSCP analysis can be a powerful tool for assessing the complexity of PCR products.

Heteroduplex Analysis (HA)

It refers to DNA polymorphisms produced by separating homo-duplex from heteroduplex DNA using non-denaturing gel electrophoresis or partially denaturing high performance liquid chromatography.

Single-base mismatches between genotypes produce hetero-duplexes; thus, the presence of hetero-duplexes signals the presence of DNA polymorphisms. Heteroduplex analyses can be rapidly and efficiently performed on numerous genotypes before specific alleles are sequenced, thereby greatly reducing sequencing costs in SNP discovery and SNP marker development.

Advantages

It is a powerful method for SNP discovery. Automated HA can be performed using HPLC. Most heteroduplex markers are co-dominant and locus specific. HA can be performed using very small DNA samples (typically 10 to 50 ng per individual). HA markers are easily shared between laboratories.

Disadvantages

Requires special equipment. One protocol may not be sufficient for heteroduplex analyses of different targets via HPLC.

Uses

Heteroduplex analysis has been mostly used in human genetics to screen disease genes for DNA polymorphism. In plant breeding, it is used for detection of pathogens which are in latent stage and thus useful in selection of disease free plants. It is also useful in the discovery of single nucleotide polymorphism.

Single Nucleotide Polymorphism (SNP):

The variations which are found at a single nucleotide position are known as single nucleotide polymorphisms or SNP. Such variation results due to substitution, deletion or insertion. This type of polymorphisms has two alleles and also called bialleleic loci. This is the most common class of DNA polymorphism. It is found both in natural lines and after induced mutagenesis. Main features of SNP markers are given below.

1. SNP markers are highly polymorphic and mostly bialleleic.
2. The genotyping throughput is very high.
3. SNP markers are locus specific.
4. Such variation results due to substitution, deletion or insertion.
5. SNP markers are excellent long term investment.

6. SNP markers can be used to pinpoint functional polymorphism.
7. This technique requires small amount of DNA.

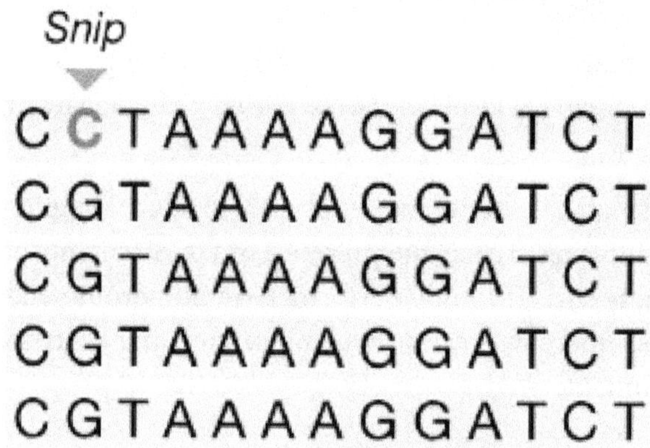

Fig. 5. Single Nucleotide Polymorphism

Advantages

SNP markers are useful in gene mapping. SNPs help in detection of mutations at molecular level. SNP markers are useful in positional cloning of a mutant locus. SNP markers are useful in detection of disease causing genes.

Disadvantages

Most of the SNPs are bialleleic and less informative than SSRs. Multiplexing is not possible for all loci. Some SNP assay techniques are costly. Development of SNP markers is labour oriented. More (three times) SNPs are required in preparing genetic maps than SSR markers.

Uses

SNPs are useful in preparing genetic maps. They have been used in preparing human genetic maps. In plant breeding, SNPs have been used to lesser extent.

Expressed Sequence Tags (EST)

Expressed Sequence Tags (ESTs) are small pieces of DNA and their location and sequence on the chromosome are known. The variations which are found at a single nucleotide position are known. The term Expressed Sequence Tags (ESTs) was first used by Venter and his colleagues in 1991. Main features of EST markers are given below:

1. ESTs are short DNA sequences (200-500 nucleotide long).
2. They are a type of sequence tagged sites (STS).
3. ESTs consist of exons only.

Advantages

It is a rapid and inexpensive technique of locating a gene. ESTs are useful in discovering new genes related to genetic diseases. They can be used for tissue specific gene expression.

Disadvantages

ESTs have lack of prime specificity. It is a time consuming and labour oriented technique. The precision is lesser than other techniques. It is difficult to obtain large (> 6kb) transcripts. Multiplexing is not possible for all loci.

Uses

ESTs are commonly used to map genes of known function. They are also used for phylogenetic studies and generating DNA arrays.

Sequence Tagged Sites (STS)

In genomics, a sequence tagged site (STS) is a short DNA sequence that has a single copy in a genome and whose location and base sequence are known. Main features of STS markers are given below.

1. STSs are short DNA sequences (200-500 nucleotide long).
2. STSs occur only once in the genome.
3. STS are detected by PCR in the presence of all other genomic sequences.
4. STSs are derived from cDNAs.

Advantages

STSs are useful in physical mapping of genes. This technique permits sharing of data across the laboratories. It is a rapid and most specific technique than DNA hybridization techniques. It has high degree of accuracy. It can be automated.

Disadvantages

Development of STS is a difficult task. It is time consuming and labour oriented technique. It require high technical skill.

Uses

STS is the most powerful physical mapping technique. It can be used to identify any locus on the chromosome. STSs are used as standard markers to find out gene in any region of the genome. It is used for constructing detailed maps of large genomes.

STR: Short Tandem Repeats

The short tandem repeats/ simple sequence repeats are also called a microsatellite. The microsatellite is the shorter repeats sequences present into the genome thousands or millions of times.

In other words, we can say that the STRs are tandem repeats but are much shorter than the VNTR. The repeats in STR are 2 to 6 base pairs per repeat.

The STR is highly polymorphic, the number of repeats differs between individuals even though the loci are homologous. Because of this reason, the STR marker is the first choice in forensic analysis.

Due to the difference in the repeat number each individual has a different STR profile.

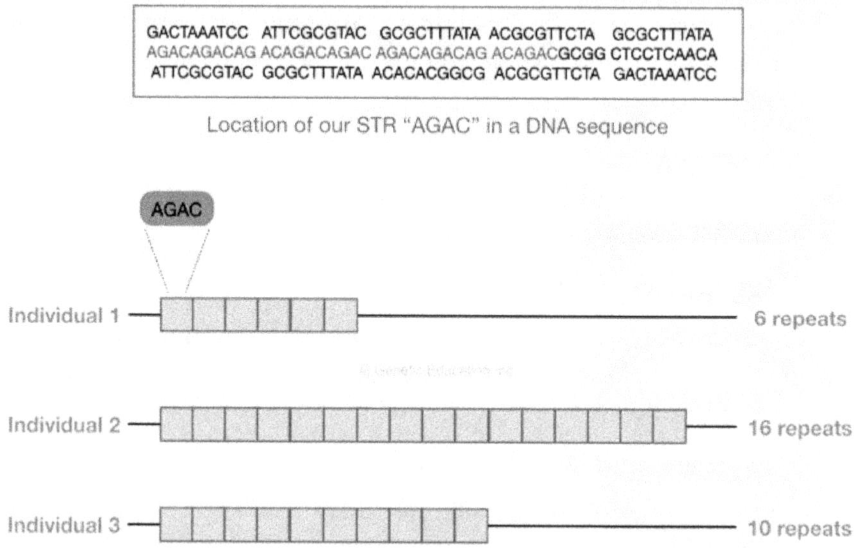

Fig. 7. The graphical illustration of the STR: short tandem repeats.

PARENTAGE DETERMINATION

Technical Aspects of Parentage Assignment

Aquaculture is now the fastest growing animal production worldwide, and provides half of the fish for human consumption worldwide. Such an important sector would be expected to use the best knowledge-based improvement methods, amongst which selective breeding is of paramount importance. However, researchers estimated that only 10% of aquaculture production worldwide is based on genetically improved stocks. There may be several reasons for this, but one clear technical weakness of aquaculture regarding the development of optimized selective breeding schemes is the fact that pedigree information is difficult and costly to obtain.

The Basic Reason is Rather Straightforward: farmed aquatic animals are all too small at hatching (from a few micrograms in mollusks and crustaceans to ca. 100 mg in salmonids fishes) to be physically tagged.

Parentage Assignment Methods

Basically, two computation methods are used for parentage assignment, exclusion-based methods and likelihood-based methods. Exclusion is very simple and makes no hypotheses other than Mendelian segregation of alleles, but is very sensitive to genotyping errors. When error rates are moderate and theoretical assignment power is high, however, genotyping errors can be dealt with by allowing a limited number of allelic mismatches between an offspring and its parents alleles, and exclusion remains the gold standard of parentage assignment. Exclusion programs used in aquaculture are PROBMAX, VITASSIGN, and FAP.

Likelihood methods use a different approach, with probabilities. In this case, the most likely couple is chosen as the true one (eventually integrating a genotyping error rate), but the decision rules rely on hypotheses on allelic frequencies. Likelihood methods generally give higher assignment rates than exclusion with low power marker sets, but sometimes give inconsistent results. Using sibship information in calculations can greatly improve the efficiency of the likelihood methods. Likelihood programs used in aquaculture are CERVUS, PAPA, and PARENTE.

A specific question is also the assignment of polyploids, especially sturgeons or induced polyploids, and specific packages have been developed for tetraploids (wHDP), diploids to octoploids (VITASSIGN-OCTO), as well as a general method to transform polyploid genotypes to pseudo-diploid dominant genotypes.

A crucial Issue: The Assignment Power of Markers Used

However, whatever the method used, the first requirement to be able to use parentage assignment in practice is to obtain high levels of unique assignments, which primarily depends on the assignment power of the marker set used. It depends on the exclusion probabilities of the markers used and on the size of the problem to be solved, the total number of putative parents having an exponential effect on the proportion of unassigned individuals. Overestimation of the assignment power of markers is very frequent, and can be explained by Hardy–Weinberg disequilibrium, sampling variance and relatedness of parents, incomplete genotypes, genotyping errors especially caused by stuttering or size-shift, and null alleles. In some species groups like mollusks, null alleles may be extremely frequent and problematic, but the main cause of overestimation of the theoretical assignment power is a widespread inappropriate calculation method. Typically, assignment power >0.99 can generally be obtained by 8–15 microsatellite markers in fish crosses involving a few tens or hundreds of parents, and a reasonable option when designing a marker set is to include a few more markers than theoretically needed. This then spares a lot of time by providing easy assignment even if small problems of genotyping errors, inbreeding or null alleles appear. High quality genotyping is also essential, and a recent review gives very useful insights to this question.

Microsatellites and SNPS for Parentage Assignment

Microsatellites, due to their high number and high variability, are the markers that allowed the development of efficient parentage assignment methods. Today, however, SNPs (single nucleotide polymorphisms) use is growing exponentially, but not yet in parentage assignment. It was estimated that ~6 SNPs give the same assignment power as 1 microsatellite. Empirical studies tend to suggest that the adequate number of SNP for an efficient panel would be in the 100–450 range. With such numbers, the classical requirement of unlinked markers within a panel cannot be met, thus lowering the real assignment power. SNPs are individually less expensive to genotype than microsatellites, but multiplexing decreased the cost of microsatellites genotyping, and for the moment SNPs remain more expensive due to the number required, but technology is rapidly evolving for SNPs and not for microsatellites. Empirical studies also sometimes reveal quite a high number of genotyping error in SNPs and the necessity to test a higher number of SNP markers than expected to select the appropriate ones. However, prospects for development of genomic selection with low-marker density may imply genotyping of a few hundred to several thousand SNPs per fish, which in this case would be sufficient to provide parentage assignment at no additional cost. The recent shift to SNP markers was, however, efficient to improve assignment at least in some mollusks species which suffered from high numbers of null alleles with microsatellites.

Implementation of Parentage Assignment in Aquaculture

Inbreeding control

Mass selection is the simplest way to improve traits such as growth or morphology, but bears a high risk of rapid genetic loss, with highly unbalanced families, which was revealed by parentage assignment mostly in mass spawning species, but also in controlled artificial reproduction systems.

The impact of different factors (mating design, mating ratio, number of parents per generation, selection pressure, trait heritability, grading practices) were simulated to improve inbreeding control and optimize genetic progress, including optimal contribution selection, which requires pedigree knowledge.

Estimation of Genetic Parameters

Estimation of heritability and genetic correlations allows to evaluate expected genetic gains and to design breeding programs. This is maybe where the possibility to access pedigree information by genotyping gave the most important and fruitful contribution to date to aquaculture genetics.

Optimization of mixed family designs for genetic parameters estimation was done by for strain effects, for heritability, and for genotype by environment (G × E) interaction. After several feasibility studies with few families, heritabilities were estimated in a growing number of fishes species for growth, processing traits, flesh color, muscle fiber diameter, deformities, disease resistance or sex ratio, and in shrimps and mollusks for growth or meat yield in mussel.

Heritabilities obtained in mixed family rearing are often higher than those recorded in separate rearing, which may be linked to the general absence of between-family environmental variance due to family mixing, although non-genetic maternal effects may persist in mixed family rearing, with possible upward biases on heritability estimates. Explicit comparisons of the same families in mixed or separate rearing design concluded that separate family rearing induced much higher levels of between-families environmental effects.

Mixed family rearing also allowed the estimation of G × E interactions between rearing systems, density or rearing temperatures, plant-based vs marine feeds, and even separated vs mixed fish family rearing designs.

A limiting factor of such studies is that as fish are generally tagged to maximize individual information collection, individual performances are not available before physical tagging, thus limiting genetic studies on early stages. However, recent advances allow individual tagging at 200–400 mg, which should change this matter of fact.

Implementation of breeding programs

Concepts Used and Implementation

The first proposal to use parentage assignment in breeding at an acceptable cost was an improvement of within-family selection called "walk-back selection". A two-step process of assignment was suggested and tested to achieve a minimal number of selected candidates per family).

Since this date, public organizations and breeding companies initiated selection programs using parentage assignment in sturgeons (France, USA), Atlantic salmon (Ireland, Norway, Scotland), tilapia (Philippines), halibut (Norway and Scotland), rainbow trout (France), cod (Norway), gilthead sea bream (France, Greece, Spain), turbot (France), European sea bass (France, Greece), meagre and red drum (France), Asian sea bass (Singapore, Indonesia, Australia), and shrimps (Australia, Thailand, Mexico, Equator, Central and South America). This list may be incomplete and represents the present informal expert knowledge of the authors. Little information is publicly available in these programs but mass selection, family-based selection (often BLUP: best linear unbiased prediction) or a combination of both are used to improve growth, processing yields, quality traits and disease resistance according to different schemes.

Key parameters to choose to develop a breeding program using parentage assignment are not only the genotyping cost (12–20 Euros per individual), but also the capacity to produce a large number of families in one batch to avoid tank effects, the true assignment efficiency, as well as the availability of tools such as rapid mass genotyping capacities (specially for species with short generation interval), individual tagging to improve traceability and facilitate data collection, automated database systems to collect, store and link performances to tags, DNA samples and pedigrees, optimized genetic softwares to rank and mate candidates to maximize genetic progress and minimize inbreeding. Use of parentage assignment is not only "genetic tagging," but requires a complete re-optimization of breeding programs.

Optimization of Breeding Schemes Using Parentage Assignment

One main target for optimization has been the limitation of numbers genotyped, using two-way nested models for partial pedigrees, or extreme phenotypes with family effect considered as a fixed effect. BLUP selection normally requires the knowledge of performance and pedigree on all candidates. In these cases, the loss of selection efficiency (compared to BLUP with pedigree known on all candidates) depends on selection intensity and genetic parameters. In addition, issues linked to mixing of families were studied, such as methods to limit non-genetic maternal effect in salmonids, effect of grading practices to limit cannibalism on family contributions in barramundi, and the importance to consider male maturation status to estimate heritability of growth more accurately. and compared expected genetic gains with different systems of evaluation (mixed/separate families and impact of G × E interactions), while proposed application of ultrasound tomography to predict processing yields on live candidates to limit the use of slaughtered sibs.

GENETIC MAPPING

Introduction

Genetics is defined as the study of heredity and embodied in the definition of heredity is the transmission of genetic characters from parents to offspring. Among the several definitions that exist for a map, the most appropriate is the concept that a map is a two-dimensional reconstruction of a part of the earth or heavens. In its totality, a genetic map does portray the ordering and landmarks of the physical entities we know as chromosomes. But in and of itself any map, including a genetic map, can only direct the observer to a specific location or place. It cannot convey to the observer the nature of that location or the activities that occur at that location. To attain an appreciation for the location we recognize that is necessary to actually visit and observe the location. As a biological metaphor genetic maps are only the genesis for understanding the evolutionary complexities of any species.

Genetic maps have been used successfully to find the gene responsible for relatively rare, single-gene inherited disorders such as cystic fibrosis and Duchenne muscular dystrophy. Genetic maps are also useful in guiding scientists to the many genes that are believed to play a role in the development of more common disorders such as asthma, heart disease, diabetes, cancer, and psychiatric conditions.

To produce a genetic map, researchers collect blood or tissue samples from members of families in which a certain disease or trait is prevalent. Using various laboratory techniques, the scientists isolate DNA from these samples and examine it for unique patterns that are seen only in family members who have the disease or trait. These characteristic patterns in the chemical bases that make up DNA are referred to as markers.

DNA markers don't, by themselves, identify the gene responsible for the disease or trait; but they can tell researchers roughly where the gene is on the chromosome.

This is why: when eggs or sperm develop, the paired chromosomes that make up a person's genome exchange stretches of DNA. Think of it as a shuffling process, called recombination. The single chromosome in a reproductive cell contains some stretches of DNA inherited from the person's mother and some from his or her father.

If a particular gene is close to a DNA marker, the gene and marker will likely stay together during the recombination process, and they will likely be passed on together from parent to child. If each family member with a particular disease or trait also inherits a particular DNA marker, it is very likely that the gene responsible for the disease lies near that marker.

The more DNA markers there are on a genetic map, the more likely it is that at least one marker will be located close to a disease gene-and the easier it will be for researchers to zero in on that gene. One of the first major achievements of the HGP was to develop dense maps of markers spaced evenly across the entire human genome.

Gene mapping is a hierarchical process and utilizes a number of different biological techniques to gain information on the genome. In this paper, some of the techniques that are currently used in the study of genomics will be outlined, and an overview of various types of genetic maps will be presented with consideration given to the source material and markers that are needed to construct these maps. Based upon the knowledge of genetic marker map orders, it is possible to relate physical chromosome morphologies to the linear order of markers on the chromosomes via gene-centromere mapping distances. This methodology will be discussed in relation to its utility in defining the location of centromeres on genetic maps.

In the latter sections of the paper, an overview of the applications of genetic mapping will be given. Included in those sections will be a discussion of gene transmission dynamics not only within species, but also between the sexes within a species. Highlighted, will be the use of a genetic map and its representative markers in investigating and understanding the process of adaptation, survival, and evolution (i.e., the raison d'etre of genetic maps).

Map Construction

Genomes may be mapped at three different levels ranging from physical maps that localize large DNA segments onto the cytological karyotype of a species, to genetic maps that depict linear marker orders along a chromosome, and finally, by cataloging and sorting raw DNA sequence along a chromosome. Genetic maps are the most common and are typically composed of linear marker arrangements depicting various linkage groups. Such marker linkage maps are generally produced using inbred/ semi-inbred or outcrossed lines in an intercross or backcross design, where F1 parents are produced by mating two strains that are divergent for some character of interest in a quantitative trait locus (QTL) study.

Haploid and diploid gynogenetic or androgenetic family lines have also been used for mapping in fishes. Some of the programs also perform QTL analyses on the same data set but require that the phase of the markers be specified in the mapping families. When access to fully inbred lines is not available then the method of producing doubled haploid progeny provides an efficient way to generate a genetic map using fully homozygous progeny lines. By producing mitotic diploids from two lines and then hybridizing the lines, the F1 may be used to re-establish haploid progeny lines in the following generation. Linkage assignments will be completely in phase as only homozygous progeny genotypes are produced and the allelic origin can be traced to either donor strain (if the alleles differ between the strains).

Most mapping studies in fishes are currently focused at the second level of genetic map organization (i.e., the construction of marker linkage maps). A marker linkage map may be defined as a landmark map that is composed of anonymous DNA markers (e.g., most microsatellite markers, AFLP (amplified fragment length polymorphism), RAPD (randomly amplified polymorphic DNA), SNP (single nucleotide polymorphism) markers), or known gene sequence regions marking either introns or exons (e.g., SNP marker derived from an EST (expressed sequence tag) or designed from partially homologous DNA sequence information). In general any marker that identifies a known expressed gene region (e.g., insertion/deletion (INDEL) polymorphism, microsatellite repeat, SNP) may be defined as an expressed sequence marker polymorphism (ESMP). If the marker identifie s a known gene using the polymerase chain reaction (PCR), amplifying DNA through a known intron-exon boundary with subsequent detection of an ESMP, the marker may also be classified as an EPIC (exon primed intron capture) marker. Markers may also be developed as STS (sequence tagged sites) from cloned or PCR amplified DNA.

By reference back to our original analogy, we must recognize that a marker linkage map is largely incomplete. It can give us a reference to landmarks along the physical surface of a chromosome, but it is largely characterized by unlinked sequence data spanning 1 cM (a centiMorgan) or more. A centiMorgan is defined as 1% genetic recombination between a pair of markers, and the actual number of base pairs embodied in this distance is dependent upon the species being examined, and even the linkage group being examined.

Certain regions of chromosomes may have higher recombination rates than others, and this will make the actual translation of estimated recombination distances into physical map distances variable throughout the genome. A sequenced linkage map of overlapping cloned fragments (contigs) that is characterized as containing large tracts of assembled sequence is generally referred to as a sequenced genome rather than a map.

Therefore, the term map will be used to refer to a collection of genetic markers. The attainment of even a draft status of a sequenced genome requires a large investment of time and money to perform the necessary multiple sequencing runs on a large number of cloned fragments from the genome.

Genetic maps are simply built on the examination of recombination differences among pairs of genetic markers transmitted by a given parent. Implicit in this is the use of doubly heterozygous markers that uniquely mark four chromosome segments in the genome. Each genetic marker position will of course identify one homologous pair of chromosome segments that possess a different allelic marker on each chromosome arm.

If the other pair of marker alleles being compared are also derived from the same set of homologs (i.e., from an identical linkage group) then the recombination levels between any pair of linked markers will be reduced inversely to the physical distance they are separated on a chromosome arm.

This will be detected as a significant deviation in the expected 1:1:1:1 segregation pattern of genotypic combinations, with parental coupling alleles being in excess to non-parental recombination alleles. In the absence of information on the phase of the markers transmitted by a parent, the phase may be inferred by comparing the genotypic combinations that occur in the highest frequency (i.e., if the genotypes being compared are Aa at locus 1 and Bb at locus 2, then either AB and ab genotypes must occur in the highest frequency, or Ab and aB gametes must occur in the highest frequency). Theoretically, only those markers that are physically close enough to one another to escape recombination via chiasmata cross-overs will map syntenically. Chiasmata cross-overs involve exchange between only one pair of non-sister chromatids that are present at meiosis I.

Markers that are physically close to one another have a reduced probability of having parental phases disrupted during meiosis, and thus, are expected to transmit a high frequency of parental (phase intact) genotypes to their progeny. In contrast, markers that are located far from one another on a linkage group may experience a cross-over event in 100% of the meioses generated. However, if chiasma interference is complete during meiosis (i.e., the formation of one chiasmata junction inhibits the subsequent formation of other junctions) then only two of the four chromatid strands present during meiosis I will experience a cross-over event. Thus, 50% of the potential gametes formed at meiosis II should contain only non-recombinant parental marker orders. Assuming that cross-over gametes are transmitted in equal frequency to non-cross gametes, then total map length is expected to be fairly accurately represented by additive pairwise recombination distance estimates.

For sets of markers that are located less than 50 cM from one another (i.e., less than 50% recombination for any pairwise combination) map distances are expected to be additive for any three point recombination estimate (e.g., A " B + B " C = A " C, if B is the intercalary marker).

Double cross-overs involving the same pair of non-sister chromatids may also occur during meiosis I, which will disrupt a linear map order by shortening the recombination distance estimates among proximally and distally located markers along a linkage group. Non-random or preferential transmission of parental phase chromatids compared to recombinant chromatids will also shorten the apparent map distances between distally located markers and may mimic the results obtained from double crossover events (see section on Distribution of cross-overs along rainbow trout chromosomes).

Mapping functions such as Haldane's and Kosambi's mapping function (see Ott, 1999 for their formulations) account for the shortened map distances that may occur between any pair of linked markers due to double cross-over events. These mapping functions readjust observed recombination distances upward, thus generating a larger estimate of total genome size than would be obtained by directly summing recombination intervals among linked markers.

The Haldane mapping function gives larger interval readjustments than the Kosambi function. Dense marker maps saturated with markers spanning only 1–2 cM intervals are largely equivalent in terms of their genome size estimates regardless of the mapping functions used.

Regional differences in recombination rates are known to bias the direct extrapolation of recombination distances to physical map distances. For example, recombination rates appear to be suppressed around the centromere and increase towards the telomeres. Also shorter chromosome arms appear to have proportionately greater map distances and this appears to be a function of maintaining chiasmata junctions in all paired chromosome arms. By virtue of their smaller physical length, short chromosome arms may experience a greater cross-over rate (per physical DNA length) than longer arms.

Genetic map construction using information from the segregation of genetically variable markers is essentially a four stage process. The first stage is composed of scoring progeny genotypes in a source mapping family, and then entering the genotypes into a computer program for analysis of the segregation patterns obtained. The second step is to ascertain the linkage arrangements among the markers tested. Implicit in this analysis is the fact that only one contributing parent is scored for segregation and linkage.

The use of androgenetic or gynogenetic haploids or diploids makes this a requisite, and when inbred lines are used, F1 hybrids between the inbred lines may be backcrossed to one of the homozygous inbred parents resulting in scored segregation of heterozygous loci from the single hybrid parent. Hybrids may also be intercrossed, although the phase assignment of heterozygous progeny genotypes in such crosses is not possible (i.e., the parental origin of the alleles cannot be ascertained). In outcrossed families, however, both parents may be heterozygous for different alleles at marker loci making phase assignments possible.

When groupings of genetic markers have been statistically identified the markers belonging to any one group may be tentatively regarded as belonging to a linkage group within the species. It is not unusual to initially have more linkage groups identified in the species than may be accommodated by the actual known haploid number of chromosomes in the species. This is due to the fact that clusters of 2 or 3 markers may occur that remain unlinked due to low marker density. These clusters may occur at opposite ends of larger linkage groups, and they will eventually be joined as intervening markers in the linkage group are genotyped.

Once a tentative linkage group has been identified it is necessary to ascertain the map order of the markers along the linkage group. For this analysis, markers are sorted into separate linkage group files, and pairwise recombination distances between the markers in the linkage group may be used to sort the markers into their linear sequence along the linkage group.

For outcrossed families, this needs to be done separately for each parent. Once the map order is ascertained, the map distances between the markers on the linkage group can be estimated using either the raw observed recombination levels or adjusted levels based upon a given mapping function (i.e., Haldane or Kosambi). Some of the existing computer programs (e.g., CARTHAGENE, JOINMAP, and LINKMFEX may also facilitate the construction of genetic maps generated from multiple mapping families or between mapping parents, provided that two or more common markers are used to 'anchor' linkage orientations within either source linkage group.

Radiation hybrid maps may also be used to construct gene maps (e.g., in zebrafish. Typically small fragments of the genome of any species are established in cell lines. These cell lines are then screened for the presence or absence of specific markers. The absence of co-expression for any given pair of markers is taken as evidence for the localization of these markers in different linkage groups (or more distal regions of a linkage group) compared to markers that are co-expressed in the same hybrid cell line. The map distance used is the cR (or centiRad) which denotes the percentage of cell lines where any two markers are not co-expressed. Radiation hybrid maps have proven to be somewhat more inaccurate in the placement of markers on human linkage groups compared to assignments made using classical genetic map.

Physical mapping involves in situ hybridization of cloned segments of genomic DNA to their physical location on a chromosome. Although at the lowest level of genomic discrimination, physical mapping is extremely useful in terms of giving researchers insights into structural arrangements within the genome (i.e., localizations of repetitive DNA (e.g., distribution of specific SINE (short interspersed nuclear elements) families among chromosomes, identification of multi-gene family locations in the genome (e.g., rDNA clusters), localization of linkage group specific repetitive clusters (e.g., chromosomal paint probes)). In situ hybridization methods can also be of great use in verifying the relative location of specific candidate genes that have been localized to large cloned DNA fragments (e.g., BAC clones). Through the integration of genetic map information with the localization of selected markers from the genetic map, the confirmation of genetic map orders to specific chromosomal landmarks can be confirmed (i.e., the localization of terminal markers in a genetic map linkage group to telomeres and markers with small gene-centromere recombination distances to their relative location on the chromosome). Markers with apparent small gene-centromere map distances may in fact prove to be variable in their placement along chromosome arms dependent upon the frequency and localization dynamics of chiasmata formation among different linkage groups.

IDENTIFICATION OF GENETIC SEX

A sex-determination system is a biological system that determines the development of sexual characteristics in an organism. Most sexual organisms have two sexes. In many cases, sex determination is genetic: males and females have different alleles or even different genes that specify their sexual morphology. In animals, this is often accompanied by chromosomal differences. In other cases, sex is determined by environmental variables (such as temperature) or social variables (the size of an organism relative to other members of its population). The details of some sex-determination systems are not yet fully understood

The XX/XY sex-determination system is the most familiar sex-determination systems. In XY/XO system, females have two copies of the sex chromosome (XX) but males have only one (X0). The ZW sex-determination system is found in birds and some insects and other organisms many other sex-determination systems exist. In some species of reptiles, including alligators, some turtles, and the tuatara, sex is determined by the temperature at which the egg is incubated. Some species, such as some snails, practice sex change: adults start out male, and then become female. In tropical clown fish, the dominant individual in a group becomes female while the other ones are male, and blue wrasse fish are the reverse. In Bonellia viridis, larvae become males if they make physical contact with the female and females if they end up on soil.

Some species, however, have no sex-determination system. Hermaphrodites include the common earthworm and certain species of snails. A few species of fish, reptiles, and insects reproduce by parthenogenesis and are female altogether.

In some arthropods, sex is determined by infection, as when Bacteria of the genus Wolbachia alter their sexuality; some species consist entirely of ZZ individuals, with sex determined by the presence of Wolbachia.

Temperature-Dependent Sex Determination

Temperature-dependent sex determination (TSD) is a system in which the temperature the eggs experience determines the sexes of the organisms that hatch. It is most prevalent and common among amniote vertebrates that are classified under the reptile class, but is also used among some birds, such as the Australian Brush-turkey. It also happens in some fishes such as salmon, gold fish, tilapia etc. It differs from the chromosomal sex-determination systems common among vertebrates. It is a type of environmental sex determination (ESD); in other ESD systems, some factors such as population determine the sex of organisms.

The eggs are affected by the temperature at which they are incubated during the middle one-third of embryonic development. This critical period of incubation is known as the thermosensitive period (TSP). The specific time of sex-commitment is known due to several authors resolving histological chronology of sex differentiation in the gonads of turtles with TSD

A variety of sex determination systems exists in reptiles and fishes, including temperature-dependent sex determination (TSD). While the precise physiological basis of TSD is unknown, numerous past studies have provided a general characterization of this phenomenon. The period of thermosensitivity encompasses the middle one third to one half of embryonic development. It has been shown that in several species of fishes this period starts prior to (histologically defined) gonadal differentiation and extends to a time when sex specific changes are becoming evident in the gonads

Studies on the mechanism of sex determination of the commercially important tilapia, Oreochromis niloticus, have demonstrated that this species exhibits a predominantly monofactorial genotypic system with male heterogamety and female homogamety. The effects of environmental conditions during the period of sex differentiation on the sex ratio of the Nile tilapia, Oreochromis niloticus.

Different sex genotypes were exposed to varying temperatures putative all-female, all-male and all-YY males.and salinities putative all-female progeny only. For a minimum period of 21 days after first feeding and were on grown prior to sexing by gonad squash. The majority of the putative all-female progeny exposed to high temperature 36.54°C. produced significantly higher percentages of males compared to controls reared at ambient temperature 27.87 °C Similarly, at high temperature, some of the all-male and YY male progenies had significantly lower percentage of males compared to controls. Sex differentiation in YY males appears to be more labile than in normal XY males although this could possibly be attributable to different levels of inbreeding. Low temperature 25.78°C.and varying levels of salinity 11.30 to 26.65 ppt. did not significantly affect sex ratios.

Evidence of the effect of environment on sex ratio has been observed in a number of fish species. A number of studies have reported on the varying degrees of temperature dependent sex determination and its adaptive significance in the Atlantic Silverside, Menidia menidia.

A number of studies have also demonstrated effects of temperature on sex ratio in tilapia. In an early study, researchers observed that at a cold temperature 19°C.the sex ratio of O. mossambicus in one experiment had a significantexcess of males 89%. compared to the control reared at 28°C mechanism.

The effect of temperature on sex differentiation is variable among species. Fish species that do not possess a clear distinct chromosomal sex, show strong temperature dependent sex determination/differentiation (TSD).

In gold fish sex differentiation is affected during a specific period of larval development, and that the timing of the sensitive period is itself dependent on temperature. Sex differentiation depended on the initial temperature until Day 17 in the group reared at 30 °C and Day 22 in the group reared at 23 °C. The sex ratio was greatly affected by a high temperature regime (0-7.7% females at 30 °C). At temperatures between 17 and 30 °C, the percentage of males increased with temperature. In conclusion, water temperature can play an important role in the process of sex differentiation in goldfish.

A temperature sensitive period (TSP) has also been demonstrated in several species of fish they show that Temperature is responsible for sex determination in the European sea bass, Dicentrarchus labrax Thermolability in phenotypic expression of sex was examined in hime salmon, a land-locked type of sockeye salmon Oncorhynchus nerka, thus produce all-female populations in practice to develop effective salmon aquaculture in a safe, simple, environmental friendly.

8

SELECTIVE BREEDING AND HYBRIDIZATION

INTRODUCTION: METHODS OF GENETIC IMPROVEMENT OF FISH

Genetically engineered (also called transgenic) fish are those that carry and transmit one or more copies of a recombinant DNA sequence (i.e., a DNA sequence produced in a laboratory using in vitro techniques). Because genetic engineering is defined by the technology that is used to create and transfer the DNA sequence, and not the source species of the donor DNA, even fish that are engineered with DNA derived entirely from fish species are considered to be genetically engineered. Currently, no genetically engineered fish has been approved for food production in the United States. To date only one company, AquaBounty, has publicly announced that it has requested FDA approval to market a genetically engineered food animal, a growthenhanced Atlantic salmon that is capable of growing 4 to 6 times faster (but not larger) than standard salmon grown under the same conditions.

Although genetic improvement of common carp probably started several thousand years ago, the application of genetic principles to most aquaculture species is a relatively recent phenomenon. Thus, the majority of farm-raised aquatic animals and plants are very similar to the wild forms. Genetic improvement programmes are beginning to be applied to more and more aquatic species, but when compared to the levels of domestication in livestock and crops, the aquatic sector is very far behind.

In 1984 aquaculture production data from the Mediterranean were reported to the Food and Agriculture Organization of the United Nations on only six species, mostly shellfish and a few mullet; culture of seabream and seabass was just beginning. Today there are between 20 and 30 Mediterranean farmed species for which data are reported; many important species cultured in the region are high value marine species. As the husbandry improves for more and more species, so will the application of genetic improvement technologies.

The growth of the aquaculture sector has raised concerns on its risk to the environment and native species through the large-scale release or escape of farmed animals and their subsequent breeding with, competition with, or predation on local species; farmed aquatic animals may also be a vector for disease transmission to wild stocks. Aquaculture has been recognized as the primary reason for the purposeful movement (introduction) of aquatic species and experience has shown that cultured species usually escape into the wild. Hatchery or farmed fish have been assumed to represent a risk to the native gene pool by introducing maladapted genes, i.e., genes adapted to the farm and not to the wild. However, researchers pointed out that direct evidence for adverse effects resulting from the mixing of hatchery and wild stocks is difficult to come by. As genetic techniques increase in their frequency of application and in their alteration of the phenotype, there may be more concern for environmental biosafety.

GENETIC IMPROVEMENTS

Genetic technologies can be utilized in aquaculture for a variety of reasons, not just to improve production. Improvements in marketability, culturability, and the conservation of natural resources can be facilitated by the appropriate genetic technology. Genetic improvement programmes can be used to provide short-term or long-term gains. The short term gains are usually immediate, within 2 generations, and generally not cumulative (unless combined with other long-term programmes), whereas the long term programmes such as selective breeding produce small gains that accumulate each generation. In general, genetic improvement technologies have not been applied to marine species to the same extent as they have been applied to freshwater and anadromous (salmonid) species. Thus, as more marine species become farmed, genetic improvement can be expected to play an increasingly important role in improving production.

Long Term Strategies

Domestication and the full potential for the utilization of aquatic genetic resources will undoubtedly only be realized by long-term breeding programmes utilizing additive genetic variance. Of the few breeding programmes applied in aquaculture, the Atlantic salmon breeding programme in Norway is an example of a long-term approach to breed improvement that has been copied in other parts of the world, e.g., in the Philippines for the improvement of tilapia and in Chile for coho salmon. Selective breeding for increased growth in red sea bream in Japan has been ongoing for over 30 years (7 generations) and resulted in a shorter time required to produce a market sized (1 kg) fish.

Traits considered for improvement should be of economic interest to the aquaculture industry. Although growth rate is the character most often to be improved in selective breeding programmes, other traits have been shown to have additive genetic variance and therefore, amenable to improvement. Disease resistance is one character that is extremely important for some aquacultured species, but improving this trait by specific selective breeding programmes may be problematic and due more to the problems of experimental design (are fish that do not get a disease resistant or have they merely not encountered the disease?) than to problems with additive variance for the trait "disease resistance". Resistance to polluted waters is also desirable in some areas, e.g., heavy metal tolerance in tilapia, but care should be taken not to use this as a justification not to restore degraded habitats. Recently, due to the high prices that high quality products reach in the market, other traits, such as flesh quality, are being considered by the industry.

Selection efficiencies may be enhanced by selecting for a particular gene loci, a process called marker assisted selection, if the proportion of additive genetic variance related to a locus or loci exceeds the heritability of a trait. However, identification of useful oci is still difficult in aquatic species. Micro-satellites probes reveal high levels of genetic variation and can help identify useful loci and help establish pedigrees in mixed family groups.

Therefore, marker assisted selection may be practical for commercial or small scale farms because no- special rearing facilities are needed, daily routine of fish farming is not interrupted and fish did not need to be marked; genetic analysis could be performed on small tissue sections or fin clips that would not detract from the marketability of the fish.

In general, care has to be taken in order to limit the negative effect of inbreeding when designing and running selective breeding programmes in aquaculture and genetic markers such as micro-satellites can help determine family relationship.

Production of transgenic fish is considered here to be a long term breedimprovement strategy because of the time required to identify a useful gene and its promoter, to insert them into the host animal and then to conduct the screening required to confirm the stable inheritance of the transgene. Several commercially important species have received genes from other species.

For example, coho salmon with a growth hormone gene and promoter from sockeye salmon grew 11 times (0-37 times was range of effects) as fast as non-transgenics. Private industry is promoting transgenic fish and testing transgenic Atlantic salmon that grow 400% faster than normal (during first year) at an inland aquaculture facility in Scotland.

Short-term Strategies

Short-term genetic improvement programmes may not require the same level of record keeping nor management as long-term projects and can impart significant gains with simple technologies in a short period of time. That is not to say that everyone will view the technologies as "simple" or that proper broodstock management and accurate record keeping are not important, they will be necessary to maintain healthy and genetically appropriate stocks for manipulation.

Table 1. Genetic improvement strategies

Genetic manipulation	Improvement
Long-term strategies:	
Selective breeding for:	
Growth rate	As high as 50% increase after generations in coho salmon (Hershberger et a/., 1990); gilthead sea bream mass selection gave 20% increase/ generation; mass selection for live weight and in Chilean oysters found gain in one generation.
Body confirmation	High heritabilities in common carp, catfish.
Physiological tolerance (stress)	Rainbow trout selected for high response showed increased levels of plasma cortisol levels
Disease resistance	Increased resistance to dropsy in common carp.
Pollutant resistance	Tilapia progeny from lines selected for resistance to heavy metals Hg, Cd, and Zn survived 3-5 times better than progeny from unexposed lines
Maturity and time of spawning	60 day advance in spawning date in rainbow trout
Gene transfer	Coho salmon with a growth hormone gene and promoter from sockeye salmon grew 11 times (0-37 range) as fast as non-transgenics. Atlantic salmon grew 400% faster than normal during first year.
Short-term strategies	
Intra-specific crossbreeding	Heterotic growth seen in and 22% of channel catfish and rainbow trout crosses, respectively. Chum salmon and largemouth bass showed no heterosis.
	Heterosis for wild X hatchery S. aurata; crossbreeds of channel catfish common carp showed 30-60% heterosis.
Sex reversal and breeding	show improvements in yield of almost 60% depending on farming system
Sex reversal and breeding	All male tilapia show improvements in yield of almost 60% depending on farming system and little unwanted reproduction and stunting. All female rainbow trout grew faster and had better flesh quality
Chromosome manipulation	Phagrus major triploids had similar growth rate to diploids at 10 months of age, but were smaller and presumed tobe sterile at18 months. Dicentrarchus labrax triploids showed inconsistent growth in relation to diploids and had lower GSI.
	Improved growth and conversion efficiency in triploid rainbow trout, channel catfish, at plaice flounder hybrids; triploid Nile tilapia grew 66-90% better than diploids and showed decreased sex-dimorphism for body weight, but other studies found no advantage. GxE interactions also influence performance.
	Triploid Pacific oysters show 13-51% growth improvement over diploids at 8-10 months of age and better marketability due to reduced gonads; triploid Sydney rock oysters showed 41% increase in body weight at 2.5 years.
	Polyploidization makes certain interspecific crosses viable.

Hybridization has been undertaken to combine favourable qualities from two genetically different groups and to take advantage of hybrid vigour (heterosis). Sterile or non-reproducing groups may also be produced through interspecific hybridization, but fertile hybrids also exist for several aquaculture species (Table 1). Usually, it is only the generation that is of interest to aquaculturists, therefore the pure parental lines must be maintained and managed by the breeding centre. Several large groups of importance to aquaculture do not produce useful hybrids, for example salmonids and peneaid shrimp.

As more and more marine hybrids are produced, their culture performance is and should be evaluated. The Kuwait Institute of Scientific Research has produced a hybrid bream, *Acanthopagus latus* (Sheim) × *Sparidentex hasta* (sobiaty), that is currently being investigated. It has good growth and body quality and appears to be fertile.

Evaluation should note the importance of the sex of interspecific crosses. In hybrid striped bass, the sunshine bass, a hybrid between male *Morone saxatilis* X female M, chrysops, is superior under culture to the Palmetto bass, the reciprocal hybrid. Approximately 80% of the culture of Clariad catfish in Thailand involves a *Clarius macrocephalus* and *C. batrachus* hybrid with the preferred female being the Thai catfish, *Clarius macrocephalus*.

Many aquatic hybrids are fertile and may be bred together or backcrossed to parental lines. Hybrid red tilapia (Red Florida tilapia × *Oreochromis urolopis honorum*) are being backcrossed to ancestral lines in efforts to improve growth and body shape in Venezuela.

Manipulation of chromosome-sets (polyploidization) has been accomplished for many aquatic species through thermal and chemical shocks to developing embryos. Triploid organisms are particularly interesting because they should be sterile and therefore be able to put more energy into the growth process rather than into maturation and reproduction. However, this assumption does not hold for all species and triploids demonstrate varying levels of reproductive activity and growth difference from diploids.

The performance in culture of triploids in relation to diploids seems to be species specific. Inconsistent performance of triploids within rainbow trout and Atlantic salmon has also been noted; in coho salmon triploids, poorer growth and survival was observed). In oysters improved growth and marketability resulted from triploidization, and the practice is common in Pacific oyster production in the USA. Researchers demonstrated that triploid oysters produced by mating tetraploids with diploids had higher survival and larger size than triploids produced by chemical shock and were larger than diploid controls.

The production of mono-sex groups is another short-term strategy that takes advantage of sexual dimorphism in important traits or when reduced chance of reproduction is desired. For example male tilapia grow faster than females and a single sex population of tilapia would not be prone to reproductive stunting, whereas female rainbow trout are desirable because of faster growth rate and delayed maturity. In species where fish roe is valuable, e.g., caviar from sturgeon, the production of all female groups would be an obvious benefit. Fish may be directly "sex reversed", by administering androgens or oestrogens to early life history stages, thus producing a group of all one sex.

Although the use of hormones early in the culture of aquatic animals is becoming widely accepted due to the fact that no residue remains when the organism reaches market size, certain areas still have restrictions on the consumption of fish that have been treated with hormones. This problem may be solved by the appropriate sexreversal and breeding of broodstock, such that single-sex progeny are produced without ever coming in direct contact with the hormone.

FISH SELECTION

Several selection strategies are available for genetically improving the performance of Tilapias. The strategies differ in terms of which kind of relative provides information to estimate the breeding value of the selection candidates. The choice of selection strategy depends on the nature of traits and the heritability of traits (i.e. the relationship between an individual's phenotype and its breeding value).

Individual Selection

Individual selection, or 'mass selection', is a strategy where the breeding candidates are selected based on their own phenotype, that is, on their own performance. It has been the most frequently applied selection strategy in fish due to its simplicity. However, it can only be used to improve traits that are recorded on the breeding candidates while they are still alive (e.g. growth, shape, color), and it is not efficient to improve discrete and lowly heritable traits such as survival rate.

If practiced without constraints, the accumulation of inbreeding can be a major problem of individual selection in fish. The selected broodstock can become more and more related to each other. If large families are naturally produced or stocked in the test environment, the wide genetic variation and common non-genetic effects (e.g. maternal and age effects) may easily result in a disproportionately high contribution of individuals from just a few parents to the next generation, and a consequent accumulation of inbreeding. Inbreeding reduces the genetic variation and the potential for further genetic improvement in the breeding population.

Generally, it also lowers the performance of fish, due to the phenomenon known as inbreeding depression. To prevent inbreeding, only a restricted number of individuals from each full and half-sib family should be tested as breeding candidates when applying individual selection. However, this requires that each family be produced and kept separately from each other before testing.

Family Selection

Family selection is a strategy whereby the breeding candidates are ranked based on records taken on their full and half-sibs, that is, individuals that have genes in common with them. This selection strategy can be applied to improve traits that cannot be recorded on the breeding candidates while they are alive (e.g. carcass quality traits), traits that can only be recorded for groups of fish (e.g. feed utilization), and discrete traits at high or low incidences (e.g. survival rate, age at sexual maturity). Furthermore, family selection is more effective than individual selection when the heritability of a trait is low. However, family selection requires that each family be produced and kept separately and that all fish are tagged before testing in common test environments.

Combined Selection

When traits can be recorded on the breeding candidates, family selection can be combined with individual selection to increase the accuracy of the estimates of breeding values, and thereby achieve a greater selection response. Combined selection increases the rate of selection response relative to the earlier mentioned methods. It optimally combines all available sources of information used in the estimation of an individual's breeding value. With fish, most often this means information from full and half-sibs. More generally, it may mean information from parents, offspring, or any other group of relatives.

Combined selection may take the form of an index, in which individual, full-sib and half-sib information are included and weighted. It may also be implemented using Best Linear Unbiased Prediction (BLUP) procedures, which allow the simultaneous estimation of breeding values and adjustment for fixed effects. For relatively balanced designs, one may expect good agreement between the selection index and BLUP approaches.

Within Family Selection

This is the opposite of the previous method. The selection is based on the deviation of each animal from the average of your family. There is the choice of animals of all families, then the method requires identification of the families. This may be achieved by maintaining them in separate tank, cages, hapas or any other means of containment, without necessarily tagging the fish. The full-sib groups reared in unreplicated hapas or any other form of containment fall into this category.

Under such circumstances selection between families would be misleading from a genetic viewpoint because of the confounding between genetic merit and common environmental effects. If replacements are chosen so that every family contributes the same number of individual to the next generation (chose one female and one male from each family) the effective population size is twice the actual.

The criterion of selection is the deviation of each individual from the mean of the family to which it belongs. This type of selection is the reverse of family selection; the family means being given zero weight in the selection decision. The chief condition under which this method has an advantage over the others is when there is a large component of environmental variance common to members of a family. Selection within families would eliminate this large non-genetic component from the variation operated on by selection. So, the within-family selection is especially advantageous when there is a large component of environmental variance common to members of the same family.

However, not all the additive genetic variance is available for selection, but only a fraction equal to the coefficient of relationship, but only a fraction equal to the coefficient of relationship among the family relatives in question will be available (0.50 and 0.25 for full-sibs and half-sibs, respectively). The lower within-family heritability can be compensated for by the high within-family selection intensity that can be applied without increasing the rate of inbreeding. The selection intensity within families will be limited only by the number of individuals tested per family.

The number of families involved in the programme will determine the lower limit of inbreeding, which can easily be controlled by applying a rotational mating system such as that earlier suggested for selection within cohorts. An important practical advantage of selection within families, especially in laboratory experiments, is that it economizes breeding space, unlike family selection. The method reduces the need for tagging large numbers of individuals. They estimate that the implementation of a selection method that entailed the individual identification of large numbers of fish and a period of communal rearing would be more expensive and difficult to implement. Within family selection has low efficiency compared to most other selection methods

THEGENETICIMPROVEMENTPROGRAMSINAQUACULTURE

Most animal breeding programs have focused on cumulative short-term genetic changes of production traits, because breeding optimization has to a very large extent been based on market economic values. Sustainable genetic improvement by animal breeding is a long-term and complex process. Farmed fish is at an early stage of domestication and selection, but rapid selection responses for growth rate are already documented for several species. Precautions should be taken to avoid the same unwanted side effects, such as increased frequencies of diseases (e.g. mastitis in dairy cattle) and leg problems, which we have often seen in breeding programs for agricultural livestock.

Careful monitoring of possible correlated responses is needed. More basic knowledge of, e.g. animal welfare and behavioural needs of fish may also be required, as we know very little about its perception of pain and satisfaction, and what it prefers.

Starting a selective breeding program should then focus on expected benefits of, e.g. faster growing fish with higher survival rates and the costs, by improving the aquaculture species in question. Because of the great fecundity of most aquaculture species, a single, well-organized breeding nucleus may supply a huge market with continuously upgraded seed by delivering selected brood stock to a number of multiplier farmers after each generation of selection.

Consequently, the running costs of the breeding programs will be more or less constant and independent of the volume of the target industry. The benefit/ cost ratio will then often depend more on the volume of the target industry than on the price of the end product, both from the consumers' point of view and from a private business point of view. The total impact of an aquaculture breeding program on food supply and resource efficiency for the society will usually be greater for a large volume species.

Genetically improvement strains are essential to aquaculture development. The application of proven quantitative genetic theory should continue for relevant species. There is an ample proof of the success such programmes can have. Greater emphasis is required in the areas of dissemination of the improved stock to farmers to ensure impact at the production level. However, consideration must be given to the potential impacts of fish escapees and the conservation of wild fish genetic resources. So, genetic management is necessary for both genetic improvement and conservation of genetic variation in cultured fish populations and should be emphasized in future capacity development initiatives

From an economic viewpoint, investment appraisal studies indicate very favorable benefit-to-cost ratios for genetic improvement programmes in *Nile tilápia* and common carp. This was show to be so even for situation in which there was genotype by environmental interaction and a single program had to service more than one environment. It is reasonable to think that these results can be generalized to other, similar, aquatic animal species.

Limitations and constraints during the implementation of genetic improvement programmes in fish often occur, particularly in developing countries. Issues are commonly related to financial resources, and to a paucity of human capacity in this field. Other constraints include technical issues related to individual identification of fish but these can be overcome with relatively small investment.

Before initiating a selective breeding programme, a farmer must ask the following question: Is a selective breeding programme appropriate or necessary? The adoption of a selective breeding programme should be restricted to farmers who are capable of keeping and maintaining records, who are good managers, who adopt and maintain new technologies and who are willing to incorporate some long-term planning into their management programmes.

In order to control fertilization, hatching and first feeding, it is essential that the entire reproduction cycle for a species is controlled in captivity. Furthermore, an inexpensive marking or tagging system is advantageous and a necessity for a breeding program based on testing and selection within and between families. In order to select the best individuals for breeding, we need records of the traits in the breeding objective or traits correlated to the breeding goal. The records should be obtained for many animals at a reasonable cost during a limited time interval. Knowledge and methodology to optimize selection by predicting and using breeding values is also essential.

The programmes starting with a population with ample genetic variation is a trademark of successful fish genetic improvement programmes. Although this in itself is not a sufficient condition for success, it is indeed a necessary condition. The failure of some attempts to achieve genetic improvement with aquatic animals may have been due more to weakness in the base population than to be selection method utilized. Irrespective of the method of choice, continued genetic improvement will hinge upon the adequate balance between high selection intensity and the maintenance of low inbreeding rate.

Before the selection is made, it is preponderant that the beginning of an improvement program originates from a base population with great additive genetic variability. In this way, it is important that the nucleus is formed by lots from several genetic sources, such as several sites or collection points of a particular river. Examples are the Norwegian Atlantic salmon program, which originated from 41 wild genetic sources and the GIFT program for Nile tilapia originated from the diallele crossing of 8 strains of the species.

The success of an aquaculture breeding program critically depends on the way which the base population of breeders is constructed. In particular, all the genetic variability that can be used for selection on the traits initially included in the breeding objective is that found in the original breeders. Also, the decisions taken when creating the base population will have consequences on the genetic progress for any other additional trait that may be part of future breeding goals, whatever production or fitness related traits.

Strains originating from the base population may be selected to serve crops in different environments, since there may be genotype-environment interaction, as evaluated for the characteristic weight at Nile tilapia farms cultivated in tanks-net and nurseries. However, until then, little or no interaction was found in rainbow trout, Atlantic salmon and catfish, suggesting that it is not necessary to select several lineages for this species. However, there is always a need for previous studies to confirm the lack of genotype-environment interaction, mainly due to the great variation between and within existing production systems in fish farming.

The decision on the type of selection to be adopted will depend on the program's long-term objectives, available resources, the nature of the goal and the value of heritability. Ideal program designs maximize the likelihood of correct ordering of organisms as to the true genetic value for the target in question. However, perfect programs are rarely possible, restricted by the limitation of individual identification and the physical separation of families to communal storage Mair.

The major goal to most selective breeding programmes that are going to be used to improve quantitative phenotypes is to improve growth rate, and this will also increase yield. There are two basic approaches to improving yields. One is to use a breeding programme to improve the fish that are being cultured. The other is to improve the improvements in management are often inexpensive.

For that, they must to conduct an assessment to determine its usefulness, needs to quantify his goals and know how the phenotype(s) will be measured, to determine the type of selective breeding programme he will use, needs to know how the results will be evaluated and must plan an annual evaluation and reassessment of the programme.

The overall aim of all selection schemes is to maximize the probability of correctly ranking all potential breeders with regard to their breeding value. An animal's breeding value can be defined as the average performance of an infinite number of its progeny or from a practical points of view its ability to produce good or bad progeny.

Selection that combines the individual information of the animal and its relatives. More efficient method, but more expensive. From this method, it is possible to predict the genetic values of the animals by BLUP (Best Linear Unbiased Predictions). Method adopted for the Norwegian National Program for Genetic Improvement of Atlantic salmon and the GIFT program for Nile tilapia. With the information of pedigree, it becomes possible the direct selection of characteristics of meat quality and survival to diseases.

One of the major difficulties of these programs in fish breeding is the mark used for the identification of animals. Most available and accessible such as numbered ribbons that span the musculature or nylon strands with small numbered labels that are also affixed in the same way, have high loss rates.

Thus, more animals must be identified to compensate for information losses, increasing program costs. The emergence of identification tags improved pitfalls considerably (almost 0%), but it is the most expensive option on the market.

For each selection methods than can be chosen, always have the side of the advantages and disadvantages. That choice will be influence for a given particular breeding scenario. However, for most methods, there is a theoretical maximum value of the correlation between the true and estimated breeding values, given different heritability for the traits in question.

ARTIFICIAL SELECTION AND ITS FORMS

Selective breeding (also called artificial selection) is the process by which humans use animal breeding and plant breeding to selectively develop particular phenotypic traits (characteristics) by choosing which typically animal or plant males and females will sexually reproduce and have offspring together. Domesticated animals are known as breeds, normally bred by a professional breeder, while domesticated plants are known as varieties, cultigens, cultivars, or breeds. Two purebred animals of different breeds produce a crossbreed, and crossbred plants are called hybrids. Flowers, vegetables and fruit-trees may be bred by amateurs and commercial or non-commercial professionals: major crops are usually the provenance of the professionals.

In animal breeding, techniques such as inbreeding, linebreeding, and outcrossing are utilized. In plant breeding, similar methods are used. Charles Darwin discussed how selective breeding had been successful in producing change over time in his 1859 book, *On the Origin of Species*. Its first chapter discusses selective breeding and domestication of such animals as pigeons, cats, cattle, and dogs. Darwin used artificial selection as a springboard to introduce and support the theory of natural selection.

The deliberate exploitation of selective breeding to produce desired results has become very common in agriculture and experimental biology.

Selective breeding can be unintentional, e.g., resulting from the process of human cultivation; and it may also produce unintended – desirable or undesirable – results. For example, in some grains, an increase in seed size may have resulted from certain ploughing practices rather than from the intentional selection of larger seeds. Most likely, there has been an interdependence between natural and artificial factors that have resulted in plant domesticat

Selective breeding in aquaculture holds high potential for the genetic improvement of fish and shellfish. Unlike terrestrial livestock, the potential benefits of selective breeding in aquaculture were not realized until recently. This is because high mortality led to the selection of only a few broodstock, causing inbreeding depression, which then forced the use of wild broodstock. This was evident in selective breeding programs for growth rate, which resulted in slow growth and high mortality.

Control of the reproduction cycle was one of the main reasons as it is a requisite for selective breeding programs. Artificial reproduction was not achieved because of the difficulties in hatching or feeding some farmed species such as eel and yellowtail farming. A suspected reason associated with the late realisation of success in selective breeding programs in aquaculture was the education of the concerned people – researchers, advisory personnel and fish farmers. The education of fish biologists paid less attention to quantitative genetics and breeding plans.

Another was the failure of documentation of the genetic gains in successive generations. This in turn led to failure in quantifying economic benefits that successful selective breeding programs produce. Documentation of the genetic changes was considered important as they help in fine tuning further selection schemes.

Quality Traits in Aquaculture

Aquaculture species are reared for particular traits such as growth rate, survival rate, meat quality, resistance to diseases, age at sexual maturation, fecundity, shell traits like shell size, shell colour, etc.

- **Growth rate** – growth rate is normally measured as either body weight or body length. This trait is of great economic importance for all aquaculture species as faster growth rate speeds up the turnover of production. Improved growth rates show that farmed animals utilize their feed more efficiently through a positive correlated response.

- **Survival rate** – survival rate may take into account the degrees of resistance to diseases. This may also see the stress response as fish under stress are highly vulnerable to diseases. The stress fish experience could be of biological, chemical or environmental influence.
- **Meat quality** – the quality of fish is of great economic importance in the market. Fish quality usually takes into account size, meatiness, and percentage of fat, colour of flesh, taste, shape of the body, ideal oil and omega-3 content.
- **Age at sexual maturation** – The age of maturity in aquaculture species is another very important attribute for farmers as during early maturation the species divert all their energy to gonad production affecting growth and meat production and are more susceptible to health problems (Gjerde 1986).
- **Fecundity** – As the fecundity in fish and shellfish is usually high it is not considered as a major trait for improvement. However, selective breeding practices may consider the size of the egg and correlate it with survival and early growth rate.

FINFISH RESPONSE TO SELECTION

Salmonids

Researchers showed that selection of Atlantic salmon (*Salmo salar*) led to an increase in body weight by 30% per generation. A comparative study on the performance of select Atlantic salmon with wild fish was conducted by AKVAFORSK Genetics Centre in Norway. The traits, for which the selection was done included growth rate, feed consumption, protein retention, energy retention, and feed conversion efficiency. Selected fish had a twice better growth rate, a 40% higher feed intake, and an increased protein and energy retention. This led to an overall 20% better Fed Conversion Efficiency as compared to the wild stock. Atlantic salmon have also been selected for resistance to bacterial and viral diseases. Selection was done to check resistance to Infectious Pancreatic Necrosis Virus (IPNV).

The results showed 66.6% mortality for low-resistant species whereas the high-resistant species showed 29.3% mortality compared to wild species.

Rainbow trout (*S. gairdneri*) was reported to show large improvements in growth rate after 7–10 generations of selection. Researchers showed that growth gains by 30% could be achieved by selectively breeding rainbow trout for three generations. A 7% increase in growth was recorded per generation for rainbow trout by researchers.

In Japan, high resistance to IPNV in rainbow trout has been achieved by selectively breeding the stock. Resistant strains were found to have an average mortality of 4.3% whereas 96.1% mortality was observed in a highly sensitive strain.

Coho salmon (*Oncorhynchus kisutch*) increase in weight was found to be more than 60% after four generations of selective breeding. In Chile, researchers conducted experiments on early spawning dates in coho salmon. After selectively breeding the fish for four generations, spawning dates were 13–15 days earlier.

Cyprinids

Selective breeding programs for the Common carp (*Cyprinus carpio*) include improvement in growth, shape and resistance to disease. Experiments carried out in the USSR used crossings of broodstocks to increase genetic diversity and then selected the species for traits like growth rate, exterior traits and viability, and/or adaptation to environmental conditions like variations in temperature. Researchers selected carp for fast growth and tolerance to cold, the Ropsha carp.

The results showed a 30–40% to 77.4% improvement of cold tolerance but did not provide any data for growth rate. An increase in growth rate was observed in the second generation in Vietnam. Researchers showed positive results when selecting for slower growth for three generations compared to selecting for faster growth. Researchers showed resistance to the dropsy disease wherein selected lines suffered low mortality (11.5%) compared to unselected (57%).

Channel Catfish

Growth was seen to increase by 12–20% in selectively bred *Iictalurus punctatus*. More recently, the response of the Channel Catfish to selection for improved growth rate was found to be approximately 80%, i.e. an average of 13% per generation.

SHELLFISH RESPONSE TO SELECTION

Oysters

Selection for live weight of Pacific oysters showed improvements ranging from 0.4% to 25.6% compared to the wild stock. Sydney-rock oysters (*Saccostrea commercialis*) showed a 4% increase after one generation and a 15% increase after two generations. Chilean oysters (*Ostrea chilensis*), selected for improvement in live weight and shell length showed a 10-13% gain in one generation. Bonamia ostrea is a protistan parasite that causes catastrophic losses (nearly 98%) in European flat oyster *Ostrea edulis* L. This protistan parasite is endemic to three oyster-regions in Europe. Selective breeding programs show that *O. edulis* susceptibility to the infection differs across oyster strains in Europe. A study carried out by Culloty et al. showed that 'Rossmore' oysters in Cork harbour, Ireland had better resistance compared to other Irish strains. A selective breeding program at Cork harbour uses broodstock from 3– to 4-year-old survivors and is further controlled until a viable percentage reaches market size.

Over the years 'Rossmore' oysters have shown to develop lower prevalence of *B. ostreae* infection and percentage mortality. Researchers selectively bred the eastern oyster, *Crassostrea virginica*, for resistance against co-occurring parasites *Haplosporidium nelson* (MSX) and *Perkinsus marinus* (Dermo). They achieved dual resistance to the disease in four generations of selective breeding. The oysters showed higher growth and survival rates and low susceptibility to the infections. At the end of the experiment, artificially selected *C. virginica* showed a 34–48% higher survival rate.

Penaeid shrimps

Selection for growth in Penaeid shrimps yielded successful results. A selective breeding program for *Litopenaeus stylirostris* saw an 18% increase in growth after the fourth generation and 21% growth after the fifth generation. *Marsupenaeus japonicas* showed a 10.7% increase in growth after the first generation. Argue et al. (2002) conducted a selective breeding program on the Pacific White Shrimp, *Litopenaeus vannamei* at The Oceanic Institute, Waimanalo, USA from 1995 to 1998. They reported significant responses to selection compared to the unselected control shrimps.

After one generation, a 21% increase was observed in growth and 18.4% increase in survival to TSV. The Taura Syndrome Virus (TSV) causes mortalities of 70% or more in shrimps. C.I. Oceanos S.A. in Colombia selected the survivors of the disease from infected ponds and used them as parents for the next generation. They achieved satisfying results in two or three generations wherein survival rates approached levels before the outbreak of the disease. The resulting heavy losses (up to 90%) caused by Infectious hypodermal and haematopoietic necrosis virus (IHHNV) caused a number of shrimp farming industries started to selectively breed shrimps resistant to this disease.

Successful outcomes led to development of Super Shrimp, a selected line of *L. stylirostris* that is resistant to IHHNV infection. Researchers confirmed this by showing no mortalities in IHHNV- challenged Super Shrimp post larvae and juveniles.

Aquatic Species Versus Terrestrial Livestock

Selective breeding programs for aquatic species provide better outcomes compared to terrestrial livestock. This higher response to selection of aquatic farmed species can be attributed to the following:

- High fecundity in both sexes fish and shellfish enabling higher selection intensity.
- Large phenotypic and genetic variation in the selected traits.

Selective breeding in aquaculture provide remarkable economic benefits to the industry, the primary one being that it reduces production costs due to faster turnover rates. This is because of faster growth rates, decreased maintenance rates, increased energy and protein retention, and better feed efficiency. Applying such genetic improvement program to aquaculture species will increase productivity to meet the increasing demands of growing populations.

INBREEDING: DEFINITION AND METHODS OF ITS DETERMINATION

Definition

Inbreeding is simply the mating of closely related individuals, nothing more nothing less. Inbreeding is neither good nor bad; it can like any breeding program, be used either wisely or stupidly. Inbreeding does not change gene frequencies. It changes genotype frequencies by increasing the homozygotes at the expense of heterozygotes.

The mating of individuals or organisms that are closely related through common ancestry, as opposed to outbreeding, which is the mating of unrelated organisms.

Inbreeding is useful in the retention of desirable characteristics or the elimination of undesirable ones, but it often results in decreased vigour, size, and fertility of the offspring because of the combined effect of harmful genes that were recessive in both parents.

Inbreeding is also one of the breeding programme that can have a tremendous impact on productivity. As recently as 200 years ago there were laws forbidding consanguineous matings even in livestock, because it was considered immoral and against the laws of God and Nature. But according to plant breeders, inbreeding is one of the most important breeding techniques; without its use, agricultural productivity would decline.

- **Inbreeding is simply the mating of related individuals.** Inbreeding does not imply, nor does the definition mention, anything about viability, growth, or productivity. (A mating between relatives is often referred to as Consanguineous mating).
- Inbreeding is neither good nor bad.
- Genetically, all inbreeding increases homozygosity in the offspring (this means inbreeding also decreases heterozygosity in the offspring by an equal amount).
- The increase in homozygosity occurs because related fish share alleles through one or more common ancestors, i.e. the parents may carry a copy of an allele that both inherited from a common ancestor.
- When relatives mate, the alleles that they share because of their common ancestor(s) can be paired in their offspring. This produces offspring that are more likely to be homozygous at one or more loci.
- The mating of unrelated fish also produces offspring that have homozygous genes. Additionally, an inbred fish looks the same as one with no inbreeding; there is no distinguishing mark that separates fish into inbred vs non-inbred categories.
- The two types of homozygosity are identical; there is no genetic difference.
- The only difference between inbred and non-inbred fish is that the homozygosity of each is created in different ways, but this distinction is most important.

- Inbred fish are homozygous because they have genes where the two alleles are identical by descent ; non-inbred fish are homozygous because they have genes where the two alleles are identical because they are alike in kind.
- Inbred fish are homozygous at a locus because they inherited identical copies of an allele from both parents that the parents, in turn, inherited from a common ancestor. Non-inbred fish are homozygous at a locus because they just happened to inherit an identical pair of alleles from their parents.
- This difference in the type of homozygosity is not physical or chemical. The only difference is the paths that the alleles took before they were paired to create the homozygosity ,i.e., how the alleles were inherited.
- The mating of relatives produces offspring that tend to be more homozygous than the population average.
- This increase in homozygosity is called "inbreeding", and the coefficient of inbreeding (F) is a measure of how much more homozygous a fish is than the population average.
- The coefficient of inbreeding does not measure how many homozygous loci the fish has; it simply quantifies the percent increase in homozygosity.
- On a gene-by-gene basis, **F** is the probability that the two alleles will be identical by descent. Since **F** measures the percent increase in homozygosity, the same level of inbreeding can produce different amounts of homozygosity in different populations, depending on the level of homozygosity that has occurred over a specific time interval (number of generations) due to identity by descent (due to the mating of relatives).
- Additionally, F is based on probability, so a given value of F is an average value; consequently, a given value of F means more homozygosity will have been produced in some fish, while less will have been produced in others.
- Because inbreeding increases homozygosity, it changes genotypic frequencies, increasing the percentage of homozygous genotypes, while shrinking the percentage of heterozygous genotypes.
- If regular systems of inbreeding are conducted, the breeding programme creates a series of inbred families which do not interbreed. This subdivides a population into numerous lines, which further increases genotypic variance.

The effect of inbreeding on genotypic and gene frequencies.

- In the P_1 generation, all fish are heterozygotes (Dd). In every generation, the following matings are made: DD x DD; Dd x Dd; dd x dd.
- These matings reduce the percentage of heterozygotes and increase the percentage of homozygotes.

- In the F_3 generation, only 12.5% of the fish are heterozygotes, while 87.5% are homozygotes. Eventually, there will be no heterozygotes. Although this mating pattern changes genotypic frequency, the frequencies of the two alleles do not change.
- While inbreeding changes genotypic frequency, it does not change gene frequency.
- Selection, genetic drift, migration, and mutation are the evolutionary and breeding forces that change gene frequency. Inbreeding itself does not alter gene frequency, but by altering genotypic frequency inbreeding can, theoretically, accelerate selection.

During an inbreeding programme, genotypic frequencies can be drastically altered if relatively few inbred families are maintained (a type of selection). When this occurs, gene frequencies will also change. In many domesticated animals it is possible to estimate the level of inbreeding in an individual if the pedigree of the animal is known.

In most fish species this is almost impossible because of the numbers involved, the difficulties in identifying broodstock and the expense involved in keeping different families of fish separate. However, as a simple basic, it is possible to calculate an average level of inbreeding from year to year for any hatchery stock.

USES OF INBREEDING

Although inbreeding in general represents a harmful phenomenon, it can also be extremely useful in fish selection. This usefulness stems primarily from the stabilization of selection traits due to increased homozygosity and the augmented expression of several of them.

Inbreeding is a breeding programme that can be used to produce superior animal and plant brood stock, and it can also be used to produce genetically improved animals and plants for grow-out. Inbreeding is the breeding programme that is often used to create new breeds or varieties that breed true for "type"; i.e., a particular body conformation or set of qualitative phenotypes. Linebreeding is a form of inbreeding that is used to increase an outstanding animal's contribution to a population. The use of herd bulls is a form of inbreeding that is used to quickly improve a population. Inbreeding can be used as a type of progeny testing to create defect-free animals. Inbreeding can be used to improve response to selection when h^2 is small by combining inbreeding with between-family selection.

Inbreeding is a breeding programme that can be used

- To produce superior animal and plant brood stock,
- To produce genetically improved animals and plants for grow-out,
- To create new breeds or varieties that breed true for "type"; i.e., a particular body conformation or set of qualitative phenotypes.
- One of the major uses of inbreeding is line breeding.

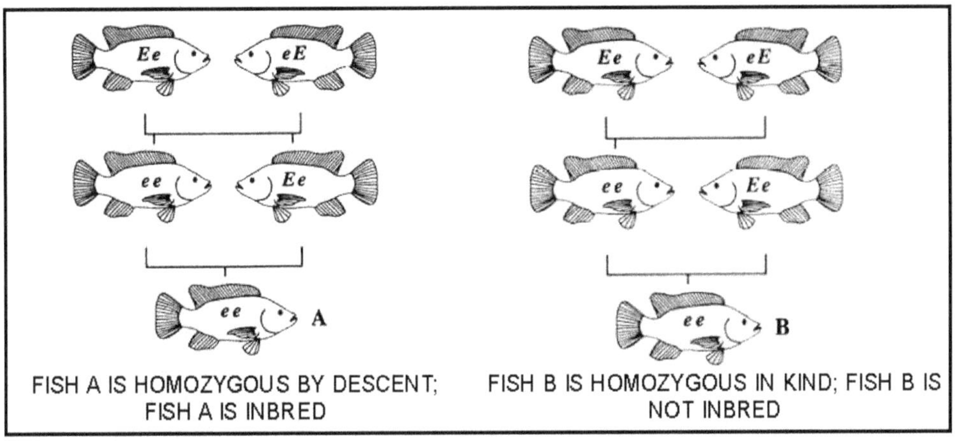

Fig. 1. Pedigrees of two fish with identical genotypes (ee), one of which is inbred and one of which is not. Fish A is inbred because its parents are related (brother and sister); it is homozygous by descent. Fish B is not inbred, because its parents are not related; it is homozygous in kind.

- A major use of inbreeding is the creation of inbred lines that can be hybridized to produce F_1 hybrids for grow-out.
- Here, two or more selected lines are inbred to fix certain alleles.
- When the inbred lines are mated, the hybrids will be identical at the desired loci and will be uniform.
- Thus inbreeding in two or more lines followed by hybridization is the classic way of producing uniform progeny for grow-out.
- Though inbreeding depression is reported, sometimes outstanding inbred animals can be valuable as brood stock because they breed true for many phenotypes and will not perpetuate undesired phenotypes.

EXPRESSION OF INBREEDING DEPRESSION IN FISH

Rainbow trout is a species of general importance in Iranian Aquaculture. Iran has many rivers and springs which provide sources of cold water, favorable for the growth of rainbow trout (*Oncorhynchus mykiss*). Using these natural advantages, commercial trout farming increased in recent years and the production was about 34,000 ton in 2005. Production of this species expanding very fast and most of the farmers try to use the fry from their own brood stocks. Rainbow trout have advantage for use in breeding, notably: with the small number of breeding individuals all the progeny of succeeding generations can be produced but this leads to increased inbreeding.

Moreover, the practice of mating unequal number of males and females, usually in this case, a smaller number of males than females, also increased the probability of inbreeding. Inbreeding is mating between individuals which are related to each other by ancestry. Inbreeding depression is the effect of inbreeding measured as the decline in performance of a quantitative trail. It has been investigated in most domestic and laboratory animal species including fish.

Studies to measure depression at a series of inbreeding levels in rainbow trout populations using the method of inbred-outbred half-sib families are discussed. While the actual depression estimates varied widely between populations and inbreeding levels, significant levels of depression were found in many hatchery, field and brood stock performance traits after only one generation of brother-sister mating.

Therefore, the high level of abnormal fish especially spinal malformation appearing in hatchery fish that is an important problem in many private farms. Malformation often is associated with growth depression, leading to high mortality rates at early fry stage. Researchers investigated the effect of three levels of inbreeding (F = 0.25, 0.375 and 0.50) on survival and growth, found that inbreeding caused significant depression in survival of eyed eggs, alevins and fry, in growth rate of adults but not in growth rate upto 160 days after first feeding. The source and the rate of abnormality still is a problem for many hatchery farms.

The objective of this study was to investigate the magnitude effect of the inbreeding on survival rate, body weight and abnormalities in brother-sister mating of rainbow trout in one of the hatcheries in Iran.

The rainbow trout breeders were obtained from a captive brood stock, fed special (Ecogen 13, Biomar, Fr.) pellets contained, crude protein 50%, crude fat 13%, ash 9.5%, fiber 1.9% and P total 1.3%. The breeders maintained by Kelardasht salmon propagation center (Iran), one pair from Yasoj fish farm (Center part of Iran-group A) and one pair from Kelardasht salmon propagation center (North part of Iran-group B). The breeders were in the same age group (3 years) and were selected based on similar growth rate, 2.8 kg. The experiment conducted from 2001-2005.

A random block design was done between A and B groups using 4 randomly chosen sires and 4 randomly chosen dams from each groups. In second experiment a full factorial diallele cross was done between A and B groups using 4 randomly chosen sires and 4 randomly chosen dams from each group, thus theoretically producing (4+)(4+4) = 64 half-sibs and full-sibs families. All dam and sire age were equal, thus making the dam and sire effect possible. Sixteen full-sib and half-sib mating in experiment one and the 64 mating in experiment 2 were individually incubated in a fully randomized design of eighty standard California incubators. Incubation went on at 8±0.5°C until hatching around 420° days (°D) and then until swim-up at 750° days (°D).

After hatching the fish were transferred to nursery tanks, one tank for each full-sib family of 150 fry (1 g L^{-1}). The tanks were rectangular of 2 x 0.5 x 0.2 m, receiving about 12 L min^{-1} of fully oxygenated water. Feeding was initiated when the fry showed signs of swimming. Fifty days of age, the fish were transferred to a set of 1.5x1.5x1 m concrete pond, with a volume of 2000 L receiving about 30 L min^{-1} of fully oxygenated water. Feeding was provided by hand feeding at a rate of 1.5% b.wt. day^{-1}. Water temperature ranged from 8-12°C with an average 11±1.41°C.

Data Collection

The following measurements were taken.

- Fertilization rate
- Survival from fertilization to eyed stage and from eyed stage to swim-up
- Hatchability rate
- Percentage of swim up fry
- Percentage of active feeding stage
- Growth at different period of time

For estimate the fertilization and eyed eggs rate 100 eggs were tested. Data of survival rate were estimate by the daily decrease in the number of larvae during experiment period (300 days).

Abnormal Larvae and Juvenile

The percentage of deformed specimens was estimated by counting 100 larvae for each experimental period. In progressive stages normal and lordotic (n = 60) juveniles were anaesthetized, weighted and radiographed using a medical x-ray system. The system was adjusted for fish at the size of 30-60 g. The x-ray pictures were used for examination of the skeleton.

The lordosis was evaluated by measuring the place and number of vertebrate at which defect occur.

Statistical Analysis

Statistical comparisons between survival data were performed with one-way factorial analysis (ANOVA), with confidence level of 95%, followed by the fisher, Protected Least Significant Difference, (PLSD) multiple comparison test using the program SPSS 11.5 for Windows.

The weight gain was measured for 10 fish in each treatment. The percentage of deformed larvae was estimated by counting total larvae for each experimental period.

The analysis of variance showed that the effect of inbreeding was not significant on fertilization, eyed stage hatching stage, neither at the swim-up nor at active feeding stage ($p>0.05$).

Analysis of variance revealed that effect of inbreeding on the weight of juvenile was not significant at 150 days and 225 days weight giant of fishes but a highly significant effect was at 300 days of rearing ($p<0.00$). The difference between inbreeding depression for male and female half-sib group was non significant ($p>0.05$). The differences were not significant of sire line ($p<0.111$) and dam line ($p<0.819$) too.

In second series of experiment sampling at fertilization, active feeding stage and weight gain at 30, 50, 70, 90 and 105 were done.

The accumulative effect of inbreeding depression of fry survival and 150 days weights on resulting total weight and total number of fish produced at 150 days of age, was calculated by using the mean inbreeding effects measured at the $F = 0.25$ level of inbreeding in Table 3.

The combined effects of the remaining two traits at an inbreeding level of full-sib mating reduced the total weight of fish produced at 150 days of age 5.83 and 8.49% in two groups.

Abnormal Larvae and Juvenile

Result on the effect of inbreeding on growth related variable at active feeding and weight at 150-300 days showed non-significant difference between percentage of abnormal larvae and juvenile between half sibs and full fish mating (pe”0.05).

The highest mortality occurred during one month at the time of hatching up to feeding stage. Many abnormal larvae could not rupture the eggs shell and died before hatching (18.4%). A few hatched but was not able to swim; they stayed at bottom of incubator (10.3%). Others reabsorbed the yolk sac and start internal feeding.

On the 30th day, the percentage of larvae survival was 57.8%; this figure decreased to 54.9%, five month later, when larvae weighted 4.6 g. The most frequent observed deformations at this time were deformation of vertebral column in tail (up to 92% of abnormal fry) and other abnormalities observed were operculum atrophies (5%), cranial abnormalities (1%) and abnormal shape.

About 10% of the newly hatched larvae observed, displayed clearly distinguishable skeletal deformities. Six days after hatching, yolk sac reabsorption occurred; larvae become active and commenced feeding. The number of deformed larvae progressing decreased from this period and at the end of larval period, they represented about 3% of survival larvae. Most of these deformed animals had deformation in the vertebral column.

Abnormal specimens showed unbalanced growth, which started to be clearly noticeable 3-4 week after hatching (1 g). The difference progressively increased and, 150 days after hatching, the juveniles measured 14.1 cm in length, while normal specimens measured 12.4 cm. X-ray examinations shows the place of defect (Fig. 1) and some juvenile showed V-shaped vertebral column. In some juveniles a skeletal sign of deformity more than one point of curvature was identified, whereas others showed such slight undulating curvatures of the vertebral column, that externally they were recognized only by the irregular trajectory of this lateral line. Juvenile salmo trout have 63 spinal vertebral (21 cervical, 21 abdominal and 20-22 caudal). The frequency of affected spinal vertebrae was estimated and the vertebrae between 38-48 were found to be most frequently affected with a maximum frequency at the 43rd vertebra. The lowest frequency was found to correspond to the vertebrae behind the head.

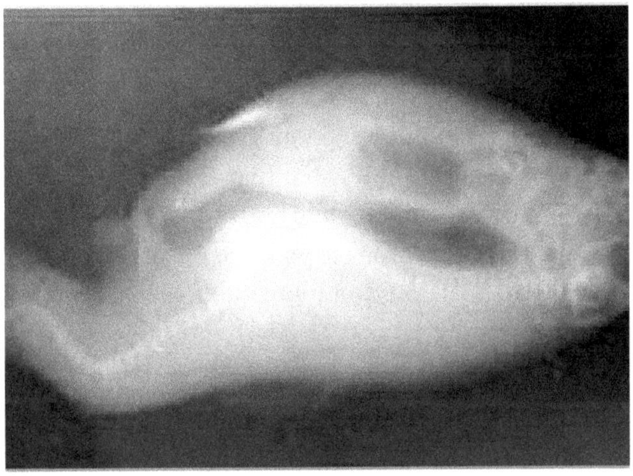

Fig. 1: Deformation in vertebral column of rainbow trout is shown by radiograph assay

In the present investigation, environmental conditions seemed to have caused higher rate of abnormalities than inbreeding as the period of incubation was very long and the eggs of cold water fish are more sensitive to environmental variables.

In reared rainbow trout, inbreeding was thought to involve two conditions, abnormality and reduced growth rate. The accumulation of inbreeding is a natural consequence for closed population undergoing artificial selection.

Critical values for inbreeding level have not been established in fish, but researchers suggested using F = 5% as a conservative estimate and F = 10% as a tolerate estimate. In the present study, detectable harmful effects on important traits such as fecundity, survival and growth at an inbreeding level of 0.25 have been investigated. For this purpose the effect of inbreeding was estimated in male and female half-sib families.

This experiment demonstrated reduced survival at eyed stage, swim-up and at feeding stage in male and female half-sib, but difference in survival rate observed was not significant as was reported by for Rainbow trout. During the rearing period from 150 to 300 days (rearing at 10°C) and from 30 to 105 days (at 18°C) at advanced stage of rearing weight reduction due to inbreeding was significant.

This trend was expected considering that these measurements were taken at a series of age intervals on the same families. Since growth occurs as a function of body mass at the beginning of the particular growth period, any deviation between the inbred and outbred, weight at the end of one growth period would be expected to increase, in each subsequent growth period (Kincaid, 1983). Therefore at progressive growth the mean depression in weight showed significant difference among male and female half-sib with full-sib mating. The cumulative effect of inbreeding depression of fry survival and weight at 150 days, showed an average 7.16% reduction in full-sib groups. This reduction evident at the fingerling stage (4.5 g) was found to be more pronounced and significant at the juvenile stage. Inbreeding depression in body weight showed a clear trend to increasing trend with age from 150 days to adult stage. It may be due to the cumulative effect of inbreeding as the body of the fish successively grows larger as stated by the researchers.

Inbreeding depression could not be found in body weight at early stages before 256 days age, as stated by researchers. They stated that at relative low level of inbreeding (an average up to 19.5%) inbreeding depression in body weight at early age stages is not serious enough to be detectable. It is in accordance with the result obtained at inbreeding level of F = 0.25. Researchers found no significant inbreeding depression in growth of fingerlings even at high level of inbreeding (F = 0.25-0.50) and they assumed that the observed inbreeding depression of survival counteracted inbreeding depression in body weight because mortality may preferentially act on the smallest individuals in the population.

The effects of inbreeding of the maternal and paternal family groups could be markedly different or even significant. Maternal effect was expected to be appreciably larger than paternal effect during early period of life because of the influence of embryo size and the quantity and quality of yolk materials, which are almost entirely determined by the female during egg formation. The formation of maternal half-sib families from a random half of the eggs of a single female maximized the uniformity of the eggs used in each of the female mating.

In parental families, however, the range in size and general quality of the eggs varied at random, depending on the particular females selected. Inbreeding would be expected to have a depression effect on maternal influence where the female was herself inbred. Using large population of breeders (more than 250 pairs) in Kelardasht fish farm and choosing randomly few breaders, no inbred male or female would be expected in half-sib mating or control groups.

In other hand as researchers reported, maternal effects should not be large for salmonid because the offspring are not nurtured. The most likely source of maternal effect would be difference in egg size and egg quality. However as reported by researchers, in most studies with rainbow trout, any correlation observed between egg size and early growth disappears soon after first feeding. Therefore full-sib family effects, as observed in the present study, most likely reflect only common environmental effects.

The same statement was obtained by researchers in their study in rainbow trout. Existing studies shows dam effect on survival at eyed stage, whereas swim–up and active feeding stage reveal no significant effect. Egg size may show significant effect in salmonid as observed in many studies.

However as shown by researchers, egg size had no effect on the survival from fertilization to the eyed stage. A review of studies of the effect of sire or dam revealed that survival is related to environmental factors or at least can be considered to have a low additive genetic component.

In a review of several reports, researchers found that heritabilities estimated from dam components were a bit higher than those estimated from sire components of variance, suggesting strong common environmental and/or non-additive genetic effect.

The full-sib family effects on body weight at two experiments (30-105 and 150-300) showed two differences in trends. First, the percentage inbreeding depression did not increase as age increased from 30 to 105 days. This trend is an anomaly largely explained by the higher fry mortality observed in inbred lots, which lowered fish density and thus reduced competition for the available space and food, even though thinning was done randomly, by culling extremely small fish, resulting in a decrease in the variance within full-sib families.

The growth of the remaining fry reared at a lower fish density would give the appearance of decreasing effect of inbreeding depression, as suggested by the trend seen in this experiment. Second, markedly evident of increasing depression with advancing age was seen from 150 to 300 days. A relatively large and significant increase of this effect of depression was found at 300 days.

Inbreeding depression for body weight was relatively high at 300 days of age. Inbreeding effect as well as inbreeding rate was high for body weight at harvest time. This may be due to some families responding more severely to inbreeding or selected future parents might have come from related families.

As we stated earlier, the high level of spinal malformations appearing in hatchery rainbow trout is an important problem in fish farms. Abnormal fish have an unfavorable shape, decreasing the production and increasing the cost.

The etiology of these syndromes in fish has been study by many researchers and nutritional, pathogenic, environmental and genetic causes have been cited. Nutritional factors such as deficiencies in ascorbic acid, vitamins and triptophem in the diet of parental fish and larvae have shown different skeletal deformities. Pathogens such as the mynosporidian *Myxosoma* sp.

Researchers may induce congenital or postnatal skeletal deformation in fish. Environmental factors such as density of eggs, mechanical or thermal shocks, pollution, radiation, salinity, oxygen deficiency of water and light intensity have also been reported to cause aberrations in development.

In the present study no significant difference in the number of abnormal fish at active feeding stage at the end of yolk resumption or in the growth at 150, 225 and 300 days was observed. Any markedly significant difference between different mating to confirm genetic causes of variability were also not observed during the investigation. In spite of high percent (about 10%) of larvae showed abnormalities and the rate decreased to 3% on the 300th day with most of the abnormal juveniles exhibiting vertebral abnormalities in the tail region. Survival of the larvae with vertebral abnormalities was found to be higher compared to other more serious types of abnormalities.

Therefore, in contrast to the other abnormalities, vertebral abnormality in larvae does not totally interfere with feeding, but during culture they are collected and culled at different ages. In these respect economical losses of vertebral abnormality is higher than those which lead to larval mortality at the time of hatching. To avoid losses and to achieve optimum production and income it is suggested that rainbow trout culture should be carried out only by standard farms using good quality of water, high value nutrition, large number of breeders and avoiding of full or half-sib mating through identification of individual families.

Simple measures to avoid inbreeding

- The broodstocks in a hatchery should be particularly replaced periodically. Exchange of brood stocks between the local hatcheries can be done.
- Brood stocks of different age groups should be bred together. This help in reducing the chance of loss of some valuable alleles due to genetic drift.
- Natural stocks may be inducted in to the hatchery periodically to increase the heterozygosity.
- The pedigree record of brood stock should be maintained to avoid the mating of close relatives.
- Cultured populations should be identified using a proper marking system. Females and males have to be originated from two different lines.
- Individual fish with poor constitutional conditions or anatomical abnormalities should be culled.

INBREEDING DEPRESSION

As inbreeding increases, it often causes a decrease in productivity which is termed "inbreeding depression".

- Inbreeding depression is a decrease in growth rate, fecundity, etc. that is observed in the inbred group when it is compared to a control population where three is no inbreeding.
- The severity of inbreeding depression depends on the level of inbreeding, the phenotype in question, and the population.

There are several explanations for why inbreeding depression occurs, and what are all contribute to inbreeding depression.

Qualitative genetic explanation

i. The most common explanation is that inbreeding depression occurs because of the pairing and expression of detrimental recessive alleles.
   - Although the terms "dominant" and "recessive" do not meant that one allele is good and the other undesirable, most of the alleles that produce abnormal phenotypes or that lower viability are recessive.
   - Even though mutation rates are usually only one in 10,000 to one in 100,000 replications per gene, because each fish can produce many gametes (up to 500,000,000 sperm per spawn) and because each fish has tens of thousands of genes in its genome, each fish produces dozens to hundreds of gametes that contain a mutant copy of one or more genes.
   - Most of these mutant alleles are recessive.
   - Since recessive alleles can be expressed only when a fish is homozygous, these mutations tend to accumulate in a population.
   - Consequently, many fish carry "hidden" copies of these mutant alleles in the heterozygous state.
   - Each fish carries a number of these detrimental recessive alleles, and inbreeding uncovers them.
   - When relatives mate, they produce offspring with an increased level of homozygosity; consequently, some of the detrimental recessive alleles that the parents carry in the unexpressed heterozygous state are paired and expressed in the offspring.
   - Detrimental recessive alleles can also be paired and expressed when unrelated fish mate. The difference is one of magnitude, and that is what causes inbreeding depression.
   - It has been estimated that each individual carries dozens of mutant alleles that lower viability.
   - Most of these mutant alleles will produce only a small reduction in fitness, but several will produce phenotypes so abnormal that they are lethal or cause premature death.
   - Even though the probability of producing a defective offspring is the same for each recessive allele (25% if both parents are heterozygote), if both parents carry a large number of identical defective recessive allele, the chance of producing defective offspring increase dramatically.
   - The more closely related the parents, the more alike they are genetically, so it is more likely that defective offspring will be created.

ii. A second explanation for inbreeding depression is that by decreasing heterozygosity, inbreeding reduces what is called "over dominance".
    - Over dominance occurs when the heterozygous genotype produces a phenotype that is superior to the two homozygous loci.
    - The best known example of over dominance occurs in humans. In many populations, the gene which produces haemoglobin exists in two allelic forms: one produces normal haemoglobin, while the other produces sickle-cell haemoglobin.
    - Individuals who are homozygous normal are susceptible to malaria and get quite ill and can die when infected. The heterozygote are superior, even though they are slightly anaemic, because they are resistant to malaria.

Quantitative genetic explanation

Since inbreeding increases homozygosity, it also decreases heterozygosity. By doing this,

- Inbreeding affects the interactions between alleles at each locus, which means that it affects V_D, and dominance effects can play a major role in the production of many quantitative phenotypes.
- When h^2 is small (<0.15), as it often is for viability and fecundity, dominance effects can be quite significant.
- The creation of excess homozygosity tends to disrupt dominance effects, which tend to be more unidirectional than the additive effects.
- Inbreeding can be considered to be opposite of crossbreeding, which is used to exploit V_D.
- Consequently, if crossbreeding maximizes exploitation of V_D, inbreeding minimizes exploitation, which means that inbreeding makes phenotypes controlled by V_D worse (inbreeding decreases phenotypic mean).

This genetic explanation makes sense, because inbred lines are often used in crossbreeding programmes. The creation of inbred lines is done to produce opposite types of homozygosity that will be maximized when hybrids are created, which is how V_D is exploited. This explanation is further supported by the fact that crossbreeding, the breeding programme that is used to exploit V_D, can also be used to eliminate inbreeding and inbreeding depression.

Simple measures to avoid inbreeding

- The broodstocks in a hatchery should be particularly replaced periodically. Exchange of brood stocks between the local hatcheries can be done.
- Brood stocks of different age groups should be bred together. This help in reducing the chance of loss of some valuable alleles due to genetic drift.

- Natural stocks may be inducted in to the hatchery periodically to increase the heterozygosity.
- The pedigree record of brood stock should be maintained to avoid the mating of close relatives.
- Cultured populations should be identified using a proper marking system. Females and males have to be originated from two different lines.
- Individual fish with poor constitutional conditions or anatomical abnormalities should be culled.

INTERSPECIES HYBRIDIZATION

Hybridization is a conventional technique commonly used in fish breeding. Hybrids are defined as the products of crossing distinct species. A considerable number of hybrids have been produced between different fish species. The occurrence of natural hybridization in fishes is known since long. It is more common in freshwater fishes than in marine.

- The purpose of hybridization is to bring out or combine useful traits by crossing fish with different phenotypes and genotypes.
- The F_1 hybrids will have new phenotypes and genotypes and usually exhibit **heterosis**.
- Hybrids with obvious heterosis in the first generation can be used directly in commercial fish production, and also in fish breeding.
- Hybridization can be used in one of several ways to improve productivity. This can be incorporated into a selection programme as a final cross to produce animals for grow-out. In this case, the fishes are selected in two lines that have been shown to produce good hybrids; after the selection programmes, the selected lines are hybridized.
- Interspecific hybridization was successfully obtained in many fish and shellfish genera and/or families (carps, catfishes, cichlids, moronids, salmonids, sparids, sunfishes, oysters, crayfish) as a means of improving production traits (e.g., growth rate, survival, disease resistance) as well as to manipulate sex ratios.
- Sometimes an interspecific hybrid does not exhibit heterosis for any trait, but is still quite important for aquaculture as it expresses a good combination of beneficial traits from both parent species.
- Most of the hybrids evaluated did not prove beneficial for aquaculture. However, a few valuable hybrids are commercially exploited. Hybridization in fishes has the following application in aquaculture.

INTERSPECIFIC HYBRIDS

(Hybrids produced by crossing between different species)

- O_+ *Labeo calbasu* x *L. rohita* (male) (calbasu - rohu). Hybrid fertile and has smaller head. Other morphometric characters are intermediate between calbasu and rohu. Hybrid grows faster than *L. calbasu*.
- O_+ *Labeo– rohita x L. calbasu* (male) (rohu - calbasu). Hybrid also fertile. Superior in growth rate to calbasu. Both the above hybrids can be used in place of calbasu in aquaculture.
- The sunshine bass, a hybrid between white bass (*Moronechrysops*) female and the striped bass (*M.saxatilis*) male grows faster and has better overall culture characteristics (good osmoregulation, high thermal tolerance, resistance to stress and disease, high survival in culture, and ability to utilize soy beans as protein source) than either parent species.
- This hybrid is commercially cultured mainly in the USA,and in Israel and Taiwan.

INTERGENERIC HYBRIDS

iii. O_+ *Labeo rohita* x *Catla catla* (male) (Rohu–Catla hybrid)

iv. iiO_+*Catla catla* x *Labeorohita* (male) (Catla-Rohu hybrid)

- Of all the hybrids rohu x catla is the most promising hybrid in terms of growth, food and feeding habits, viability and higher flesh content.
- Both hybrids are intermediate in most of the taxonomic characters.
- The most distinguishing feature of both hybrids is that in rohu x catla pectorals almost reach the ventral base, whereas in catla - rohu they do not reach the ventral base and leave a wide gap.
- In both hybrids, the head is smaller than catla and flesh more than either of the parents.
- Growth of both hybrids is faster than rohu but slower than catla.
- Even F_2 fry of rohu - catla hybrid grew faster than rohu but slower than catla in 6 months when reared in 0.1 ha pond. Rohu - catla hybrid attains maturity in 3 years.
- Rohu x catla hybrid when compared to equal sized parent contains more quantity of flesh (54 percent) than catla (44 percent) and rohu (48 percent).
- Catla-calbasu hybrid grows faster than calbasu, has a smaller head than that of catla, accepts artificial feed more readily (unlike catla which does not accept artificial food readily).

INTERSPECIFIC HYBRIDIZATION OF THE TIGER PRAWNS

Interspecific hybridization can increase productivity through

- Hybrid vigour,
- Produce animals that are sterile, or
- Combine desirablecharacteristics found in one species with those of another.
- Attempts to produce hybrid crustaceans are few, largely because control of reproduction and mating in crustaceans was limited.
- Interspecies hybrids of penaeid prawns have been successfully achieved for *P. setiferus* x *P. stylirostrus, P. setiferus* x *P. schmitti* and reciprocal crosses of *P. monodon* and *P. penicillatus* although spawn rate, hatch rate, and the survival of hybrid progeny to post-larval stages were low compared with the intraspecific matings.
- No successful matings or induced maturation of hybrid females by eyestalk ablation has been successful, although sperm and ova were observed in mature hybrids from *P. setiferus* x *P. schmitti* crosses.
- The hybrids of *P. monodon* x *P. penicillatus* reciprocal crosses, were recorded as growing faster than either of the parental species.
- The several interspecies hybridization attempts with penaeids have successfully produced larvae is promising, but the low hatch rates in the hybrids, lack of hybrid vigour, sterility in hybrids suggest that immediate commercial application are unlikely.

HYBRIDIZATION AMONG FRESHWATER PRAWNS

A cross was made between a male *Macrobrachium malcolmsonii* and a female *M. gangetium*; these two species are considered alternative or additional candidates to *M. rosenbergii* for aquaculture in India. Hybrids obtained did not attain maturity within the first year of rearing but gonad maturation, breeding and spawning were observed after 14 months. F_1 and F_2 hybrids were also obtained. Earlier attempts to produce inter specific hybrids between *M.rosenbergii* and *M. malcolmsonii* suggested the presence of in complete reproductive barriers between the species. *Macrobrachium* species have been hybridized by natural service matings, but the development of artificial insemination techniques, and *in-vitro* fertilization, allowed inter species crosses where behavioral or anatomical differences prevented natural service. Hybrids that were obtained crossing female *M. nipponse* with the male of *M. hainanense* had a hatching rate of over 90 percent and a survival rate (to PL) of 20-60 percent.

9

CHROMOSOME SET MANIPULATION AND SEX CONTROL

INTRODUCTION

Chromosome-set manipulation techniques of sperm chromosome inactivation (with radiation or chemicals) and suppression of cell divisions (with heat shock, cold shock, or pressure) can be readily applied to fish to produce gynogenetic and polyploid individuals. Gynogenetic individuals have all their chromosomes from the female parent and should all be females in species with XX females. Polyploids include triploids that are expected to be sterile, and tetraploids that have the potential of being fertile and producing sterile triploids when crossed to normal diploids. Partially inbred gynogenetic diploids and triploids may be produced by treatments causing retention of the second polar body of the egg. Completely homozygous gynogenetic diploids and tetraploids may be produced by treatments blocking the first mitotic division.

Chromosome set manipulation methods include induced gynogenesis, androgenesis, and polyploidy. Artificial sex regulation in common carp, as well as in other fish species, is connected mainly with elaboration and application of the hormonal sex reversal method. Common carp is a traditional subject for chromosome set manipulation studies. It is a popular aquaculture species with established methods of artificial spawning and rearing. Common carp is a convenient subject for these types of investigations since it has high fecundity and unique morphological genetic markers such as genes for scale cover types and color mutations. The development of techniques for production of gynogenetic, androgenetic, polyploid, and monosex progenies in common carp is described in this review from a chronological perspective.

Before presenting information on common carp, a brief general description of each method is provided. More detailed general information may be found in reviews on chromosome set manipulation and sex control in fish.

INDUCED GYNOGENESIS

Gynogenesis is embryo development under control of only maternal heredity after activation of eggs by insemination. Natural gynogenesis has been described in several allfemale forms of fish. Gynogenetic progenies may be obtained in species reproducing by the normal sexual mode. In this case, the main cytogenetic features of gynogenesis, inactivation of male chromosomes and prevention of female chromosome set reduction, are achieved by special experimental treatments.

Inactivation of male chromosomes is induced by irradiation of sperm. X-rays, gamma rays, or ultraviolet (UV) irradiation may be used for this purpose. This method of chromosome inactivation is based on different sensibilities of chromosomes and cytoplasmatic structures of spermatozoa. At high dosage of irradiation, chromosomes are inactivated but spermatozoa maintain ability for movement and insemination of eggs.

After insemination of eggs with genetically inactivated spermatozoa, gynogenetic haploids are produced. Haploids in fish are morphologically abnormal ('haploid syndrome') and die before or soon after hatching. In order to obtain viable diploid gynogenetic fish, the haploid female chromosome set must be doubled. A female chromosome set may be doubled in two ways: either by suppression of the second meiotic division in eggs (meiotic gynogenesis) or by suppression of the first mitotic division in haploid embryos (mitotic gynogenesis). For artificial suppression of meiotic or mitotic divisions, strong physical treatments (shocks) are applied to the embryos. Most commonly used treatments are either low or high temperatures (cold and heat shocks) or hydrostatic pressure.

For suppression, shock is usually applied at anaphase of the corresponding division. Under the influence of this treatment, spindle fibers are destroyed and, as a result, division stops and daughter products are fused. When the second meiotic division is suppressed, the second polar body is not extruded; it is fused with the haploid female pronucleus. During suppression of the first mitotic division in haploid embryos, two haploid nuclei are united to form a diploid nucleus. After the next mitotic cycle two diploid blastomeres are formed. With low frequency (usually less than 0.1%) viable gynogenetic diploids may appear without shock application due to sporadic spontaneous suppression of the second meiotic division in eggs (meiotic gynogenesis).

INDUCED ANDROGENESIS

Androgenesis is embryo development under the control of only paternal chromosomes. Genetic inactivation of female chromosomes for induction of androgenetic development in fish may be achieved by irradiation of eggs. After insemination of eggs with genetically inactivated chromosomes by intact spermatozoa, androgenetic haploids are produced. To produce viable diploid androgenetic fish, the paternal haploid chromosome set must be doubled. This may be achieved by suppression of the first mitotic division in the haploid embryos.

Diploid androgenesis in common carp has been reported by researchers. Female chromosomes were inactivated by irradiation of eggs with X-rays; diploidy was restored by suppression of the first mitotic division in androgenetic haploids with heat shock. Researchers succeeded to induce androgenetic development in common carp by UV-irradiation of eggs stirred in a synthetic ovarian fluid; the diploidy was induced by heat shock. In further studies viable androgenetic YY males were identified, that after crossing with normal females (XX), produced all-male progenies. Androgenesis has been used for obtaining of clones in common carp.

INDUCED POLYPLOIDY

Induced polyploidy is artificial production of individuals, with an increased number of haploid chromosome sets. In aquaculture and fisheries, this method is primarily used for production of triploid fish, i.e. fish whose karyotypes contain three haploid chromosome sets. Triploid fish are genetically sterile, i.e. they are not capable of producing viable progeny. Also triploids are characterized by complete or partial reduction of gonads. Abnormalities in development of reproductive systems in triploids are caused by the presence of a third, 'additional' haploid chromosome set, which disturbs the normal process of conjugation and disjunction of homologous chromosomes.

There are several methods for producing triploid fish. The simplest and better developed method is based on suppression of the second meiotic division in eggs after insemination by intact spermatozoa. Production of triploids by this method has been elaborated along with the technique for inducing meiotic gynogenesis, since both methods use similar shocks for doubling of the female chromosome set. Shock application should provide a high frequency of triploids, but at the same time does not decrease drastically embryo survival.

Cold shocks were used frequently in initial studies on induced triploidy in common carp; in further investigations heat shocks were mainly applied. From a practical point of view, application of heat shocks appears to be more convenient for mass production of triploid progenies since duration of heat shock is much shorter than duration of cold shock. Recoubratsky et al. (1989, 1992) described the method for mass production of triploid common carp using heat shock. The following parameters were recommended as optimal: timing - 0.2 s0 after insemination (about 6 min at pre-shock temperature 20 °C), shock temperature and duration - 40 °C and 2 min, or 41 °C and 1.5 min. These treatments provided 80–100% triploids in progenies with embryo survival 50-70% of controls.

As already mentioned, triploid fish are characterized by a reduction in gonad development. This makes triploids potentially valuable for commercial rearing. It was speculated that, during the rearing of triploids, there would not be retardation in somatic growth, usually recorded in diploid fish during the period of gonad maturation. Theoretically, this advantage should be especially profound when sex maturation is reached before fish reach a marketable size. Researchers assessed triploid common carp for culture in such conditions. A reduction in gonad development (especially ovaries) in triploid fish was observed and verified in both studies. However, the reduction in gonad development in triploids did not result in an increase of somatic growth rate.

Moreover, triploid fish grew slower than diploid fish in almost all comparative trials. These data cast doubt on the potential of triploid common carp for commercial rearing.

The second method of production of triploid fish is the crossing of previously obtained tetraploid fish with normal diploid fish. Tetraploid fish can be obtained by suppression of the first mitotic division in normal diploid embryos. This method of triploid fish production was realized in rainbow trout.

Researchers have reported results of successful experiments on mass production of tetraploid common carp larvae by application of heat shocks; in some trials the frequency of tetraploids in progenies was up to 100%. However, attempts to raise common carp tetraploids have demonstrated their very poor survival. Only three tetraploid fingerlings have been reported.

PLOIDY MANIPULATION

'Ploidy' means a set of chromosomes or genome and 'manipulation' means to increase or decrease. So, ploidy manipulation means the change (either increase/decrease) in the number of set of chromosome of an organism.

For example, fishes are diploid organisms; by ploidy manipulation we can produce triploid or tetraploid fish.

A number of techniques have been developed to produce haploid, triploid and tetraploid fish and even to produce fish whose chromosome come solely either from their mother (Gynogen) or from their fathers (Androgens).

Chromosome manipulation can be also defined as the alteration of chromosome set to N, 3N or 4N by applying temperature, pressure or chemical shocks.

Significance of ploidy manipulation:

Ploidy manipulation is very much important in aquaculture because triploid, teraploid, gynogens and androgens are produced through ploidy manipulation are given bellow:

1. Triploids have an increased DNA content, nuclear size, number of chromosomes and cellular size compared to diploid, and also triploids do not divert energy for growth, to gamete production, reproduction or care of young. So, they show higher growth.
2. To provide monosex male or female population.
3. To discover the sex-determination system.
4. To create highly inbreed lines.
5. To save the endangered species.

Why chromosome set manipulation?

The basic goal governing the research and technology of chromosomoe manipulation is the control of sex and fertility. The ability to direct sex or create sterile fish is done

1. To prevent early maturation
2. To prevent reproduction during grow-out
3. To produce larger fish
4. To improve carcass (skeleton) quality
5. To grow the sex that has the greater market value and
6. To utilize exotic species both in fish farming and in fisheries management

The main key topics is –

- Polyploidy
- Gynogenesis
- Androgenesis
- Sex-reversed and
- Production of monosex population.

Methods of ploidy manipulation:

The methods of ploidy manipulation are mainly divided into three types-

- Temperature :
- High temperature /heat
- Low temperature /cool
- Pressure
- Chemicals

Triploids

Triploids are produce by shocking newly fertilized eggs (E) to prevent extrusion of the second polar body (SPB). The fertilized egg thus contain an egg haploid nucleus (EN), a sperm haploid nucleus (S) and a second polar body haploid nucleus which fuse to form a triploid nucleus and create a triploid zygote.

Triploids are those fish that have three sets of chromosomes. Triploidy has been induced in a variety of fish species and is of interest to aquaculturist because it results in functional sterility and increased growth.

Production of Triploids

Triploids are created by shocking newly fertilized eggs, shortly after fertilization. Fish eggs do not extrude the second polar body until they are fertilized. Because of this, if a newly fertilized egg is shocked, the shock prevents the second polar body from leaving the eggs. As a result, the fertilized eggs contain three haploid nuclei, one from the eggs, one from the sperm and one from the second polar body. The three haploid nuclei will fuse to from a triploid zygote nucleus which creates a triploid.

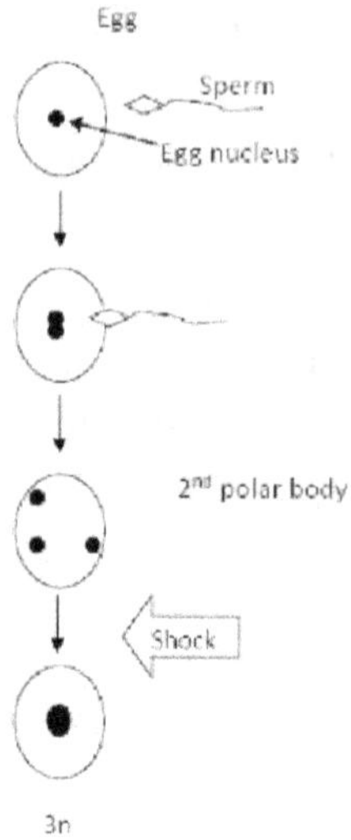

Fig. 1: disruption of cellular processes associated with meiosis-II, causing the retention of second polar body.

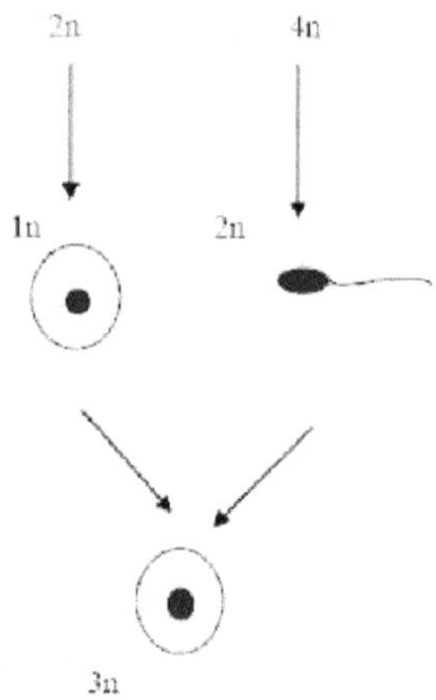

Fig. 2: The production of interploid-triploids is achieved by fertilization of a haploid egg form diploid female with a diploid sperm from tetraploid male.

For triploidy the retention of second polar body causing the formation of nucleus with three sets of chromosomes.

Triploidy can also be produced by crossing between a tetraploid and diploid. Tetraploid rainbow trout have also been crossed with diploid to produce triploids. These progeny are referred to as interploidtriploids. Because no shocks are applied, this method appears to be more reliable and avoids detrimental side effects associated with conventional methods of triploidy induction.

Why triploids are produced?

Triploids are created for two reasons:

Increased growth

Triploids have an increased DNA size, nuclear size, number of chromosomes and cellular size compared to diploid. Beside these, the nucleus of triploids contains more than 33% allele for growth and triploids don't divert energy for growth to gamete production, reproduction or care of young. So, triploids have faster growth.

Sterility

More importantly, triploids are created to produce sterile fish so that species that might cause adverse environmental impacts can be cultured on farms or used in natural resources management. In triploid organisms, gametic incompatibility occurs during meiosis because triploids have three sets of chromosomes.

Due to presence of three sets of chromosomes, the chromosomes cannot properly pair and align themselves, which make it difficult to undergo meiosis, because the chromosomes cannot be divided into two equal haploid sets. Gametes that are produced usually aneuploid and they contain unbalanced number of chromosomes.

Aneuploid gametes usually produce abnormal and sub-viable offspring that do not live long. So, triploids are sterile.

Examples of triploid production in fishes

Triploids have been created in many species including common carp, grass carp, bighead carp, rohu, *T. aurea*, *T. nilotica*, channel catfish, Asian catfish, African catfish, European catfish, Atlantic salmon etc.

Significance of triploids production

- Triploidy produce sterile fish, so the species that might adverse environmental impacts can be cultured on farms or used in natural resource management.
- The production of sterile fish can reduce the risks of exotic introduction and contamination of wild gene pools.
- It can be safe guard against unauthorized perpetuation of transgenic genomes and project intellectual property.
- The development of sterile stocks for food fish production can be used to control precocious spawning in fishes that reach sexual maturity before they reach market size and may improve growth.

Tetraploidy:

Method 1:

Tetraploids are organisms/fish that have four sets of chromosomes instead of normal two sets. Tetraploids can be produced by shocking the diploid zygote after the extrusion of the second polar body but before the first cell division. Application of shocks just prior to the first mitotic division causes the breakdown of actin filaments and prevents the formation of the cleavage furrow. So, the shock prevent nuclear and cell division. So the resulting zygote nucleus will contain four sets of chromosomes instead of two.

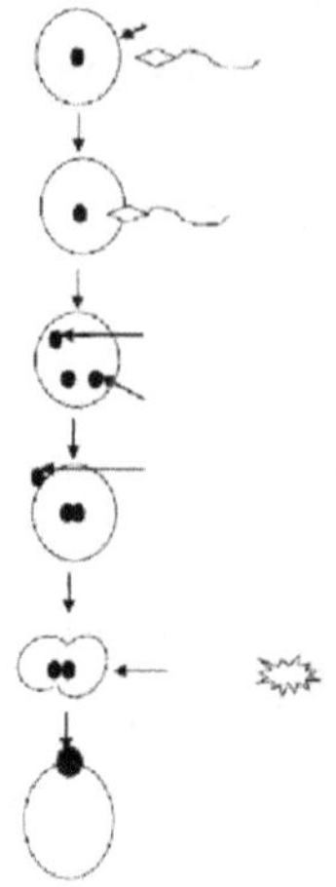

Fig. 3: disruption of cellular processes associated with meiosis-II, causing the retention of second polar body.

Method 2:

Tetraploids can also be produced by crossing between two tetraploids organisms. The mating between two tetraploid rainbow trout can produce second generation tetraploids (Fig. 4).

Uses

Tetraploids could be used to produce interploid triploids, which would eliminate the need to continually create triploids manually. For this a tetraploid female is mated with a diploid male. The female should be tetraploid because the micropyle of eggs from diploid female might not be large enough to admit sperm from tetraploid males (Fig 5) .

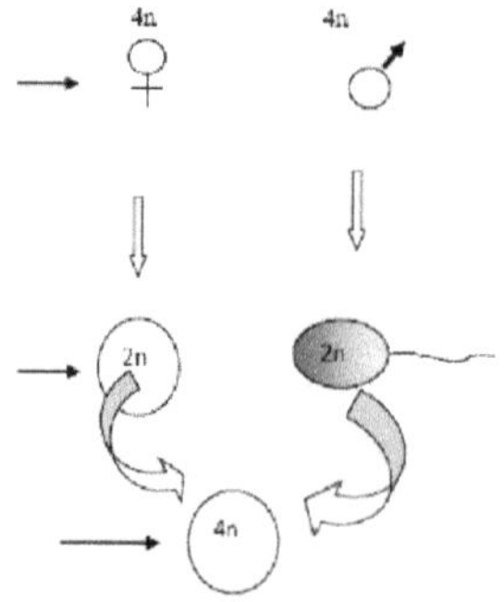

Fig. 4 Maintaining tetraploid lines by fertilizing diploid eggs from tetraploid female with diploid sperm from tetraploid male.

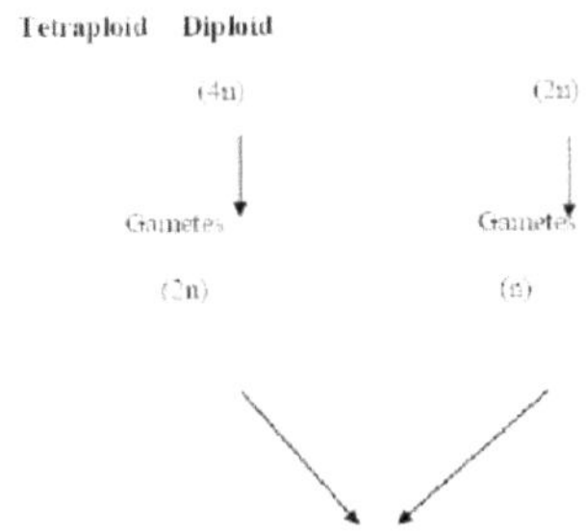

Fig. 4: 'Triploids'

Gynogenesis

Fish that contain both sets of chromosomes from the maternal parents are called gynogens, and gynogenesis is the process of producing gynogens by fertilizing the eggs with sperm which have had their genetic material destroyed by X-rays, gamma-rays, or UV-irradiation.

Significance of gynogenesis

- To produce monosex female population.
- To discover the sex-determination system. It is possible to differentiate whether a female is homogametic; gynogenesis will produce only females, and when the female is heterogametic, gynogenesis will produce equal numbers of males and females.
- To create highly inbreed lines.
- To create homogenous lines, if it is coupled with sex reversal.
- To study gene mapping.
- To create clone lines.

Types of gynogenesis:

In a fish, a mature egg completes meiosis-I, before fertilization. After fertilization, meiosis-II is resumed and the separation of the two sets of chromosomes occurs. One set forms polar body and other set forms the egg pronucleus.

Generally gynogenesis is **two** types:

1. Meiotic gynogenesis
2. Mitotic gynogenesis

Meiotic gynogenesis:

It is a process of producing gynogens by preventing the extrusion of second polar body during meiosis-II, after the fertilization of eggs with inactivated sperms. In this process, the produced zygote receives both sets of chromosomes from the mother.

Mitotic gynogenesis:

It is the process of producing gynogens by suppressing the first cleavage of embryo/zygote. Here, the first step is to activate the eggs with the sperms which have had their genetic materials destroyed by irradiation. This will produce gynogenetic haploid. When this gynogenetic haploid undergoes first cleavage, the embryo is shocked and this prevents cell division. Two haploid nuclei fuse to form a 2n zygote and both sets of chromosomes come from mother.

Techniques of gynogenesis

There are three ways to produce gynogens

1. The first techniques:

The first techniques is to produce activate the eggs with sperm which have their genetic material destroyed by x-rays, gamma- rays, or uv–irradiation. Shortly after activation the egg is shocked to prevent the second polar body nucleus. This all fuse to form a 2n nucleus and both sets of chromosomes originated in the mother (Meiotic gynogenesis).

2. The second technique:

The second technique begins the same way; the first step is to activate the eggs with the sperm which have their genetic material destroyed by irradiation. This will produce a gynogenetic haploid. When the haploid zygote undergoes first cleavage, the embryo is shocked and this prevents the cell division. The two haploid (n) nuclei fuse to form a 2n zygote, and both sets of chromosomes come from the mother (Mitotic gynogenesis).

3. In the third techniques:

In the third techniques, the gynogens are produced by fertilizing eggs from tetraploid females with sperm who's DNA has been destroyed by irradiation. Tetraploid females produce diploid eggs. So, once they are activated, they don't have to be shocked to produce a diploid chromosomal component.

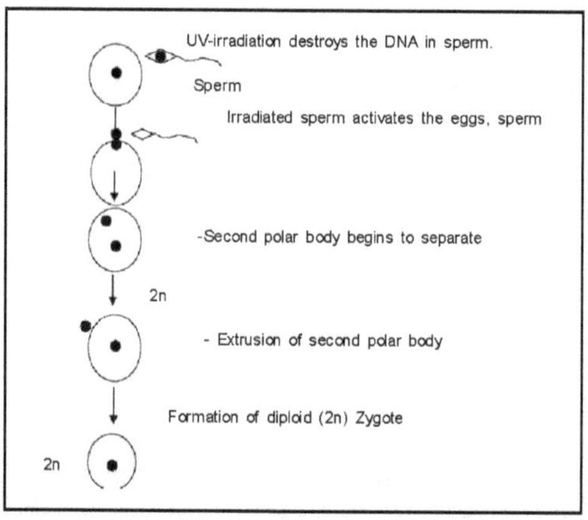

Fig. 6: method of gynogenesis by fertilizing eggs from tetraploid females with sperm which DNA has been destroyed by irradiation.

Adrogenesis

Androgens are fish whose chromosomes come from father but they receive mitocondrial DNA from mother, which gives some phenotypic variation. Androgenesis is the process of producing androgens (males) by fertilizing inactivated eggs with the normal sperms.

Importance of androgenesis:

- To produced highly inbred lines
- It can be used to produce super male (YY). For this purpose, the species which have XY-sex determination system can be used
- By using super male, we can produce all male population
- To produce clone lines
- To conserve the indigenous species through cryopreservated sperm
- To discover the sex-determination system
- To observe the effects of mitocondral DNA.

Types/ Techniques of androgenesis:

The androgens can be produced by one of two methods, or there are two types of androgenesis:

Androgenesis is a genuinely weird form of quasi-sexual reproduction of some eukaryote species in which a male's gamete develops to produce a new male, a clone or sub-clone of his father. In its original definition, androgenesis referred to when a sperm or pollen nucleus fuses with an egg nucleus, but the nuclear genome of the mother is eliminated by the paternal genome (Fig. 7). Alternative mechanisms of androgenesis are, however, recognized, particularly the development of an embryo from a male gamete in an egg that lacks a maternal nucleus. When such a 'non-nucleate' egg is fertilized by sperm or by pollen it develops into an offspring whose entire nuclear genome is of paternal origin. Both types of androgenesis (with and without selfish elimination of the female genome driven by the male) require diploidization of the paternal genome, or the ability of the gamete to develop and survive as a haploid male. As we explain in the following sections, whether and how diploidization occurs, and under which form of sex determination, greatly affects the predicted fate of androgenetic lineages.

Androgenesis is a developmental process, facilitating the inheritance of an exclusively paternal genome. It involves two steps:

- Elimination or inactivation of the egg's genome and
- Monospermic or dispermic activation of embryonic development by the haploid or diploid gamete;

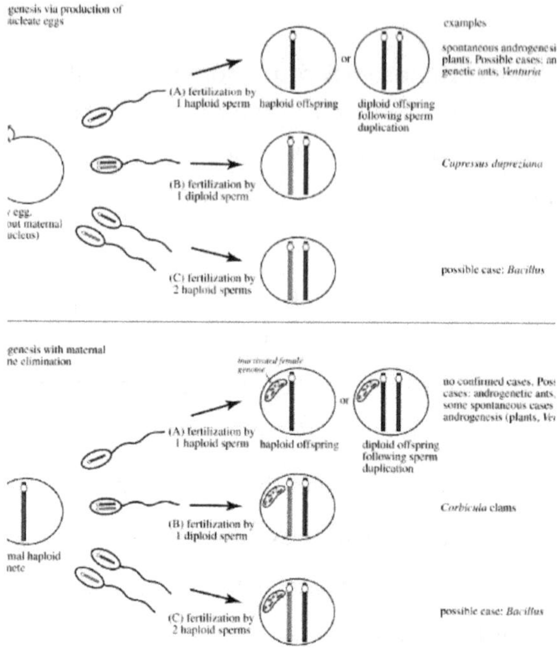

Fig. 7. Different forms of androgenesis.

Haploid monospermic activation requires restoration of diploidy by suppression of the first mitotic cleavage.

- Reliable protocol for inducing androgenesis, requiring the suppression of the first mitotic cleavage, however, is available for less than a dozen species only.
- Due to irradiation damage suffered by eggs,homozygous expression of lethal genes and damage inflicted by thermal shock treatment to suppress the first mitotic cleavage, survival of the androgenotes is very low.
- Androgenesis is induced in fish by fertilizing irradiated eggs (gamma or X ray or U-V ray) with normal spermatozoa.
- U-V (254 nm) irradiation is easy, inexpensive and can be easily set up under laboratory condition.
- Diploid androgenetic individuals can then be produced by various treatments (thermal, pressure or chemical shock) to suppress the first cleavage division yielding a homozygous diploid.
- Theoretically 50% XX and 50% of YY homozygous individuals will be produced.
- The resultant YY males, if viable and fertile, can be crossed with females to produce XY isogenic (genetically uniform) all-male progeny.
- Androgenetic trout was produced using sperm from tetraploid males and found that these males showed better viability than androgenetic diploid.

- Among the aquarists the male fishes are often found more colourful and beautiful thus commanding a greater demand. It would also be advantageous for species in which the male shows better characteristics for culture, like tilapia.
- In androgenesis, polyploidization by inhibition of meioticphase I and II is impossible due to the distruption of the female nucleus.
- Androgenesis results in a rather low rate of development compared to gynogenesis since the egg cytoplasm as well as the egg nucleus is affected by radiation treatment.
- Androgenesis has not been widely investigated in fish as gynogenesis.
- Like gynogenesis,androgenesis can be used to generate clonal lines.
- An advantage of androgenesis is the option of storing and regenerating lines from cryopreserved sperm by androgenesis; eggs have not yet been successfully cryopreserved in fish and this precludes the storage of gynogenetic lines in that way.
- Androgenetic fishes were successfully produced in tilapia, rainbow trout and Siberian sturgeon. Androgenetic tilapia can beuseful for monosex fish culture.

Androgen from tetraploid males

Androgenesis is a uniparental developmental mode, in which the offspring only inherit the paternal nuclear genome making them clones of the father]. This mode of reproduction has been successfully carried out in several fish species that adopt external fertilization, such as *Cyprinus carpio*, *Oncorhynchus mykiss*, *Oreochromis niloticus*, and *Misgurnus anguillicaudatus*. Androgenesis is one of the best chromosome manipulation techniques available and is useful in constructing pure genetic lines, generating monosexual fish, restoring endangered or extinct species from cryopreserved spermatozoa, and evaluating how the mitochondrial genome affects development.

In general, embryonic development in teleost fish is activated by fertilization. In androgenesis, an egg whose nuclear content is destroyed serves as the vector; it is then fertilized with an untreated native sperm to form a haploid embryo. The egg provides the functional cytoplasm, but does not contribute to the nuclear genome of the offspring during development.

Subsequently, to obtain diploid embryos, the first cell division is inhibited by either heat or pressure shock. After the first cell division blockage, the genome is duplicated to form a diploid. There are two physical treatments in the entire process. This first is inactivating eggs by ultraviolet irradiation or gamma ray, the second is preventing first mitosis by heat or pressure shock. Researchers have compared the effects of both treatments on embryo survival rates.

Thorgaard et al. found that the survival rate of androgenic progenies with diploid sperm is much higher than those with haploid sperm, which indicates that pressure shock treatment and/or homozygosity may be responsible for the low survival rate of androgenic double haploids; while the ultraviolet irradiation inactivation method did not pose much of a problem. Researchers found that both heat and pressure shock induced developmental disorders during embryonic development, in particular they suppressed dorso-ventral differentiation. To date, few examples of successful androgenesis with diploid sperm have been reported for limited kinds of fertile tetraploid fishes.

Researchers have reported the first allotetraploid hybrids (4n = 200, abbreviated as 4nAT) derived from crossing between red crucian carp (*Carassius auratus* red var.) (2n = 100, abbreviated as RCC) (F) and common carp (*Cyprinus carpio* L.) (2n = 100, abbreviated as CC) (M). Since then, a consecutive allotetraploid line has been established from generations F_3 to F_{24}.

Recently, autotetraploid crucian carp (4n = 200) have been obtained by distant hybridization between RCC (2n = 100) and blunt snout bream (Megalobrama amblycephala) (2n = 48, abbreviated as BSB) (F); the autotetraploid crucian carp line was formed from generations F_2 to F_8. Both tetraploid lines are useful for genetic research and breeding, especially androgenesis, because tetraploids can produce many diploid spermatozoa. Researchers induced androgenesis in diploid 4nAT sperm and successfully yielded bisexual fertile diploid AT-ag. Unfortunately, all of the male AT-ag fish obtained by researchers were heterozygous for both the sex ratios of male to female fish in F_1 ag × ag and AT × AT-ag did not differ significantly from the expected ratio of 1:1.

We used the diploid spermatozoa of autotetraploid crucian carp to artificially induce androgenesis to screen homozygous super male fish. In the androgenetic offspring, a male fish with white sperm was selected. The following characteristic analysis revealed that the YY crucian carp was diploid and produced unreduced diploid sperm. Furthermore, self-crossing and hybridization of the YY crucian carp with diploid female RCC and 4nAT sired all male triploids and tetraploids, respectively, which proves that the male fish selected was a homozygous super-male YY crucian carp. This is the first report in which a viable homozygous super-male fish with diploid sperm was obtained by distant hybridization and androgenesis. This fish will benefit the study of the sex determination mechanism in teleost fish, and producing all male fish making it important for genetic breeding in aquaculture.

The 2nd technique that can be used to produce androgens is to fertilize the eggs, whose genetic material has been destroyed by irradiation, with normal sperm from tetraploid males. Tetraploid males produce diploid (2n) sperm; consequently the sperm pro-nucleus is diploid rather than haploid which means the resulting zygote will also be diploid (2n).

Meiotic androgenesis is impossible:

In fisheries, a mature egg completes meiosis-1, before fertilization. Upon activation, meiosis-2 is resumed the separation of the two sets of chromosome occur. One set from the polar body which is normally extruded from the egg and other set from the egg's pronucleus. If meiotic androgenesis is possible then shock treatment is applied to prevent the extrusion of second polar body. But during androgenesis, eggs are irradiated which destroys the DNA not only of eggs but also of second polar body and sperm does not produce any second polar body. So, meiotic androgenesis is impossible.

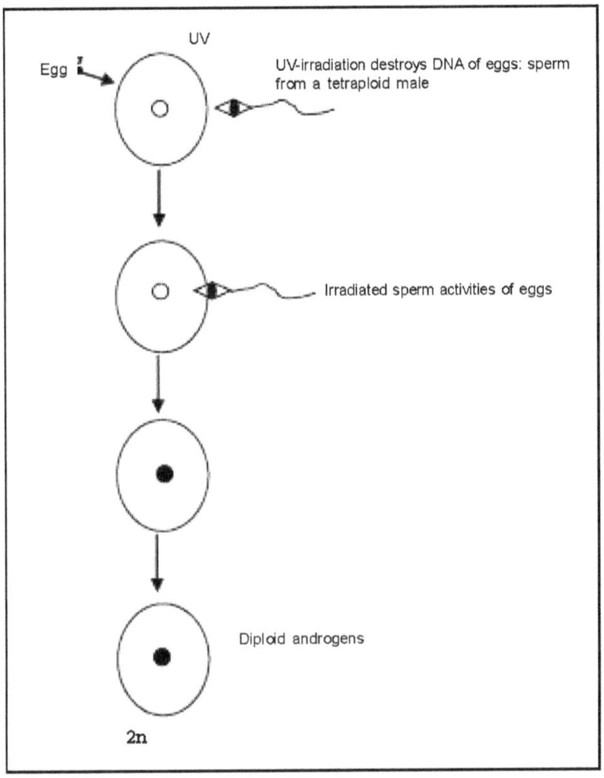

Fig. 9. Androgens from tetraploid males

HORMONALSEXREVERSALANDGENETICSEXREGULATION

Gonochoristic fish display an amazing diversity of sex determining systems, ranging from a strict genetic control (genetic sex determination, GSD) to an environmental control (environmental sex determination, ESD) or the joint action of both factors. GSD was originally reported in medaka but has been shown to be widespread within gonochoristic teleost species. In most fishes, sex is determined at fertilization by sex chromosomes with a monofactorial system often based on a XX/XY male heterogamety or less frequently a ZZ/ZW male homogamety. Sex in some fishes as well as in a few other animal species relies upon a polyfactorial system where the sum of various independent loci determines the sexual phenotype and birds, where the sex chromosome pair can be identified by its heteromorphy with a decayed 'heterogametic sex chromosome', in most of the poikilotherm vertebrates including fish, sex chromosomes are homomorphic and poorly differentiated.

Because they cannot be directly identified in fish through karyotype analysis, indirect approaches have been developed to search for the sex chromosomes of various species in order to analyze or control sex determination. More recently, molecular sex-linked markers have been identified in some model fish and aquaculture species. All-male or all-female populations (also known as monosex) have been produced for various aquaculture species to benefit from the best sex-linked dimorphisms (e.g., growth, age at puberty).

The sexual phenotype in fish is a plastic trait that can be easily manipulated by hormonal treatments (induction of sex reversal) during critical stages. Complete and functional sex reversal was produced by oral administration of androgens to masculinize or estrogens to feminize medaka fry. The use of Y-linked color genes brought evidence that part of the phenotypic males from the masculinization groups were neomales (genetic females masculinized by the treatment = mXX). Similarly, in the feminization process, part of the phenotypic females were neofemales (genetic males feminized by the estrogens = fXY).

By mating mXX with classic fXX females, all-female progenies have been obtained (mXX × fXX → fXX). Conversely, by mating fXY with mXY, viable and fertile YY males (mYY) were obtained (mXY × fXY → 1 fXX + 2 mXY + 1 mYY), and these YY males when mated with classic fXX breeders sired only male progenies (mYY × fXX →mXY) . Following this pioneer work, numerous studies have challenged the hypothesis that estrogens and androgens could be the natural 'gynoinducers' and 'androinducers', respectively. These studies mainly relied upon in vivo treatments using steroids, steroid enzyme inhibitors, or steroid receptor antagonists, *in vitro* steroid metabolism, steroid enzyme immunodetection, and gene expressions, concluding that aromatase and estrogens are the ones which play a key role in fish sex differentiation.

All the hormonal sex reversal treatments that have since been applied to researchers, this plasticity of sex differentiation can also lead to xenobiotic-induced sex reversal in fish due to pollution. Since these endocrine disrupting effects have been recently reviewed, they will not be discussed further in the present review.

Identification of an ESD system is more recent in gonochoristic fish species. ESD implies that the environment experienced by the offspring during early ontogeny determines the sex of a progeny. Temperature was the first environmental factor demonstrated to be able to determine the sex of an Atlantic silverside *Menidia menidia* progeny through a temperature-dependent sex determination (TSD).

A temperature-induced sex reversal (TISR) was first evidenced in the Nile tilapia *Oreochromis niloticus*, with high temperatures being able to masculinize sensitive progenies despite the fact that sex in this species is usually determined by major and/or minor genetic factors. TSD was then found in the pejerrey *Odontesthes bonariensis*. Together, these studies strongly stimulated research on TISR/TSD in various other fish species. EISR/ESR (environmentally induced sex reversal/environmental sex reversal) and especially TISR have been obtained in an increasing number of fish species under controlled conditions.

This included species with (i) a male heterogametic GSD system, such as the Japanese flounder *Paralichthys olivaceus*, the medaka *Oryzias latipes* , and the Southern flounder *P. lethostigma*; (ii) a female heterogametic GSD system, such as the blue tilapia *O. aureus*, the turbot *Scophthalmus maximus*, or the half-smooth tongue sole *Cynoglossus semilaevis*; or (iii) a polygenic GSD system, such as the seabass *Dicentrarchus labrax* or domestic strains of the zebrafish *Danio rerio*.

Coexistence of Genetic and Environmental Influences

Reviews on the environmental effects on fish sex determination/differentiation have generated debates about TSD in gonochoristic fish as well as in other vertebrates. Mainly based upon reptile sex determination, criteria have been proposed for identifying TSD in vertebrates. Amongst them, 2 main criteria have been used by researchers to reclassify 19 out of the 27 fish species previously reported to be TSD species as GSD + environmental effects. This was based on (a) the existence of sex chromosomes which was considered a signature of an exclusive GSD; and (b) for some of the fish species the efficient temperature treatments which skewed the sex ratios were considered to be outside the natural range specifically encountered by the fry or juveniles of these species during their critical period.

Concerning these 2 criteria, it is important to enrich the debate with recent data concerning GSD in vertebrates and TSD in poikilotherm vertebrates which have led to the deconstruction of some common myths about the evolution of sex determination in the animal kingdom. Mainly based upon few mammal models and to a lesser extent upon bird models, sex in animals has long been considered to be stable and determined by sex chromosomes. This predominant misconception that sex chromosomes are necessarily associated only to GSD has for long led to underestimate that at least in poikilotherm vertebrates, genetic systems of sex determination without sex chromosomes or with the presence of both major and minor genetic factors can also exist, especially in fish.

Indeed, in amphibians, reptiles, and teleosts 'sex-determining mechanisms are diverse and can evolve rapidly', with multiple evolutionary transitions between TSD and GSD having occurred in reptiles and amphibians and therefore, the line between a sex determination system relying on pure genetics or on pure environmental factor(s) is faint. Following the theoretical model, ESD and GSD have for decades been considered as 2 mutually exclusive systems.

However, based upon extensive studies in 2 lizard species, *Pogona vitticeps* and *Bassiana duperreyi*, unambiguous evidences have shown that GSD with sex chromosomes (respectively ZZ/ZW and XX/XY) and TSD can perfectly coexist. Furthermore, *B. duperreyi* has even a heteromorphic pair of sex chromosomes.

GSD systems can become sensitive to the environment and then present genotype-by-environment interactions or evolve towards TSD, allowing the coexistence of environmental and genetic influences. As hypothesized previously, the major role of sex reversal to induce this type of transition has been recently confirmed in the Australian lizard *(P. vitticeps)* by an experimentally temperature-induced rapid shift between a predominantly GSD (with heteromorphic sex chromosomes ZZ/ZW) and a TSD system using sex-reversed ZZ females identified in the wild.

Although the W chromosome was eliminated in one generation under laboratory conditions, a similar process should be much more gradual in wild populations, and therefore the transition between GSD and TSD will probably need multiple generations. Altogether, these recent data confirmed that at least in these lineages TSD and GSD are not mutually exclusive systems but the 2 extreme endpoints of a continuum, underlying the coexistence of both genetic and environmental influences in the middle of this spectrum: 'many species have differentiated sex chromosomes, but also show a temperature override, where genes and environment interact to determine sex'. In fact just recently, a three-ended continuum that can be visualized as a triangle has been proposed in which genes, environment, and random factors (i.e., any stochastic fluctuation in the expression of key genes of the cascade) could interact to determine sex.

Regarding fish, such a continuum with possible coexistence of genetic and environmental factors has also been proposed by some authors. In the Atlantic silverside, whose TSD system reached a consensus, thermal sensitivity of the wild populations varies from high to low levels within a natural thermal gradient. Because large females of *M. menidia* have a better reproductive fitness than large males, long growing seasons are more beneficial to females than to males. Therefore, at low latitudes when growing seasons are longer, early spawning (low temperatures) favors female differentiation. Temperature is a reliable cue of the future length of the growing season.

Sex determination in Southern (high thermal sensitivity with an adaptive role of TSD) or in Northern (low sensitivity) populations, respectively, relies upon a polygenic control, with both temperature and genetic factors interacting, or upon major genetic factors. Between these 2 extreme conditions, a mixture of TSD and GSD determines the sex of individuals.

Further studies analyzing 31 wild populations have demonstrated that the pattern of sex determination in the Atlantic silverside is actually non-linear, with none of the studied populations displaying a complete TSD or pure GSD, since in the first case some genetic component always exists, while in all the GSD populations some thermal sensitivity still persists. Hence, fish TSD relies upon genetic × environmental interactions with gradual shifts in sex ratio, unlike the abrupt pattern found in many reptiles.

Furthermore, the identification of a major sex determinant (the *amhy* gene) in another species, the pejerrey *O.bonariensis* with a high thermal sensitivity again underlines that in sensitive fish species, both genetic and environmental factors can determine the sexual phenotype.

Finally, it is also important to remember that differences exist in TSD traits between reptiles and fish. In reptile TSD species, especially in turtles and crocodiles, specific temperatures usually determine the sex of all individuals, producing therefore all-male or all-female progenies.

Conversely in fish with TSD, such as *M. menidia,* extreme temperatures usually do not induce 100% male or female broods. For some individuals in a progeny, sex is therefore not determined (or not only) by temperature. Similarly, apart from the pejerrey, temperature-induced monosex populations are very rarely reported in the Nile tilapia, the European sea bass, as well as in other thermosensitive species.

Besides, high between-family variance in sex ratios have been reported in tilapia, with contributions from both male and female breeders. Because skewed sex ratios are strongly heritable, they can be selected.

Moreover, in the same species, these 3 characteristics can also be applied to its thermosensitivity. Skewed sex ratios have also been shown in the rainbow trout, *Oncorhynchus mykiss*, following high temperature exposure (18°C treatments applied from 42 days post-fertilization [dpf] onwards).

In this species there is a strong genetic XX/XY sex determination where the major sex determining gene *sdY* has been identified, but the presence of a minor loci (a recessive mutation), which was called *mal,* associated or not to other loci and/or to environmental variations can lead to spontaneous masculinization. The study of this transmission was performed through several generations showing that it was a heritable trait. High temperatures of 18°C were able to masculinize XX females that carried *mal* in gynogenetic doubled haploid individuals.

In tilapia (but probably also in other thermosensitive species) thermosensitivity has been demonstrated to be highly heritable (h = 0.69), and therefore, it can be selected. In the turbot, *S. maximus*, sex determination mainly relies on a female heterogametic system (ZZ/ZW), but again, thermosensitivity and family temperature interactions have been reported at least in some progenies. These types of sex systems resemble polyfactorial systems. Until now, very few cases of 'true' polyfactorial sex determination have been reported in fish.

However, according to researchers, a polygenic sex determination is characterized by a large between-family variance in sex ratio, paternal and maternal effects on sex ratios, and a sex ratio response to selection. Interestingly, species such as rainbow trout and tilapia that are usually classified as GSD can present some of these characteristics. It is also interesting to draw a parallel between the hypothesis of an ancestral polygenic sex determination in fish and the recent demonstration that domesticated strains of zebrafish (classified as polygenic sex determination by researchers have lost the natural sex determinant present in wild zebrafish populations that rely upon a classic monofactorial GSD of ZZ/ZW type.

Interestingly, besides the ZZ/ZW system, other factors might be at play, because sex-reversed females have been reported. As suggested by the authors, domestication or/and selection or/and inbreeding of zebrafish laboratory lines have resulted in the evolution of the sex determination system possibly by using cryptic minor genetic factors. Such minor factors could coexist with the sex determinant in the wild population and partially explain why sex-reversed zebrafish can be found in both wild populations and domestic strains. These sex reversals could also be associated with zebrafish sensitivity to various environmental conditions (oxygen, density, temperature, thermocycle, poor nutrition), and/or to the unveiling of minor factors.

In haplochromine cichlids from Lake Malawi in East Africa, where multiple interacting loci control sex determination, sex-determining loci or sex determination systems can hide other loci or systems through epistatic dominance mechanisms or sex determination hierarchies. Altogether, this could explain why in tilapia but also in species considered to have a strong GSD (e.g., medaka, rainbow trout, common carp), minor factors have been detected at least under extreme genetic (i.e., clonal lines in the rainbow trout) or environmental (including those that are 'out of the normal incubation range', Ospina-Alvarez and Piferrer) conditions. Vertebrate sex-determining systems are therefore probably more complex than previously thought, especially in poikilotherms, and the hypothesis of simultaneous involvement of multiple interactive factors should not be ignored even in species where major factors have been identified. Often considered as a transient state, polygenic sex determination is nevertheless stable over time at least in the house fly, *Musca domestica*.

Regarding the second criteria that sex ratio deviation should occur within a range of temperatures naturally occurring during decisive developmental steps, recent data also bring some new perspectives. In the Californian grunion fish, *Leuresthes tenuis*, both temperature and photoperiod can influence the sex ratio.

However, because the seasonal thermal variation is very low (4°C) along the North American Pacific coast, photoperiod has become a more informative cue of season length than temperature. Longer day lengths and lower temperatures induced higher proportions of females. This is the first case of a photoperiod-dependent sex determination system in fish.

Temperature and photoperiod seem to act in synergy in this fish to determine the sexual phenotype. This suggests that (i) temperature in fish is not the only environmental factor involved in ESD systems; (ii) in some environments, photoperiod might be a more informative factor of the future environment of the fry or juveniles than temperature; and (iii) in species for which at least 2 environmental factors might have synergetic effects on sex ratios, application of a treatment of only one environmental factor probably requires higher levels of this factor to obtain a similar effect than when the 2 factors are applied. Therefore, in sensitive species the effects of additional environmental factors on sex ratios have to be searched for. If other factors (at least a second one) are also efficient to influence the sex ratio, then synergic effects will have to be searched by applying decreasing values of temperatures.

Because minor and/or environmental factors can both induce sex reversal in gonochoristic fish, we consider that their sex reversal can be defined as a mismatch between the phenotypic sex and the sex theoretically governed by the major genetic factor(s).

SOCIALLY CONTROLLED SEX CHANGE

The Indo-Pacific cleaner wrasse (*Labroides dimidiatus*) has fascinated biologists and non-biologists alike with its habit of gaining nutrition through setting up 'cleaning stations' on reefs where larger fish visit and allow the cleaners to remove parasites. In 1972, Ross Robertson showed cleaner wrasses were equally remarkable sexually when he induced socially controlled sex change in the largest females of polygynous groups by removing dominant males. Sex change had already been recognized as a valuable 'natural experiment' in reproductive biology, but his was the first demonstration of social control of the phenomenon and in the animal's natural habitat. This paper and a contribution by Fishelson on the fairy basslet (*Pseudanthias squamipinnis*) stimulated a great deal of interest in socially-controlled sex change from both behavioral ecologists and neuroendocrinologists.

The range of sexual patterns in coral reef and warm temperate reef fishes includes male-to-female sex change (protandry), female-to-male functional sex change (protogyny), bidirectional sex change, and simultaneous hermaphroditism.

Protogyny can be either monandric, where all males are sex changed females, or diandric, where individuals may mature as either males or females (the 'initial phase') and either can become a 'terminal phase' male. Sex change is observed in seven families across 27 orders of fishes and many species on coral reefs, but most of the research has focused on five families that the discussion here will be restricted to (wrasses - Labridae, parrotfishes - Scaridae, gobies - Gobiidae, damselfishes - Pomacentridae, groupers and basslets - Serranidae).

Steroid Hormones and Socially Controlled Sex Change

Gonadal steroid hormones are key regulators of sexual differentiation and reproduction across vertebrate animals. This is also true for sex change in fishes, but with some interesting variations in the key sites of steroidogenesis and especially the role of estrogen signaling. The first studies of the physiology of sex change used wrasses and hormone manipulations to determine whether protogynous sex change could be induced through administering androgens. Working with the bluehead wrasse, *Thalassoma bifasciatum*, Researchers found that androgen implants in females could induce the blue coloration characteristic of terminal phase males. A role for androgens has been supported and extended in more recent studies using both correlational and manipulative approaches as described below.

The most comprehensive studies of steroid correlates of socially-controlled sex change in a protogynous species have been in the Hawaiian saddleback wrasse, *Thalassoma duperrey*. Sex change in *T. duperrey* can be induced by housing large females with smaller female conspecifics and inhibited by also housing them with a larger individual, which can be either a larger female or a terminal phase male. These housing conditions will be referred to below as 'socially permissive' and 'socially inhibitory' respectively. Researchers compared females and terminal phase males caught and sampled directly from reefs on the island of Oahu ('natural' females and terminal phase males) to females sampled at different points during sex change induced by housing with smaller females in floating pens ('experimental females',).

Plasma levels of estradiol 17b (E_2) in experimental females in pens were significantly lower than E_2 levels in natural females from the earliest stages of sex change and comparable to the very low levels found in natural terminal phase males. Conversely, 11-ketotestosterone (11KT) levels were relatively low in natural females and elevated in natural terminal phase males. 11KT levels remained low in experimental females during sex change in pens. Interestingly with respect to patterns in tetrapods, testosterone (T) levels were not different between natural females and terminal phase males and did not exhibit significant variation across the sex change process. The lack of variation in T may be due to its being primarily a prohormone in many fishes, serving as a biochemical precursor for E_2 and 11KT synthesis.

Gonadal steroid hormone patterns in saddleback wrasses are paralleled by differences in gonadal ultrastructure and steroid hormone synthetic capacity. Females exhibited greater gonadal E_2 synthetic capacity than advanced sex changers *in vitro* when stimulated by salmon gonadotropin, while terminal phase males showed greater 11KT synthesis than females *in vitro* with advanced sex changers being intermediate. A follow up study compared terminal phase males with female-mimic initial phase males. Initial phase males are externally indistinguishable from females except for the genital papilla. This external similarity is likely important for success in 'sneaker' mating tactics. Initial phase male saddleback wrasses show both lower plasma levels and lower production of 11KT by the testes *in vitro* than terminal phase males.

Gonadal steroid hormone patterns described for other sex changing species are generally consistent with those for saddleback wrasses. Stoplight parrotfish (*Sparisoma viride*) exhibit a similar pattern of sexual phenotype differentiation characterized by diandric protogyny with colorful and territorial terminal phase males and initial phase males that are very similar to females in external morphology. As with saddleback wrasses, terminal phase male stoplight parrotfish exhibit higher 11KT levels than females or initial phase males while females exhibit the highest E_2 levels.

Researchers were also able to sample fish undergoing natural female-to-male sex change and found elevations in plasma 11KT and decreases in E_2 in these transitional individuals. Initial phase males undergoing only color change during development into terminal phase males showed increases in 11KT consistent with a role in development of the terminal phase male phenotype. The pattern of higher E_2 levels in females and higher 11KT in males is also seen in gobies such as the blackeye goby *Coryphopterus nicholsi* and a variety of grouper species.

The associations between sex and circulating steroid levels in protandrous and bidirectionally changing species are somewhat less consistent than those for protogynous species, particularly with respect to 11KT. In agreement with the findings described for protogynous species and despite the reversed direction of sex change, the protandrous anemonefish *Amphiprion melanopus* displays higher levels of 11KT in males and higher levels of E_2 in females. Testosterone levels are also higher in female *A. melanopus*, but this is likely explained by T's role as a biosynthetic precursor to E_2. Gobies able to exhibit serial, bi-directional sex change are especially interesting in this context as they maintain both ovarian and testicular tissue in the adult gonad with only one type being active at a time. Consistent with other sex changing species, higher levels of E_2 are observed when gobies are functioning as females. By contrast, Lorenzi and coworkers [105] examined excreted 11KT in the bluebanded goby *Lythrypnus dalli* and found no difference between individuals functioning as males versus females, a finding consistent with the lack of differences across functional sexes in whole-body 11KT levels in the bidirectionally sex changing *Gobiodon erythrospilus* [both *L. dalli* and *G. erythrospilus* are very small bodied, precluding plasma measurements].

The lack of sex differences in 11KT levels may be related to their ability of serial, bidirectional sex change and/or their relative lack of external sexual dimorphism. Recall that initial phase male saddleback wrasses and stoplight parrotfish also lack male secondary sexual characters and have low 11KT levels, a feature generally consistent with other examples of teleost alternative male phenotypes and gonochoristic teleosts that do not display pronounced sexual dimorphism. This relation between 11KT levels and the display of secondary sexual characters was also strongly supported in studies of the wrasse *Pseudolabrus sieboldi*, where shifts from the female-typical to terminal phase-male typical coloration of the anal fin were closely correlated to serum levels of 11KT [but not E_2].

The Use of Sex Reversal to Understand and Control Fish Sex Determination or Differentiation

Sex control is often a good answer to various aquaculture problems mainly related to sex-linked dimorphisms or for reproduction management. Depending upon the species, one sex grows faster than the other, which can be the case for males (tilapia) or for females (e.g., flatfish, sea bass, turbot). The sex-linked dimorphism can also concern the color or the body shape of ornamental fish species, such as the guppies, the Siamese fighting fish, or the platyfish.

Besides, skewed sex ratios can be a means to optimize reproduction like in tilapia where a 1:3 (M:F) ratio is usually recommended for fry production. Furthermore, in salmonids flesh quality is affected by the endocrine changes associated with sexual maturity, and consequently, it is more interesting to rely upon female production because they mature later than males. Hence, studies on sex determination and sex differentiation have been strongly stimulated in various aquaculture species due to the benefits generated by controlling sex (i.e., better management and profitability). In some laboratory models and aquaculture species, sex reversal has been used to investigate the process and mechanisms of sex differentiation and/or to develop sex control approaches.

YY Male Production

The pioneer studies of Yamamoto [1969] demonstrated that viable and fertile YY males can be produced in the medaka *O. latipes* through estrogen feminization and breeding fXY × mXY. Since then, YY 'super-males' have been developed in various species either through this hormonal approach coupled with progeny testing or through alternative procedures such as androgenesis of XY males or gynogenesis of fXY (sex-reversed females by estrogen treatment) in numerous fish species. In some cases a large scale production of YY individuals exists. Good survival rates of YY males have been reported in rainbow trout, tilapia, and yellow catfish, whereas low survival rates or non-viability of the YY individuals have been reported in the medaka and the guppy versus *Betta splendens* and *Amatitlania nigrofasciata*. In 2 important aquaculture species, the Nile tilapia and the rainbow trout, XX and YY males have been generated in order to produce female (XX) and male (XY) monosex progenies, respectively, through a genetic approach.

WW Female Production

Because of important sex-linked growth dimorphism in turbot, the use of all-female populations could enhance farming profitability. As reported in our previous results on turbot, the breeding of androgen-induced neomales (mZW) with classic females (fZW) provided some 1:3 (M:F) progenies, suggesting the viability of WW females (mZW × fZW → 1 mZZ + 2 f ZW + 1 fWW). However, due to the late sexual maturity of turbot, this 3-generation process requires at least 6 years to produce fWW (and more to also obtain mWW for WW mass production) in the absence of reliable sex chromosome markers. Moreover, turbot sex also relies upon minor genetic and thermal factors. Therefore, although this approach is technically feasible, it has not been yet developed at a commercial level neither in turbot nor in any other ZZ/ZW commercial species displaying a better female growth rate.

FISH SENSITIVITY TO SEX REVERSAL TREATMENTS

Until recently, based upon the results of researchers, it was generally assumed that effective sex reversal treatments had to be applied during a critical neutral time window which was species specific and generally different for feminization and masculinization, since ovarian differentiation happens prior to that of testis. This window precedes the onset and covers the first steps of gonadal sex differentiation. Once the natural sex differentiation process had started, a complete and functional sex reversal was considered as no longer possible. Species-specific characteristics of efficient treatments have been extensively described previously]. Our review will therefore focus on more recent aspects of gonad sensitivity to sex reversal treatments.

In cichlid and cyprinid species, where histological sex differentiation usually begins after first feeding, treatments mainly relied upon dietary administration of steroids or inhibitors of steroidogenic enzymes leading to androgens or estrogens. Interestingly, based upon the expression analysis of several candidate gene profiles, the inhibition of endogenous estrogen synthesis could even result in the production of neomales that are physiologically more similar to genetic males than phenotypic males produced by androgen treatments.

In the Nile tilapia, the standard and very efficient treatment used by most farmers to produce male monosex populations is based on a 3-4 weeks dietary administration of 17α-methyltestosterone (17MT) from the first feeding (around 10 dpf). A precocious sensitive period for sex reversal might also exist in tilapia monosex embryos, since the proportion of females or males increased (20-25%) with 17MT or EE2 immersion treatments of <12 h on XX or XY eggs. Although similar androgen 11KT (11 ketotestosterone) treatment or treatment with an aromatase inhibitor resulted in only 10% masculinization.

Uptake and clearance rates showed that EE2 levels in the eggs dropped at 4 dpf, but the whole body levels were still elevated at 10 dpf. Accumulation of hormones in the egg or vitellus could be delaying the effect on the embryo. However, this hypothesis does not explain the increase in male percentage when precocious high temperatures were applied.

Moreover, similar EE2 bio-concentrations were obtained with a single immersion treatment at 10 dpf resulting in lower female proportions. Surprisingly, no YY embryos were affected even at superior doses of 6,500 μg/l, although both XY and YY progenies were sex reversed (83.7-100%) at later stages of 10 dpf. Perhaps a Y-linked repressor or an X-linked activator is acting at this embryonic stage, which may still be partially active later since female proportions increase but are very rarely 100%.

Together, these studies suggest a precocious sensitivity to external factors in tilapia embryos just before the beginning of the morphological differentiation of the brain (31 hours post-fertilization, hpf) and also before the possible identification of the primordial germ cells (46 hpf) in the Nile tilapia. An important challenge is to compare the mechanisms of the precocious versus classic treatments, considering that the precocious doses were very elevated. We assume that both treatments probably act at different levels of the cascade leading to sex differentiation. The use of species sensitive to environmental treatments will allow bypassing the question of steroid accumulation associated with precocious hormonal treatments.

Until now, identification of the critical inversion window mainly relied upon the empiric analysis of the sensitivity of sex differentiation to external factors (hormones, xenobiotics, and environmental factors) rather than upon specific and precise developmental, histological, or molecular traits. We believe that this imprecise period has to be revisited based on the present knowledge on sex determination/differentiation and using up-to-date genomic tools available for some model species.

For example, histological sex differentiation in rainbow trout has been reported to occur around 5 weeks after hatching at 10°C (around 67 dpf) in genetic females, but the physiological or molecular sex differentiation of the gonads characterized by sexually dimorphic expression patterns begins much earlier around hatching time (around 32 dpf).

Fish, like other vertebrates were supposed to have a neutral stage during which the gonads are bipotential before they differentiate towards ovaries or testis. As reported before, sex differentiation towards a specific sexual phenotype can be easily induced in teleost fish with exogenous treatments applied at the 'critical' sensitive window covering at least the beginning of the sex-differentiating process. Gonadal bipotentiality was supposed to be lost after sexual differentiation. However, since the evidence that oocytes can be generated from frozen type A spermatogonia of rainbow trout transplanted into female recipients, it is clear that bipotentiality still persists in adult teleost gonads. Similarly in adult mice gonads, Sertoli cells and granulosa cells can also be reprogrammed in the absence of their respective transcription factors, DMRT1 or FOXL2.

Based upon the hypothesis of the possible persistence of gonadal bipotentiality, functional masculinizations have been obtained in adults of Nile tilapia and medaka through long E2 depletion treatments using an aromatase inhibitor resulting in sex-reversed males that had functional testis and sperm as well as male specific sexual behaviors. Female-to-male sex reversal was also possible in adult zebrafish with a 5-month aromatase inhibitor treatment causing retraction of the ovaries after which testes-like organs appeared, and cyst structures filled with spermatozoa-like cells were observed in sections of these tissues. Therefore, sexual plasticity still exists in adult gonads of teleosts, but functional sex reversal in adults needs longer treatments (3-6 months in tilapia and zebrafish; 2 months in medaka) than in fry (2 weeks 1 month in tilapia).

Based on these results, we assume that several steps of the gonadal development are therefore probably sensitive to external influences but also that sex reversal can be obtained either through a control of sex differentiation or through the inhibition of the maintenance of the phenotypic sex (especially femaleness).

MECHANISMS OF SEX REVERSAL

Sex reversal can be considered as a change in the sex threshold trait. In fish this threshold appears to be extremely susceptible to environmental modifiers (e.g., temperature, pH, density) shown to act at cellular and/or gene levels. Modifiers that affect the onset of ovarian meiosis, germ cell proliferation, or the number of germ cells lead to sex reversal. As previously questioned, the number of germ cells could be an important factor or even the threshold modifying the gonad sex fate at least in some species. The presence of germ cells is indeed necessary for ovarian formation and differentiation in some fish species. Loss of germ cells induced by high temperatures or by other means in zebrafish, medaka, and fugu induced sex reversal of females into males, with somatic cells seemingly predisposed to develop into male cells.

However, this was not the case in the goldfish. In the TSD pejerrey, germ cell apoptosis by high temperatures alone or coupled to busulfan did not affect the sex ratio of the sterile gonad. Likewise, researchers have shown that germ cell depletion by high temperatures does not apparently cause sex reversal in the Nile tilapia.

NEW ACTORS OF SEX DETERMINATION AND DIFFERENTIATION

The first step at which environmental factors like temperature may be overriding the genetic sex is by somehow affecting the primary sex-determining gene. We are in a fascinating era with the discovery of several different master genes in teleosts that will all be valuable and reliable sources to establish the genotypic sex of wild populations of these species in order to find sex reversals.

Most of these genes have emerged from duplicated copies of downstream sex-differentiating genes that have been recruited to become master sex determinants. *Dmy/dmrt1bY*, which arose from the testis *dmrt1* gene, was first identified in the medaka *O. latipes*. Spontaneous XY female medaka are attributed to mutations of the male-specific region, the *dmy* coding region, or a single nucleotide mutation in exon 3 resulting in a truncated *dmy*. In another medaka, *O. luzonensis*, sex is directed by the male-specific expression of *gsdfY*, a Y allelic form of the gonadal soma-derived growth factor, a member of the TGF-[2] superfamily.

Gsdf may also be the sex determinant in the sablefish *Anoplopoma fimbia*. In the Patagonian pejerrey *O. hatcheri*, maleness is directed by a Y copy of the anti-Müllerian hormone *amh* gene also belonging to the TGF-β superfamily. Despite having a strong TSD, another pejerrey, *O. bonariensis*, also expressed *amhy* at early stages of male promoting temperatures followed by the expression of the autosomal *amha* gene, showing the coexistence of both temperature and genetic sex determination in this species.

Two tandem Y copies of the *amh* gene have been identified in the Nile tilapia with a 233-bp deleted region in exon VII and a 5-bp insertion in exon VI that leads to a truncated *amh* gene lacking the TGF-β domain, defined as *amhΔY,* and another *amhy* having a missense SNP which may be determining sex in some strains, notably the Japanese strain. In several pufferfish *(Takifugu)* species an allelic variation in the *amhr2* gene, the *amh* receptor, has been shown to be responsible for maleness. Surprisingly, in the rainbow trout it is a copy of an immune gene, the interferon regulatory factor 9, which has become the master *sdY* gene].

Sex-determining genes are being searched for in several other fish species with many advances achieved in identifying linkage groups (LG) with quantitative trait loci and SNPs associated to sex using genome-wide sequencing or restriction associated DNA sequencing (RADseq). These techniques have, for instance in the Nile tilapia, narrowed down an inverted region on LG1 to 8,8 Mb that harbors a sex-determinant but not the *amh*/*amhy* and contains 8 genes amongst which is the Wilm's tumor protein homolog *wt1b* in the Manzala and Ghana strains. In tilapia, sex-linked loci seem to vary in function of the strain and population.

Environmental and parental factors are additional factors that influence the phenotypic sex of tilapia. Genotyping of functional SNPs evidenced that 2 were heterozygous for males, while females were homozygous, with misassignments corresponding to XX males [Palaiokostas et al., 2013]. Associations were confirmed in a temperature-masculinized family which had a 66% male sex ratio. Study of LGs linked to temperature-induced phenotypic male tilapias was performed by comparing a thermosensitive line versus a low or non-thermosensitive line showing a strong association of LG1, LG3, and LG23.

A strong SNP association was also shown in LG20 in another family sensitive to temperature-masculinization using RADseq. In zebrafish, sex-determining studies have followed a long windy road: RADseq revealed a polyfactorial system in domestic strains, but it found a ZZ/ZW system in the wild populations with the sex determinant on chromosome 4R. Some female-to-male sex reversals (female genotype *f*/*m*) were found but not the reverse, suggesting that either there is a dominant *f* allele which is required but not sufficient for a female phenotype, or alternatively there is a 2-dose male 1-dose female hypothesis. The authors suggest that environmental factors might be behind these female-to-male reversals affecting the meiotic oocytes by inhibiting primordial germ cell proliferation, meiosis entry, or stimulating oocyte apoptosis.

Gsdf is a novel gene, specific to fish and possibly an endogenous inducer of testis development. In medaka, *gsdf* seems to be the target of *dmy* since they co-localize in somatic cells (pre-Sertoli cells). It is detected just after *dmy*, with a downregulation seen in XY females treated with estradiol. *Dmy* knockdown with RNAi resulted in male-to-female sex reversal induced by suppressing the male pathway genes *(gsdf* and *sox9a)* and upregulating the female pathway (R-spondin1, *rspo1*).

Complete female-to-male sex reversal of the medaka *O. sakaizumii* but not of *O. latipes* stimulated *gsdf* and subsequently *dmrt1*. In tilapia, upregulation of *gsdf* transcripts were initially suggested to be at 8 dpf prior to *dmrt1* and localized also in the somatic cells surrounding germ cells. However, CRISPR/Cas9 knockouts revealed that *gsdf* is in fact expressed after *dmrt1* and an aromatase inhibitor treatment rescued the testicular phenotype of XY gsdf$^{-/-}$ gonads. *In vitro* analysis showed that *dmrt1* co-transfected with *sf1* was able to activate *gsdf* in a dose-dependent manner. G*sdf* action in XY tilapia is thought to be related to the suppression of estrogen production possibly via the inhibition of ovarian-differentiating genes.

Ontogeny expressions show that both the sex determining gene *amhy* and the autosomal *amha* gene are implicated in testis differentiation in the TSD pejerrey even in males lacking *amhy*, while in the GSD pejerrey *amha* is not implicated. In the Nile tilapia, *amhy* and *amh*Δ*Y* are expressed specifically in XY larvae from 9 dpf, but only the *amhy* knockdown resulted in male-to-female sex reversal. Overexpression of *amhy* in XX fish but not of *amh*Δ*Y* resulted in female-to-male sex reversal, suggesting that *amhy* is the sex determinant at least in the Japanese strain. A non-synonymous SNP in exon VI was found in one of the tilapia *amh* genes (but which of the 3 genes it corresponds to is unclear) associated to temperature-induced males, that could be useful to increase the proportion of males in the selective breeding of temperature-induced lines.

Amh and *amhr2* together with *gsdf* appear to play important roles in primordial germ cell proliferation which might explain why they have been recruited to become primary sex determinants. In the medaka XY *hotei*, a mutation in the *amhr2* gene causes hyperproliferation of the germ cells and male-to-female sex reversal due to lack of Amh signaling on the mitotic self-renewing germ cells. In tilapia *amhr2* knockout caused 100% male-to-female sex reversal but in contrast, only 60% were obtained with *amhy* due to F_0 generation being mosaic with the mutation varying between individuals.

Therefore, both *amhy* and *amhr2* are critical to determine sex in tilapia, but it remains to be known how the signaling occurs. Knockdown of both *amhy* and *amhr2* in XY tilapia causes upregulation of the aromatase *cyp19a1a* gene as well as higher E2 levels, but no increase in *cyp19a1a* expression was seen in *amhy*/*amhr2* overexpressed XX fish.

EPIGENETICS: A LINK BETWEEN THE ENVIRONMENT, GENE EXPRESSION, AND SEX REVERSAL

One of the underlying mechanisms leading to sex reversal in some fish species might be epigenetics. It is now evident that the environment can induce epigenetic modifications of histones affecting the compactness of the chromatin structure and DNA accessibility without causing actual nucleotide changes that play essential roles in gene regulation and ultimately modify the phenotype. One of the most studied epigenetic mechanisms is DNA methylation of cytosines (C5 position) at CpG dinucleotides because of its implication in a large number of cellular functions such as development and cell differentiation, tissue-specific gene expression, genome imprinting, X chromosome inactivation, and in disease susceptibility. Hypermethylation is associated with gene silencing, but methylation also acts at the level of transcript splicing.

Higher DNA methylation levels have been found in sea bass gonads of differentiating males in the aromatase *cyp19a1a* promoter that were inversely correlated with the *cyp19a1a* gene expression. Exposure during early life to high temperatures (>17°C) induced a male bias and also increased the methylation levels in both males and females. Interestingly, sex-specific differences were seen for the 7 CGs sites, while only 2 showed temperature-induced differences. The in vitro study suggests that the hypermethylation blocked the stimulatory effect of SF1 and FOXL2.

Opposite methylation patterns were evidenced in the half-smooth tongue sole, with males having a low methylation rate at the *dmrt1* promoter (located on the Z chromosome) associated to an elevated *dmrt1* expression. Tongue sole is a ZW/ZZ species but temperatures of 28°C lead to ~73% masculinized ZW females, and these phenotypic males show a demethylation of the *dmrt1* promoter. Genome-wide DNA methylation and the transcriptome were studied in this species, permitting a much larger analysis of the methylation status of gene regulatory elements and gene expressions. Differences in the methylation rates were found between ovaries and testes notably for the sex-differentiating pathway genes. Interestingly, higher levels of methylation were found in a 2-Mb region of the Z chromosome which differed from the W chromosome that clustered for testes, which might lead to some dosage compensation in ZW males. In addition, the tongue sole showed splicing alternatives but not silencing, for a female W gene *figla* (factor in the germ-line alpha) associated to differential expression between ZW females and ZW males.

Very recently, the global DNA methylation status was analyzed in the Nile tilapia between mature (>4 months) females, mature males, and from a matured mixed-sex group treated at high temperatures. Overall higher CpG methylated domains were found in the gene bodies of females compared to males. High temperature treatments increased the methylation levels in both males and females. The orphan nuclear receptor *dax1 (nr0b1a)* gene had higher methylation levels in induced males when compared to control females, while the steroidogenic enzyme *hsd17b8* gene as well as the *erb2* and *gsdf* genes had lower levels.

Cytosine methylation is catalyzed by the DNA methyltransferase enzymes (DNMT). In mammals, DNMT1 is involved in the maintenance of the CpG methylation pattern on the new DNA strand after replication, while DNMT3A and DNMT3B are considered de novo methyltransferases implicated in the initial CpG methylation pattern. DNMT3L is an additional member with no enzymatic activity that acts as a cofactor of DNMT3A and DNMT3B activity.

In teleosts, due to the genome duplication, up to 6 *dnmt3* genes have been identified in zebrafish which complicates the analysis of the Dnmt enzyme contribution. We have started to analyze the expression levels of Nile tilapia *dnmt* genes during sex differentiation.

Sexual dimorphism and temperature differences were observed for some of the de novo DNA methyltransferases, *dnmt3* while *dnmt1* remained unchanged which suggests that the *dnmt3a* and *dnmt3b* paralogues have specific roles during tilapia genetic sex differentiation and temperature-dependent masculinization. It might indicate differences in gene methylations occurring between genetic males and females but can also be induced by a temperature- induced masculinization.

This stage coincides with repression of the *cyp19a1a* gene in temperature-induced males. Whether its promoter is hypermethylated in tilapia males similarly to what has been seen in sea bass remains to be investigated.

DNA methylation associated with sex-determining and -differentiating genes has begun to be investigated in other fish species, but a whole avenue is now possible with genome-wide studies of the methylome associated with RNAseq which will enlighten how the methylation marks are activated or erased, implicated in transcript silencing or dosage compensation, between genetic sexes and sex-reversed fish.

Epigenetic marks other than methylation are only starting to be analyzed in fish during sex determination and differentiation. For instance microRNA, small non-coding RNAs that inhibit at post-transcriptional level target-gene expressions, have begun to be profiled between differentiating ovaries and testes in the Nile tilapia.

Integration of miRNA and mRNA was compared for early differentiating XX and XY Nile tilapia identifying 130 novel miRNA of which 49 were unique for XX and 45 solely found in XY gonads. Sex-biased miRNA were negatively correlated with genes from the steroid hormone biosynthesis pathway *(Cyp11a1, Hsd3b, Cyp19a1a, Hsd11b)* and to the ovarian or testes differentiating network genes such as *Foxl2, Amh, Star1, Sf1, Dmrt1*, and *Gsdf*.

Epigenetic mechanisms can be reversible and heritable. The study of transgenerational epigenetic inheritance is easily accessible in fish, notably those with short life stages, and a better understanding of these mechanisms will be possible by studying successive generations of sex-reversed individuals. It would be very intriguing to see how sex-reversed fish have adapted to natural high temperature regimes and how populations have been potentially affected in the wild, depending upon the presence or absence of such epigenetic marks.

HORMONAL SEX REVERSAL IN FISH

In many species of cultured fish, females grow more than males and mature later. Sex-related differences in growth and sexual maturation can be detrimental to modern aquaculture practices because they can limit or decrease production.

Controlling the proportions of sexes in production stocks to overcome these problems can be easily achieved because the process of gonadal differentiation is generally very labile in fish. This process is naturally controlled by steroid hormones, androgens, and estrogens, which the developing gonads themselves produce. It is thus possible to make a batch of fish develop only into the desired sex, usually females. In aquaculture, this is known as sex control.

Hormonal Treatment

To masculinize fish in those few cases where males are the faster-growing sex – in tilapia, for example – sexually undifferentiated alevins can be treated with androgens known as "male hormones." However, to feminize fish – which is the case in most aquacultured species for which controlling sex is advantageous (eels, carp, salmonids, catfish, sea bass, flatfish) – estrogens or "female hormones" can be used. There are two methods to produce all-female stocks: indirect feminization with androgen treatment and direct feminization with estrogens.

Indirect Feminization

In the indirect method of feminization, sexually undifferentiated fish are first masculinized with an androgen treatment. Androgens do not affect genetic males but sex-reverse genetic females. Thus, a genetic female treated with androgens will develop as a male and have testes capable of eventually producing sperm.

However, this treatment leaves the sex chromosomes unaffected. Genetic females will remain so in that their sex chromosomes will be XX, regardless of the androgen treatment. If, in the next generation, eggs obtained from normal females (which always carry the X chromosome) are fertilized with sperm from the genetic females turned into males (which also carry the X chromosome), all-female offspring will result. This is the basis of the indirect method of feminization.

The advantage of this method is that fish that reach market have never been treated with hormones – only their parents were, and for a short period of time. The major inconvenience is that setting up the methodology – masculinizing a sexually undifferentiated batch and identifying the males that are, in fact, genetic females to be used as sperm donors to produce the second generation – is costly and time-consuming. In addition, in order to work, this methodology requires the target fish to have a female XX and male XY system of sex determination, which is not the case in many species.

Direct Feminization

Direct feminization involves treating a batch of sexually undifferentiated fish with estrogens. If performed correctly, which is not as straightforward as it may sound, this results in the production of an all-female stock in a single generation. This is the most commonly used method of feminization for its simplicity. In many species, it results in increased production.

However, the fish thus produced have been treated with sex hormones, which can make producers and consumers reluctant to commercialize and consume the fish. Nevertheless, as explained below, there is scientific evidence that proves treating fish with hormones for purposes of sex control does not present any risk for the fish, the operators who administer the treatment, the environment, or the final consumers.

Natural Estrogen

The hormone used to feminize fish is the sex steroid estradiol (E_2), the major estrogen produced by all vertebrates from fish to humans. This naturally occurring estrogen is sufficiently potent to feminize many species of fish if applied correctly. Therefore, there is no need to use potent synthetic estrogens such as diethylstilbestrol (DES) or ethynylestradiol (EE_2). Their use should be discouraged or even prohibited.

Short Treatments

Effective E_2 treatments for direct feminization can be as short as hours or, at the most, three months. Taking into account that these treatments must be administered when fish are very small and still sexually undifferentiated, this means the feminization treatments are finished many months – and often years in many species – before the fish are marketed.

Therefore, not only are fish treated with a hormone they naturally produce, but the treatment is terminated so far in advance of marketing that there are no residues present. In fact, several studies have demonstrated that essentially all the administered hormone is cleared in two to three weeks after the end of the treatment.

Low Doses

Fish have a labile period during which the sexually undifferentiated gonads are most responsive to the effects of administered E_2. If this period is scientifically determined, then effective dosages for feminization can be very low.

To compare, the intake of EE_2 for women who take the contraceptive pill is in the range of 30 to 50 ìg per day. To ingest the same amount of the natural E_2, one would have to consume tens or hundreds of kilograms of treated fish every day.

Effects of Methyl Testosterone (MT)

In tilapia, sex reversal involves the treatment administration of male steroid to recently hatched fry so that the undifferentiated gonadal tissue of generic female develops testicular tissue, thus functioning reproductively as males. However, sex reversal of tilapia should consider food safety and environmental issues associated with the use of steroids. It is the obligation of the producer to ensure that the public receive the highest quantity of fish products using techniques that have minimum adverse effects on man and the environment. In some countries, restrictions exist on the sale of hormone treated fish unless it is proven that there are no risks to human health from consuming them. When such restrictions are in place, there is a problem created for the marketers of fish treated with hormone drugs. For instance marketing of treated fish is illegal in the EU countries and in India.

The hazard associated with the administration of MT in the culture of tilapia can be grouped into three:

- Risk on tilapia consumers (man)
- Tilapia culturist
- Environment

Risk of MT on tilapia consumer: Research findings have consistently shown that sex reversing tilapia with MT does not lead to accumulation of the hormone in the fish flesh after cessation of hormone treatment. Researchers revealed that whole fish body levels of MT were not detectable 100 h after withdrawal of the hormone treated diet. Researchers further reported that hormone levels in tilapia falls to normal level five days after hormone feeding was stopped. Based on these and other scientific evidences, it is clear that MT is rapidly removed from fish and will not persist after several months of culture to market size.

A conclusion can be drawn that the quantities of MT eaten by tilapia during sex inversion/reversal (fry treatment) is equivalent of 0.0-0.2 mg MT/fish represent less than 0.001% of the typical daily dosage of MT prescribed in human medicine (20-40 mg) and that even this minute quantities declines to less than 0.00001% of the daily human dosage a week after cessation of hormone treatment. It has been reported that the testes of an adult man releases about 15 mg of endogenous testosterone per day while about 10 mg of androgen are excreted daily.

The short treatment duration and rapid metabolism of MT help ensure that tilapia is free of MT before fish reach the consumer. Digested MT is rapidly metabolized and excreted. This rapid metabolism and excretion of MT by a fish treated early in its life history, combined with the extended period needed to produce a marketable size fish results in a safe consumer product.

However, as Nigeria is starting to invest in tilapia culture and the drive for sex reversal is on the top gear to increase tilapia production, there may be need for caution. Although many workers have reported that the hormone is no more detectible in the muscle of the fish after two months of cessation of the hormone treatment, it is necessary to look beyond the muscle and examine some vital organs such as the liver and gills which have the propensity to accumulate substances. Furthermore, the product of biodegradation/mineralization of MT should be assayed for possible effect on the organs and metabolic parameters.

Effect of MT on tilapia culturist: Tilapia culturists purchase commercially available androgen in powder or tablet form. This hormone is usually added to powdered tilapia feed which the culturist offers to tilapia fry many times per day during sex reversal using hormone treatment. The culturist or his workers may come into contact with the hormone in two ways: (1) When it is added to tilapia fry feed and (2) When MT treated feed is administered to tilapia fry in hatchery tanks. The hazard to fish farm worker can be minimized by following the standard for handling such substances. Use of protective surgical gloves and face mask are strongly recommended to reduce the risk of inhalation of the hormone or entry through the skin contact.

Although, the use of MT in commercial tilapia production is becoming more and more common globally, there has not been any report of adverse effect on farm workers. A more likely hazard that may arise from intentional ingestion of MT by staff is the well recognized anabolic (muscle building) and perceived libido-enhancing properties of the hormone.

Environmental concerns

Another aspect that can be of concern is the environmental impact of using sex steroids such as E_2 in aquaculture. British researchers have calculated that the total environmental input resulting from the feminization of 100,000 fish with E_2 during 30 days would be 40 milligrams per day.

Again, to compare, the daily release into the aquatic environment of the same hormone plus its derived metabolites from one single pregnant woman is in the range of 20 to 40 mg. Thus, the advantage of using the natural estrogen is that it cannot result in an input to the environment of compounds that are different from those which are normally present.

The use of sex steroid hormones for sex control in fish is a safe practice if performed correctly, and is not analogous with other uses of hormones, for example, to fatten livestock. Natural estrogen can be used for sex control in fish at low dose and duration, and the treatment is finished well before marketing, so no residues are present in the product going to market.

The use of sex steroids for sex control in fish is permitted in the European Union and some other parts of the world. However, in other areas, treatments may not be regulated and updated and, more importantly, may not incorporate the latest advancements in knowledge of sex determination and differentiation in fish.

It is anticipated that because of its inherent economic advantages, monosex (all-female) stocks will be sought in many species of aquacultured fish, and that for its simplicity the direct method of feminization will be prevalent. Dissemination of the latest advances in this field will result in the highest efficiency with the minimum exposure to steroids.

10

ENGINEERING AND PRODUCTION OF TRANSGENIC FISH

Advancements in gene technology in recent years have been driving the aquaculture industry forward. Improvements in growth performance, feed efficiency, and omega-3 content are goals of the industry that could capitalize on applications of genetic engineering. One of the major challenges in the industry is to reduce the use of fish meal and oil, to improve the environmental and eco-nomic sustainability of aquaculture. The recent development of genetically engineered feed ingredients is one potential solution tothe looming problem of fish meal and oil dependency.

Furthermore, the development of transgenic fish has potential to improve production efficiency and other future desirable characteristics that relate to feed utilization and product quality. New gene technologies are beginning to revolutionize how we produce our food, and in aquaculture, will ultimately reduce pressure on wild fish stocks, help to preserve natural aquatic ecosystems, and improve nutritional profiles of farmed fish for human consumption.

Transgenic are organisms into which transgene has been artificially introduced and the transgene stably integrated into their genomes. Transfer of transgene into the nucleus of a target cell where integration into the host genome takes place. The first transgenic fish were produced in and as of 2003, more than 35 species had been genetically engineered in research laboratories worldwide.

A number of fish species are in focus for gene transfer experiments and can be divided into two main groups: animals used in aquaculture and model fish used in basic research. Among the major food fish species are carp (*Cyprinus* sp.), tilapia (*Oreochromis* sp.), salmon (*Salmo* sp., *Oncorhynchus* sp.) and channel catfish (*Ictalurus punctatus*) while zebrafish (*Danio rerio*), medaka (*Oryzias latipes*) and goldfish (*Carassius auratus*) are used in basic research. Transgenic fish show better gross food conversion, the increase in fish weight per unit of food fed than their unmodified relatives.

Making transgenic lines of vertebrates represents a rather new technology. While transgenic fish have been introduced in the laboratory since the nid eighties, it has also becolne a reality in aquaculture since the early nineties. There are many reasons why people want to create genetically modified organisms (GMO) and there are many strategies to nlake "fish with novel genes". As such it has generated high expectations in Society and directed the interests of researchers, but has also raised concerns about environmental and consumer safety, and generated numerous ethical and legal discussions.

GENETIC ENGINEERING

The history of biotechnology in general dates back to as early as 500 BC to enhance productivity through processes such as fermentation to produce bread, dairy, and alcohol. In agriculture, selective breeding was also used as a simple form of gene selection based on phenotypic traits and was noted as a baseline component in the development of this industry. Building on the foundation of these primitive examples, the discovery and development of techniques such as gene cloning, genetic engineering, and recombinant DNA technology has enabled the modification of DNA sequences. Present day gene editing techniques are based on identifying specific genes that affect the desired phenotype of an organism, through the direct manipulation of the relationships between DNA, RNA, and the expressed proteins. New innovations in genetic technology have become highly prominent in the scientific community through the development of novel and innovative gene editing systems such as clustered regularly interspaced short palindromic repeats(CRISPR)-Cas-9.

TRANSGENIC FISH

A transgenic fish is one that contains genes from another species. A transgenic fish is an improved variety of fish provided with one or more desirable foreign gene for the purpose of enhancing fish quality, growth, resistance and productivity.

Typically, genes of one or more donor-species are isolated, and spliced into artificially constructed infectious agents, which act as vectors to carry the genes into the cells of recipient species. Once inside a cell, the vector carrying the genes will insert into the cell's genome. A transgenic organism is regenerated from each transformed cell (or egg, in the case of animals), which has taken up the foreign genes. And from that organism, a transgenic variety can be bred. In this way, genes can be transferred between distant species, which would never interbreed in nature.

The application of genetic engineering to animals, like potatoes with built-in insecticide, could provide numerous benefits, including the possibility of a safer, cheaper food supply and the creation of new sources for inadequate pharmaceutical resources.

With the advancement in the field of genetic engineering, the application of its commercial use has also increased. Aquatic animals are being engineered to increase aquaculture production.

The use of genetic engineering and rDNA technology has done miracles in medical and industrial research. The transgenic fish are being promoted as the first marketable transgenic animals for human consumption.

One of the most important aspects between fish and other terrestrial animals for cultivation and genetic improvement is that, usually, fishes have higher levels of genetic variation and hence more scopes for selection than most mammals or birds.

Using the gene transfer technology, scientists now have created a genetically engineered variety of Atlantic salmon that grow to market size in about 18 months, otherwise the fish takes about 24-30 months for becoming market size fish. It is also hoped that we can now modify a large number of fishes with fast growing characteristics and bring Blue Revolution.

Methods in the Production of Transgenic Fish

Microinjection method: Microinjection method has been used successfully in the production of transgenic fish and is a commonly used technique due to its simplicity and reliability. The use of the microinjection method results in higher survival rates for manipulated fish embryos than the electroporation method. The most established method for gene transfer in fish is microinjection. Microinjection that allows delivery of the transgene directly into the nucleus, Transgene is directly microinjected into the male pronuclei of fertilized eggs. Transgenic technology through DNA microinjection into zebra fish embryos has made great gain in the last decade. It is shown that the DNA injected into the cytoplasm of fertilized zebra fish eggs could integrate into the fish genome and be inherited in the germ line. The frequency of germline transmission of a microinjected DNA could be as a high as 20% in zebra fish.

Disadvantage

This method is not only time-consuming procedure for animals such as fish that produce a large number of eggs and labour intensive but is also limited by the physiology of fish eggs. The nuclei of fish eggs are small and difficult to visualize, the chorion, hardens soon after fertilization and in many fish species the pronuclei of fertilized eggs is not visible and transgenes are usually injected into the egg cytoplasm. This technology, however, still has as the major constrains the low efficiency generation of transgenics. To improve the efficiency of selection of transgenics, genetic markers are co-injected with the transgene to monitor for transformed zygotes. The Green Fluorescent Protein (GFP) from Jellyfish (Aequorea victoria) has been used for this purpose in zebrafish.

Electroporation method: Electroporation has been shown to be the most effective means of gene transfer in fish since a large number of fertilized eggs can be treated in a short time by this method.

Electroporation utilizes a series of short electric pulses to permeate the cell membrane that make possible the formation of temporary pores on the surface of the target cells through which the transgene is introduced into the cytoplasm where it is then delivered to the nucleus by the cellular machinery. Electroporation has been preferred in many laboratories for gene transfer in fish systems because of its efficiency, speed and simplicity.

Disadvantage

The presence of a tough chorion layer around the fish eggs reduces efficiency, removal of the chorion is a tedious procedure and introduces additional stress on newly fertilized eggs. In traditional electroporation methods, either levels of transfer have been low with few studies showing germ-line transmission and expression or the device used is no longer commercially available.

Sperm mediated gene transfer method: Spermatozoa are capable of binding DNA and carrying it into an egg. Fish spermatozoa can be stored in seminal plasma with little loss of viability for long periods. Therefore, this technique appears very promising for gene transfer in fish.

Retroviral infection method: The successful use of Retroviruses for gene delivery in fish has been reported.

Disadvantage

The preparation of retroviral particles including the transgene of interest is a very laborious process, increases costs and requires technology.

Advantages of Transgenic Fish

The growth rate of Transgenic fish can be increased by 400% to 600% while simultaneously reducing feed input by up to 25% per unit of output, thereby improving food conversion ratios.

Transgenic fish have been developed for applications such as the production of human therapeutics, experimental models for biological research, environmental monitoring, ornamental fish and aqua cultural production. Growth enhanced transgenic fish have improved feed-conversion efficiency, resulting in economic and potential environmental benefits such as reduced feed waste and effluent from fish farms. Transgenic strains provide useful model systems for studying the consequences of growth enhancement from genetic, physiological and ecological standpoints.

The trait of growth rate has received the greatest attention because of its importance in aquatic animal production systems. Two to threefold increase relative to non-transgenic fish have been reported for tilapia and Atlantic salmon and up to twofold increases in common carp. Commercial production of transgenic fish able to transmit desirable characteristics, such as enhanced growth or disease resistance to their progeny. Growth acceleration in fish has been one of the targets of gene transfer experiments.

Zebra fish are a good model for the rapid study of GH-transgenes, before selecting the construct to use in economically important species. AquAdvantage salmon, a modified

Atlantic salmon reach commercial size in one-third of the time required for nontransgenic salmon. The faster growth of each generation can lead to increased production per unit time along with savings on total food per pound of meat produced. Use of transgenic fish as bioreactors for the large-scale production of rare human therapeutic proteins or novel foods for specific dietary requirements.

Transgenic lines of tilapia engineered to produce human clotting factor VII, which is used in liver transplants and in treating injuries. Zebra fish genetically modified to manifest a defective aortic valve development has made it possible to identify the role of an enzyme, UDP-glucose dehydrogenase, in the process of the embryonic development of this valve. Transgenesis is proposed as a mean for removing allergenic substances in seafood. Fish are recognized as test organisms with distinct and superior benefits in providing insights to disease processes. The development of transgenic animal models represents a revolutionary advance in the study of a variety of disease processes.

Fish as cost-effective and important animal models in genetics, developmental biology and toxicology. Development of transgenic fish as a model in reducing or replacing selected mammals used in toxicity testing. Medaka fish has several characteristics that are especially well suited for environmental toxicology, including well-characterized histopathology. They have been used extensively in chemical hazard testing, carcinogenesis bioassays and germ cell mutagenesis studies.

Tilapia expressing extra copies of trout growth hormone gene grow much faster than their non transgenic counterparts which do not possess extra copies of the growth hormone gene. Transgenic fish have been used to study developmental regulation of genes, improve cold tolerance, increase growth rate, and improve feed utilization. Fast growing transgenic coho salmon, medaka, rainbow trout and tilapia containing novel growth hormone genes reached sexual maturity sooner than unmodified counterparts.

The efficiency of growth and feed conversion can also be increased in finfish by creating transgenic fish that incorporate a gene construct encoding growth hormone. Genetically modified fish have also been developed as experimental models for biomedical research, especially in studies involving embryogenesis and organogenesis as well as in the study of human diseases, Xenotransplantation and recombinant protein production for producing important therapeutic agents.

Disadvantages of Transgenic Fish

Inducing transgenesis is a relatively transgene. Mosaicism, the condition in which not all cells in embryos have the targeted transgene, can occur. The primary ecological concerns regarding utilization of transgenic fish are the loss of genetic diversity and loss of biodiversity, and reduction in species richness. Transgenic fish are those related to their involuntary escape into environment.

Concerns range from interbreeding with native fish populations to ecosystem and effects resulting from heightened competition for food and prey species. The sterilization of transgenic fish will help in reducing the interbreeding risks associated with the escape of transgenic fish. Introduction of transgenic fish into natural communities is a major ecological concern. Reduced sperm production in the GH transgenic Nile tilapia, Oreochromis niloticus.

Extremely fast-growing transgenic salmon and loach have low fitness and die inefficient process. Only about one out of every 100 eggs microinjected will stably incorporate the recombinant DNA sequence into its genome and subsequently transmit the transgene to its progeny. Transgenesis is unnatural in the sense that the insertion of DNA material in transgenic animals would not take place without human intervention. Transgenic fish could produce new or modified proteins that could be toxic to humans, they generally show reduced swimming ability and lower reproductive performance than non-transgenic fish.

The transgenic fish are also more active and aggressive when feeding, and more willing to risk exposure to predation. Changes in cognitive abilities or brain function and structure have occurred in transgenic fish. The transgenic salmon pose serious ecological threats to wild populations and are not engineered for natural environments. Transgenesis can result in an unpredictable number of copies and site of integration of the early.

PROPERTIES OF TRANSGENIC FISH VALUABLE FOR AQUACULTURE

Transgenesis is the introduction of exogenous genetic material (DNA) into a host genome, resulting in its stable maintenance, transmission and expression. The first successes in producing genetically modified transgenic fish occurred in 1984 and 1985 with rainbow trout and goldfish. Since then, transgenic individuals in more than 35 aquatic animal species have been produced.

Advocates of transgenesis perceive it as an effective means of improving the productivity of aquatic animals. By contrast, opponents see it as a threat to both the environment and human health. Transgenic aquatic animals are also being produced for ornamental and pharmaceutical purposes.

Production of Transgenic Fish

Fish are very tolerant of genetic manipulations, especially in the early stages of development. There are various methods for foreign gene insertion, such as microinjection, electroporation and sperm-mediated gene transfer. The process of creating transgenic individuals generally involves the introduction of a DNA construct containing genes for the desired trait(s) into the pronuclei of fertilized eggs. This is followed by in vitro or in vivo incubation of the injected embryos and subsequent maturation into a fully developed transgenic organism.

Despite the progress made, however, some problems remain. Transgenesis can result in an unpredictable number of copies and site of integration of the transgene. Mosaicism, the condition in which not all cells in embryos have the targeted transgene, can occur. Further research should work with the DNA sequences of fish, rather than mammalian origin to increase public acceptance, and avoid sequences of bacterial or viral origin.

At the present time, several transgenic lines are nearing or have reached the stage at which they could be considered for commercial production due to their alleged growth potential. These include common carp in China, hybrid tilapia in Cuba, Nile tilapia in the United Kingdom, Atlantic salmon in Canada and mud loach in Korea.

Generally, at a country level, an assessment of food safety by national regulatory bodies is carried out before granting approval to genetically modified (GM) organisms to enter the food chain. Overall, the general public has shown very limited interest in and much concern over GM-related products.

CURRENT AND POTENTIAL APPLICATIONS

Transgenesis has a number of potential applications.

Growth Rate

The trait of growth rate has received the greatest attention because of its importance in aquatic animal production systems. Two- to threefold increases relative to nontransgenic fish have been reported for tilapia and Atlantic salmon, and up to twofold increases in common carp.

Precise estimates of the comparative performance of transgenic fish and the nontransgenic strains currently used by farmers are lacking due to difficulties in making rigorous comparisons, such as the large numbers of animals needed, the risks of escape of transgenics and the absence of any systematic means of benchmarking fish strains.

Freeze Resistance

Resistance to freezing is the second major application of transgenesis to fish, but it is restricted to one species, the Atlantic salmon. The idea is to use "antifreeze" genes from fish that live in polar regions. This is still in an experimental phase, and there are fears about possible increased fitness and the invasive potential of the cold-resistant fish that may be created.

Disease Resistance

If successful, developments of increased disease resistance could result in reduced production losses and less use of antibiotics to control disease. Whereas there are promising results with channel catfish in the United States, there is still more work to be done. Some fear that disease-resistant fish could be toxic to humans or become reservoirs for pathogens.

Metabolic Modification

Some fish species used in aquaculture are highly dependent on high-protein diets. Wild-caught pelagic fish are used to manufacture their feed. If the carbohydrate metabolism of the fish could be improved, they could better use diets based on ingredients of land origin, such as soybean meal and vegetable oils. This area, too, remains in an experimental phase.

Sterility

The production of sterile fish would be an ideal means of containment of transgenic fish in production systems. Even if they escaped and became established in a water body, their presence would be limited in time since they could not reproduce. The aim is to produce fish that are deficient in a critical path leading to successful reproduction, but whose infertility could be restored by exogenous means such as hormone injection. At present, this is still a hypothetical possibility, but some progress has already been made with rainbow trout.

Biomedical Products

The generation of transgenic animals for high-value biomedical applications has gained wider public acceptance than for food production. The main application in this area includes the use of transgenic fish as bioreactors for the large-scale production of rare human therapeutic proteins or novel foods tailored for specific dietary requirements.

One example involves lines of tilapia engineered to produce human clotting factor VII, which is used in liver transplants and in treating injuries. However, the choice and modification of single genes to generate desired phenotypes and novel products remain challenging because important traits in fish are complex and controlled by multiple genes.

Objections and Risks

Varied concerns regarding transgenic organisms have been expressed by consumer and environmental groups, as well as industry and scientific experts.

Procedure is 'unnatural'

It is true that transgenesis is unnatural in the sense that the insertion of DNA material in transgenic animals would not take place without human intervention. However, public perception is not necessarily consistent, because not much is "natural" about artificial insemination, artificial incubation or farming in general, but these practices are generally accepted.

Undesirable Properties Could Appear

The possibility that undesirable properties could develop is a real risk. Transgenic fish could produce new or modified proteins that, while neutral or beneficial to the fish, could be toxic to humans. Because of the possible "position" effects of transgenes, transgenic lines produced for human consumption should undergo continuous testing to ensure there is no human intolerance to the proteins produced.

What About Escapes?

The impacts of escapes are of substantial concern. Containment of aquatic animals can be much more problematic than that of their terrestrial counterparts. If fish are not effectively contained, there could be long-term consequences through the introgression of transgenes into wild populations. Some evidence suggests that transgenic fish are generally less fit than wild fish, so the effects of escapes may not be as great as anticipated.

Even if transgenic fish were accepted from other viewpoints, rigorous containment measures must be adopted. These would vary in nature and include physical, geographical, thermal and biological containment. The possibility of producing sterile fish was mentioned earlier. Although the fish could not interbreed, they could still be viewed as alien to their environment and posing potential negative environmental impacts during the period they survive.

Animal Welfare

There is growing concern about the welfare of transgenic fish. They generally show reduced swimming ability and lower reproductive performance than nontransgenic fish. The transgenic fish are also more active and aggressive when feeding, and more willing to risk exposure to predation. It is possible that changes in cognitive abilities or brain function and structure have occurred in transgenic fish. Further research is needed to address species-specific welfare concerns.

Selective Breeding

Whereas transgenesis presents a potentially effective solution for advancing desired traits in fish, conventional selective breeding based on quantitative genetics is an approach that has been used in all genetic improvement programs implemented by WorldFish with its partners from Africa and Asia.

Because of their widespread culture in developing countries and desirable attributes, the two species on which WorldFish has predominantly focused in genetic improvement are tilapia and carp. For both, substantial improvements in growth rate have been achieved using a conventional selective-breeding approach without undesirable correlated responses. Unfortunately, no rigorous comparisons have been made between selectively bred strains and transgenics. Such comparisons would be valuable in the assessment of the potential benefits of transgenics.

Growth rate can be effectively improved by conventional means, whereas improvements in traits such as disease resistance or tolerance to environmental challenges are more difficult to achieve. Transgenesis is likely to be most useful with the latter type of trait in the future.

Whereas some growth-enhanced transgenics may soon find their way to commercial use, it is unlikely that transgenics with other, potentially more useful traits will become available for at least five years. In any case, all commercial applications will have to overcome the hurdles of consumer acceptance and demonstrating minimal impacts on the environment and biodiversity.

The following are the important points needed for genetic engineering (gene transfer) to produce transgenic fish:

1. A gene sequence is to isolate for the particular characteristics; for example, growth hormone gene.
2. These genes (gene sequence) are then inserted into a circular DNA known as plasmid Vector (enzymes endonucleases and ligases are used).
3. Plasmids are harvested in the bacteria to produced billions of copies.
4. Plasmids are introduced into linear DNA. The linear DNA is sometimes called a gene cassette because it contains several sets of genetic material in addition to new inserted gene; for example, growth hormone gene. The technology is available to integrate genes in germ line of developing individual (fish) and finally transmitted into further generations.
5. Making the cassette a permanent part of fish's genetic makeup.

DEVELOPMENT OF TRANSGENIC FISHES

Development of transgenic fish has focused on a few species including salmon, trout, carp, tilapia and a few others. Salmon and trout are cash crops while the others primarily provide sources of protein. Currently, about 40 or 50 labs around the world are working on the development of transgenic fish.

About a dozen of them are in the U.S., another dozen in China, and the rest in Canada, Australia, New Zealand, Israel, Brazil, Cuba, Japan, Singapore, Malaysia, and several other countries. Some of these labs are associated with companies that expect to commercialize their fish in a few more years.

Many of the fish under development are being modified to grow faster than their wild or traditionally bred aquaculture siblings.

Faster growth is usually accomplished by transferring a fish growth hormone gene from one species of fish into another. The faster growing fish not only reach market size in a shorter time, they also feed more efficiently. Trout growth hormone (GH) was used to produce transgenic carp with improved dressing properties. Such transgenic carp are recommended for production in earthen ponds.

Transgenic Salmon

The Atlantic salmon is engineered with a pacific salmon, growth hormone driven by the arctic antifreeze promoter gene. The rapid growth of that transgenic salmon is achieved, not so much by the transgenic growth hormone as by the antifreeze gene promoter that functions in the cool water desirable for salmon flavour.

Researchers research scientists with Fisheries & Oceans, Canada, in West Vancouver, British Columbia has modified the growth hormone gene in Coho salmon by developing a gene construct in which all the genetic elements are derived from sockeye salmon.

The transgenic Coho grew on average 11 times faster than unmodified fish and the largest fish grew 37 times faster. The growth hormone levels in the transgenic fish are high year-round, rather than falling off in the winter as occurs in ordinary salmon. The modified salmon are large enough to be marketed after one year, in contrast to standard farmed salmon that do not reach market size for at least three years.

Transgenic Tilapia

Tilapia fish, native to Africa, are cultured world-wide as "poor man's food", second only to carp as warm water food fish, and exceeding the production of Atlantic salmon (whose market value is twice that of tilapia).

Tilapia has been extensively genetically modified and promoted as a transgenic fish exclusive for isolated or contained production.

Transgenic tilapia, which is modified with pig growth-hormone, has three times larger than their non-transgenic siblings. Tilapia genetically modified with human insulin grew faster than non-transgenic siblings, and could also serve as a source of islet cells for transplantation to human subjects.

Transgenic Medaka Fish

Purdue animal scientist used tiny Japanese fish, Oryzias latipes called medaka to examine what would happen if male med akas genetically modified with growth hormone from Atlantic salmon. Inserting a gene construct consisting of the human growth hormone driven by the salmon growth promoter into medaka produced the transgenic medaka.

The viability of groups of modified and conventional fish was measured at three days of age, and 30 percent fewer transgenic fish survived to that age. The researchers calculated that large males had a four-fold mating advantage, based on observations of wild-type medaka. In another experiment Silk moth genes were introduced into Medaka fish to create resistance to bacterial pathogens.

Transgenic Zebra Fish

The tiny zebra fish (Bmchydanio rerio) that lives in aquariums, was genetically modified to produce a fluorescent red pigment, and is being promoted for sale as a household aquarium pet, the "goldfish".

The goldfish caused a stir in the United States because regulation of such transgenic pets is murky and none of the major regulatory agencies: Food and Drug Administration (FDA), United States Department of Agriculture (USDA) or Environment Protection Agency (EPA), has been willing to take the lead in regulating the goldfish (even though USDA does deal with pet animals).

These transgenic zebra fish with vivid fluorescent colours (green, yellow, red or orange) fluorescent proteins can be seen with naked eyes under both daylight and ultraviolet light in dark. The green fluorescent protein (GFP) is originally isolated from the jellyfish (*Aequorea tictoria*).

Transgenic common Carp

Researchers transferred into common carp the growth hormone DNA from rainbow trout fused to a sequence from an avian sarcoma virus.

The genetic material was injected into fertile carp eggs with microinjection. The offspring of the first generation of transgenic fish grew 20 to 40% faster than their unmodified siblings. Scientist is also developing transgenic catfish, tilapia, striped bass, trout, and flounder.

Researchers have developed transgenic carp and catfish that grow 20 to 60% faster than standard farmed varieties.

They use microinjection and electroporation to inject another copy of a fish growth hormone gene into fertile fish eggs. The growth of the resulting modified carp and catfish is stimulated by extra fish growth hormone.

In India, research in transgenic fish was initiated in Madurai Kamaraj University (MKU), Centre for Cellular and Molecular Biology (CCMB), Hyderabad and National Matha College, Kollam with borrowed constructs from foreign scientists.

The first Indian transgenic fish was generated in MKU in 1991 using borrowed constructs. Scientist in India has developed experimental transgenic of rohu fish, zebra fish, catfish and singhi fish.

Genes, promoters and vectors of indigenous origin are now available for only two species, namely rohu and singhi for engineering growth. Transgenic rohu recently produced from indigenous construct at Madurai Kamaraj University has proved to be eight times larger than the control siblings. This transgenic rohu attains 46 to 49 grams body weight within 36 weeks of its birth.

GENE TRANSFER TECHNOLOGY FOR DEVELOPMENT OF TRANSGENIC FISHES:

The most commonly used methods in fish biotechnology are chromosome manipulation and hormone treatments, which can be produced triploid, tetraploid, haploid, gynogenetic and androgenetic fish.

Other popular methods of gene transfer in fish are microinjection, electroporation of sperms, electroporation of eggs and incubation of sperms. Following are the main steps in gene transfer for development of transgenic fish.

Preparation of DNA Construct

Desired transgene should be a recombinant gene or DNA construct, which is constructed in plasmid that contains an appropriate promoter-enhancer element and a structural DNA sequence.

The foreign genes are typically introduced with strong genetic signals, promoters and/or enhancers, which enable the foreign genes to be expressed at very high levels continuously (or constitutively), effectively placing those genes outside the normal metabolic regulation of the cell, and of the transgenic organism resulting from the transformed cell.

FUTURE PROSPECTIVE OF USING TRANSGENIC FISH IN AQUACULTURE

Transgenic application for value-addition in ornamental fishes has been more useful than for food fishes in this instance because, transgenesis has long been used as a tool for increasing our understanding of developmental biology by expressing the reporter gene, green fluorescent protein (GFP) or red fluorescent protein (RFP) driven by a tissue-specific promoter . The genetic engineering technique of transgenesis could be an alternative approach for commercially viable genetic improvement to meet future food demand. With the vast diversity of fishes, transgenesis could be useful for producing new color varieties of ornamental species and to control reproductive or sexual phenotypes.

The genes of interest in transgenic fish production are growth hormone genes, metallothionein, crystalline, antifreeze protein, esterase, and regulatory gene sequences. For example, metallothionein gene promoter, heat shock gene promoter, and other tissue-specific promoter like myosin light polypeptide chain (mylz) and keratin. The GFP transgenic Medaka line has been produced for monitoring a wide-ranging endocrine disrupters (estrogenic) with a promoter derived from the vitellogenin (vtg) gene. It has been shown, tumors also can be generated via transgenic expression of oncogenes in zebrafish.

TABLE 1: TRANSGENIC FOOD AND ORNAMENTAL FISHES DEVELOPED SO FAR.

Sl. No.	Species	Target genes
1	Rohu (Labeo rohita)	GFP
2	Rohu (Labeo rohita)	Mylz
3	Channel catfish (Ictalurus punctatus)	GH
4	Zebrafish (Danio rerio)	Color genes
5	Rohu(Labeo rohita)	GH
6	Catla (Catla catla)	GH
7	Black tetra (Gymnocorymbus ternetzi)	Mylz2
8	Japanese rice fish (Oryzias latipes)	Vitellogenin
9	Coho salmon (Oncorhynchus kisutch)	GH

Sl. No.	Species	Target genes
10	Chinook salmon (Oncorhynchus tshawytscha)	GH
11	Zebrafish (Danio rerio)	Mylz2
12	Asian stinging catfish (Heteropneustes fossilis)	GH
13	Pond loach (Misgurnus anguillicaudatus)	GH
14	Nile tilapia (Oreochromis niloticus)	GH
15	Rainbow trout (Oncorhyncus mykiss)	GH
16	Red abalone (Haliotus rufescens)	GH
17	Common carp (Cyprinus carpio)	GH
18	Wuchang bream (Megalobr amaamblycephala)	GH

The metallathionin promoter an inducible promoter has been characterized from several fishes, which are used for binding heavy metals in cells with cadmium, copper, zinc, and mercury. The evidence depicted that transgenic common carp contacting growth hormone transgene is less tolerant of un-ionized ammonia, which is an important stress indicator in aquaculture. In addition, tissue-specific promoters and inducible promoters, such as heat shock protein 70 (HSP70) of Zebrafish, has been used for controlling expressing transgenes in the preferred tissues under the desired conditions.

Apart from food fishes, ornamental transgenic fishes have been developed using different color genes. The stable lines of transgenic zebrafishes expressing several living color proteins, such as GFP, RFP, YFP, BFP and CYP were successfully generated. As of now, zebrafish and medaka were model fish species in many of laboratories because they reproduce with external fertilization, have a short life span and produce transparent eggs.

The possibility of the stable expression of color genes was analyzed by zygotic injection of expressible constructs driven by the zebrafish muscle-specific promoter of the myosin light polypeptide 2 (mylz2) gene. The zebrafish (*D. rerio*) mylz2 promoter has also been used successfully to generate fluorescent white skirt tetra (*Gymnocorymbus ternetzi*), medaka (*O. latipes*) and farmed rohu carp, Labeo rohita. The GFP color can be visualized only under a fluorescent microscope for mylz: GFP transgenic rohu. The value-added aquarium fishes have already been commercialized, such as 'GloFish' with six attractive fluorescent color combinations, including Starfire red, cosmic blue, electric green, galactic purple, sunburst orange and moonrise pink.

STRATEGIES FOR TRANSGENIC FISH PRODUCTION

Techniques such as the direct microinjection, electroporation, retrovirus infection, transposable elements, and particle gun bombardment have been utilized to introduce external genetic material (DNA) into vertebrate cells. Among those methods, microinjection and electroporation have proven to be the most reliable techniques of gene transfer in fishes. Efficiency of gene transfer is determined by several factors including: hatching percentage, gene integration frequency, quantity of eggs manipulated in a given amount of time and effort required to manipulate the embryos.

Microinjection

In this method, foreign DNA has been injected into fertilized eggs. The first evidence was reported in goldfish, where foreign DNA microinjected into the fertilized eggs. This method has also been adopted in a wide range of fishes, including goldfish, common carp, atlantic salmon, catfish, medaka, tilapia, rainbow trout, and zebrafish. The earlier studies depicted the survival rate of microinjected fish embryos ranging from 35% - 80%, which was species dependent and expertise. The rate of integration of DNA ranges from 10% - 70% reported in the fishes.

The difficulties of microinjection include the tough and opaque chorion, which hinders insertion of glass micropipettes, although this has been solved by making a hole in salmonids and injection into oocyte nuclei, which is two-step injection. Dechorionation of eggs were also attempted manually as well as trypsinization and pronase treatment.

Thus, this microinjection technique in fish eggs is complicated, slow, tedious, time-consuming, requires a great deal of skills and relatively expensive equipments, and unsuitable for mass gene transfer. Also microinjection causes mechanical damage to the fertilized eggs. All these issues prevented rapid progress in transgenic research in fish.

Electroporation: Sperms vs Fertilized Eggs

Electroporation was accomplished by subjecting cell membranes to a high-voltage electric field, resulting in a shortterm breakdown and forming pores that were large enough to allow macromolecules to enter or leave the cells. Electroporation was simple, operator-independent, a rapid, potential mass method and avoided the difficulties accompanying microinjection.

Electroporation was a successful technique for transferring foreign DNA into the genome of fishes. The study showed that electroporation could be more efficient than microinjection with higher integration rates. The study depicted that the rate of DNA integration in electroporated embryos was more than of 20%. This was the popular method for introducing foreign DNA into fertilized eggs and or embryos of fish.

This was further supported in medaka and remarked higher number of transmitting ability produced by electroporation. The efficiency of the electroporation reduced because of voltage, number and frequency of pulses. The earlier study revealed that this was the successful transfer of transgenes into fish by electroporating sperm. The electroporation was, therefore, considered an efficient and versatile gene transfer technology.

The electroporation method accompanied by sperm has been used in laboratories for improving transgenic efficiency. The sperm electroporation technique was standardized to ensure 25% hatching survival and 37%, presumptive transgenic. In this method, sperm was used for delivering the foreign DNA or gene to the unfertilized eggs or oocyte during fertilization. In this method, DNA was incubated with the sperm, followed by in vitro fertilization, so as to hasten transfer of DNA. Recently, the method of electroporated rohu carp (Labeo rohita) sperm was successfully used to produce

FUTURE STRATEGY

Advances of gene editing technology: Inactivation of a gene can be accomplished by applying a gene knockout strategy or incorporating a mutated copy of the gene. The gene integration at a predetermined position mediated by the cre - recombinase with mutant lox sites successfully carried in the transgenic zebrafish.

Recent advances in gene editing technologies, such as zinc finger nuclease (ZFN), transcriptional activatorlike effector nucleases (TALENs) and CRISPR (clustered regulatory interspaced short palindromic repeats)/CRISPRassociated nucleases in combination with homologous recombination mediated transgene introduction at a predefined location of the genome. It has been possible to undertake development of model animals, including zebrafish, to knock-in or -out a particular gene of interest. The site-specific nick by nucleases can then be repaired by error-prone nonhomologous end joining (NHEJ) resulting in animals carrying deletions or insertions at the cut site and disruption of any gene of interest can be made by enforcing homologous recombination mediated repair at a specific nick site.

Along with that, CRISPR systems (Streptococcus pyogenes) has been well characterized, which includes an endonuclease Cas9, a CRISPR RNA (crRNA) and a transacting RNA (tracrRNA). It can be programmed to introduce site-specific DNA double strand breaks by providing a single guide RNA that is gRNA chimera containing fusion between crRNA and tracrRNA. The previous studies in fishes revealed heritable targeted gene disruption in zebrafish, using designed zinc-finger nucleases (ZFN), resulting in the desired effect. The knock-out of duplicated copies of myostatin gene (mstna) in yellow catfish, using zinc finger nucleases (ZFN) and TALEN, carried out the first endogenous geneknockout in farmed fish. Those investigations encouraged the exploration of creating a Cas9/gRNA-based gene mutation or modification platform for large bodied farmed carps.

The study showed mice carrying a targeted dominant mutation in MSTN gene, which suggested an increase in skeletal muscle mass as a result the combined effect of muscle hypertrophy and hyperplasia. Afterwards, skeletal muscle growth because of mutation in the MSTN gene was reported in dogs, cattle, buffalo, porcine and sheep. Thus, to improve skeletal muscle mass for increasing production, a major area of scientic research in the animals has opened. Whereas, in a fishes, the MSTN gene is well conserved throughout evolution, two distinct MSTN genes were found in some fish species.

To investigate the myostatin role in fish muscle development and growth, a transgenic approach was used to express Myostatin pro-domain in zebrafish muscle cells, thus, researchers speculated that MSTN function was inhibited.

Recently, MSTN gene characterization as well as knock-out were carried out in other fish species, such as yellow catfish, medaka, rainbow trout, lobster, chinese scallop, yellow catfish and bay scallop. Researcher has noted MSTN expressed in the liver and intestine of channel catfish (*Ictalurus punctatus*). Thus, diverse tissue distribution of the MSTN gene suggested its functional variations in different species.

Thus, isolation and analysis of MSTN genes from different species helped in the understanding of its specific role in certain species. Thus, it is likely that applications of gene-editing and gene-integration technologies, such as TALEN and CRISPR associated nucleases have the potential to incorporate indels (by NHEJ repair) at desired location(s) of the fish myostatin gene.

As shown in Fig. 1, we have depicted a strategy for generations of transgenic fishes using gene editing techniques. This could lead to loss of gene function. Alternatively, the gene of interest could be knocked-out by enforcing HR mediated repair of nick created by TALENs or CRISPR/Cas9 nucleases.

Driving Organ Specific Expression to Generate Colorful Ornamental Varieties

The development of Glofish sets a model for developing newer color varieties in ornamental fishes. This is the only ornamental fluorescent zebrafish (Danio rerio; Yorktown Technologies, Austin) were reached in the US market. Several genes, such as green fluorescent protein (GFP), luciferase, bacterial chloramphenicol acetyltransferase (CAT), neomycin phosphotransferase II (neor), and β-galactosidase (β-gal) are commonly used to produce a DNA construct with reporters. Attempts have been reported to produce novel varieties of ornamental fish for rearing as pet, and to use the change of body color as reporters for detecting environmental changes as biosensor. The U.S. Food and Drug Administration (FDA) determined there is no evidence that these genetically engineered zebrafish pose any greater threat to the environment than their widely sold unmodified counterparts.

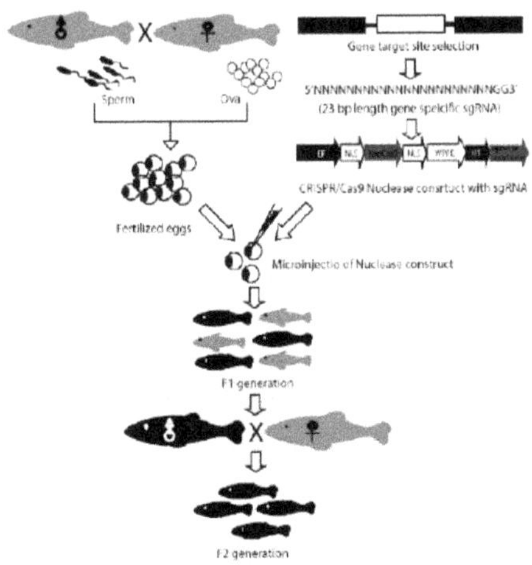

Fig. 1: General strategy for generation of transgenic fish using gene editing technology.

The reporter genes such as green fluorescence protein (GFP) and several GFP variants driven tissue specific promoters, such as skin specific keratin gene (krt8) and muscle-specific gene (mylz2) are the genes used most. The GFP, extracted from jellyfish (*Aequorea victoria*) has become convenient markers for transgenic fish because of their sensitive signal detection. With this, red fluorescence protein (RFP) reporters can be detected easily in the live transgenic fish, such as zebrafish and medaka.

The melanin concentrating hormone (MCH) was used to generate colored transgenic medaka, which was isolated from chum salmon. Till now, several colorful transgenic zebrafish displaying whole-body green, red, yellow or orange fluorescent colors was generated successfully. The mutant zebrafish and medaka with less interfering pigmentation for optical observation also have been developed. The GFP-tagged transgenic embryos, as well as fish, have become the most convenient and widely used tools in basic biological research, particularly to regulate gene expression, morphogenetic movements of cells and organogenesis during embryonic development.

The greater number of fish genes have been characterized by encoding the growth hormone. The cloning of β-actin promoter from carp has allowed the production of an all-fish construct by means of expression vectors for use in transgenic fish. The use of an 'auto transgene' construct, where the promoter or regulatory element and targeted gene derived from the same species, was shown to be helpful in transgenesis.

Studies were conducted in fish, using mammalian and avian promoters, also, sometimes, viral promoters. This requires isolation of a species-specific functional promoter capable of driving either ubiquitous or tissue-specific expression of the gene of interest. The β-actin promoter showed variable but, sometimes, ubiquitous transient expression in medaka. At CIFA, studies were carried out on characterization of rohu β-actin gene and promoter, which was an evolutionary conserved sequence. Its promoter could be used for ubiquitous expression of genes of interest. The entire rohu β-actin promoter/ regulatory region showed closest similarity to that of catla (Catla catla), among the sequences of other species available in GenBank. The potential implications in the basic research, vaccination,.

In this way, the use of an endogenous promoter constructs, where the promoter-regulatory element is derived from the same species, was shown to be useful in driving the foreign gene expression. Studies by fisheries research institutes (CIFA) depicted that the isolation of the rohu (*L. rohit*a) mylz2 promoter will have potential use in tissue-specific, desired expression of genes of interests for generating transgenic rohu.

The 5'-flanking region (1.2 kb) of a muscle-specific gene myosin light chain 2 polypeptide (mylz2) of rohu carp, was cloned successfully. Direct muscular (skeletal) injection of the construct containing the mylz2 promoter (0.6 kb) linked with GFP reporter gene marked efficient expression in *L. rohita*, confirming its functional activity, which was confirmed by observation and promoter drove GFP expression in the skeletal muscle of transgenic rohu.

The β-actin promoter may have prospective applications for value-addition in ornamental fishes as well as understanding gene regulatory network or functions. The study revealed direct muscular injection of the construct containing â-actin promoter fused to a GFP reporter gene showed efficient expression at the site of injection as well as in the areas away from the site.

About 20% of transgenic offspring showed ubiquitous expression of the reporter GFP gene and massive epithelial and/or muscular expressions, with visible green color under normal light, were documented in a few transgenic rohu.

Regulations and Policies for Transgenic Fish Production/ Promises to be Made

Now, researchers are concerned about doing research on transgenic fishes because of the safety of such work, which led to the optimizing guidelines for safe laboratory practices. The researchers or scientists are alert to assessing the Biosafety of genetically modified fish. The World Health Organization (WHO) and the Food and Agriculture Organization (FDA) have given complete international guidelines for recommendations and the safety assessment of foods derived from biotechnology, as well as safety evaluation of all genetically modified organisms. The detail policies and approval process required for transgenic fishes has been reviewed and discussed several times.

The National Institute of Health (NIH) and FDA framed different and often-revised guidelines for research, production and commercial use of transgenic organisms. Until now, twelve countries have framed their respective national policies for regulation of transgenic aquatic organisms. Out of twelve countries, only six countries have developed field-testing of transgenic fish (China, Israel, Hungary, New Zealand, UK and USA), comprising transgenic carp, salmon, nile tilapia and channel catfish. The Cartagena protocol has a detailed treatment of risk assessment towards protecting biological diversity from transgenic organisms.

Regulatory approval of the first transgenic fish for human consumption is being considered by the U.S. and Canadian governments. Screening methods, such as southern blotting and transcription analysis, can be done using Northern blotting, ribonuclease protection assays and RT-PCR for transgenic fishes. Finally, characterization should be performed using protein product and its expression, which can be correlated with any phenotype of the fishes using techniques, including Western blotting, enzyme-linked immunosorbent tests (ELISA), radio-immunoassay and immune-histochemistry.

FUTURE PERSPECTIVES

The production of transgenic food fish is promising, with techniques to meet the future food demand. But it has been held up because of regulatory issues. In this field, it is necessary to isolate and identify novel, useful genes and suitable promoters of fishorigin, which are helpful for expression of the transgene in a tissue at the desired level. Advance techniques, such as gene editing technology, could be applied to resolve transgenic efficiency issues. Continuous monitoring of integration of the transgene in further generation is needed.

Biosafety issues of transgenic fishes also need careful consideration. For framing or updating regulations affecting GMOs, accurate and current information must be used. Therefore, the challenge is great in developing GM food with no adverse effects on every single individual in the human population. Thus, accurate information about the transgenic must be spread among the general population. Producing the knock-out or -in gene inactivated transgenic fish may be more acceptable to people, who are worried about the safety of the foreign genes introduced into the fish genome.

In case of ornamental fish, it is possible to use transgenic techniques for developing new varieties with novel or brilliant colors or shapes. Glofish produced in Singapore and marketed in the USA, highlighted the new possibilities for producing new multicolored fluorescent fish, having economic benefits. The color genes, such as green (GFP), red (RFP), blue (BFP), yellow/golden (YFP) and cyan (CFP), provided avenues for value-addition to small aquarium fishes other than zebrafish and medaka.

FISH GENOMICS

Genome sequence comparisons between such distantly related organisms are highly informative due to the accumulation of neutral mutations in nonfunctional regions. Yet humans and fishes share many developmental pathways, organ systems, and physiological mechanisms, making conclusions relevant to human biology. The respective advantages of zebrafish, medaka, *Tetraodon*, or *Takifugu* have been well exploited so far with bioinformatics analyses and molecular biology techniques. However the full potential of fish genomics is about to be unleashed with the integration of more traditional disciplines such as biochemistry and physiology, with the study of additional species such as carp, trout, or tilapia and a broadening of its applications to environmental genomics or aquaculture.

While the gastronomic qualities of fish did not escape Jules Verne in his 1870 *20,000 Leagues Under the Sea*, fish are no less put to good use in twenty-first century biology. In this new context, one could easily replace fin color and flesh quality by genome size and embryo transparency in a similar enumeration of the advantages of these animals for biology in general and molecular genetics in particular. If Captain Nemo was in a position to offer such variety on his menu, it is partly because fish comprise more than 25,000 species, by far the most successful vertebrate group.

Indeed few aquatic ecosystems have eluded colonization by at least some fish species, from Tibetan streams to the abyss of the oceans via sub-zero Antarctic seas. Of these species, many have long been used as models in different disciplines of biology (Fig. 2) because of this very diversity:

The atrophy or exaggeration of important anatomical or physiological functions occur with sufficient frequency to have attracted biologists to fish models. This includes molecular genetics and genome research, for which fish also possess interesting and outstanding features, if not all-time records, among vertebrates.

Combined with large-scale mutagenesis screens initiated in the early 1990s, zebrafish filled a gaping hole in vertebrate developmental biology: the ability to study genes via their mutant phenotypes on a large scale as in *Drosophila melanogaster* or *Caenorhabditis elegans*. Since then, capital discoveries for our understanding of vertebrate development and human disease have already emerged from zebrafish studies.

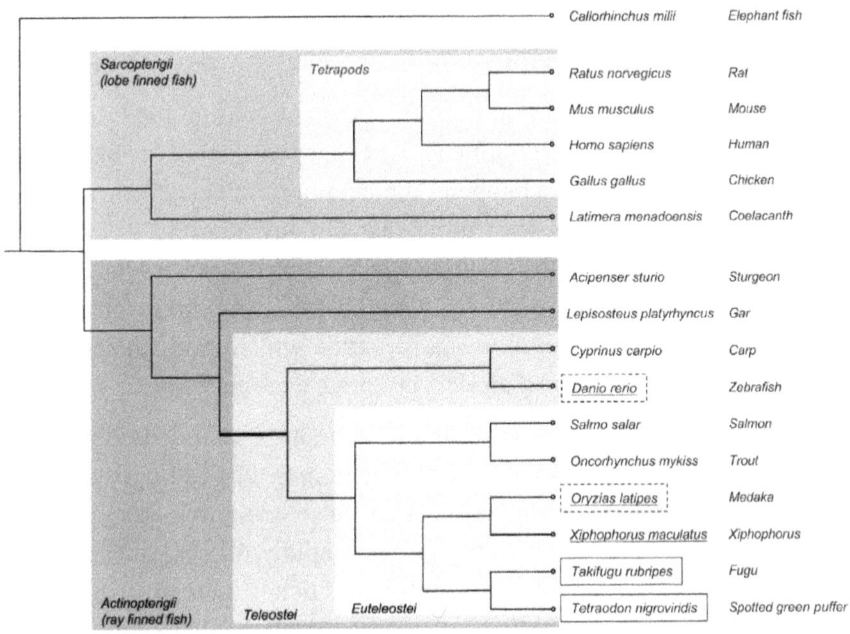

Fig. 2. Consensus phylogenetic relationships between fish and tetraopods, including fish considered as biological models (underlined), and those for which a sequence has been published (boxed, continous line) or is underway (boxed, dashed line). The thick branch indicates the most likely position of the whole-genome duplication at the root of the teleosts

However if one considers genome analysis a question of DNA sequence acquisition and "mining," then fishes really became a major player in 1993 when Sydney Brenner suggested a new species as a genome model, the marine pufferfish *Takifugu rubripes* (*fugu*). Aside from its gastronomic delicacy status in Japan and China, *fugu* possess one of the smallest vertebrate genomes. This feature, already recorded for its freshwater relative *Tetraodon nigroviridis* in 1968, is a major advantage to rapidly gain access to a large catalog of genes in a vertebrate at a cost comparatively smaller than for the much larger genome of a mammalian species. However, both pufferfish are species for which we know little in terms of physiology, reproduction, or life cycle.

Since biology is still a science largely driven by the quality and depth of the experimental data and our ability to extract meaning from it, the outcome of both pufferfish genome programs has until recently been confined to discoveries on the structure and evolution of genes and genomes, with few connections to development, cell biology, or physiology.

However this is about to change, with the emergence of networks of scientists and cross-disciplinary platforms where precise biological questions are examined with an array of tools and resources that include genome sequences and genomic techniques. Here we review the impact of the genome sequence for those fish species for which it is already available, and we examine how the combination of genomics with more traditional disciplines might pave the way for a much wider impact on biology.

Angling in the genomic aquarium

The sequence of the human genome was still a distant goal in 1993, but the project of sequencing the entire genome of a multicellular eukaryote, that of the nematode worm *C. elegans*, was well on its track. As it happens, many important eukaryotic model organisms that were already being studied with molecular biology approaches had genome sizes within the reach of sequencing technologies of the time: *Saccharomyces cerevisiae* (14 Mb), *D. melanogaster* (180 Mb), *C. elegans* (100 Mb), and *Arabidopsis thaliana* (125 Mb). But none of these were vertebrates, a situation which motivated a pilot project to evaluate the usefulness of *fugu* as a model vertebrate genome.

This influential analysis showed that the genome was indeed about 400 Mb, or eight times smaller than the human or mouse genomes. Just as important, exon–intron boundaries seemed conserved, suggesting that gene structures were likely to be very similar, albeit in a much more compact sequence. Rapidly, other studies strengthened the notion that pufferfish DNA could help identify and better understand the structure and sometimes the function of mammalian sequences.

In one such early attempt, a *fugu* sequence next to the *Hoxb4a* gene and conserved with mouse, showed enhancer activity in transgenic mice. The power of comparative genomics in vertebrates was also soon illustrated when the complete Huntington disease (HD) gene from *fugu* was sequenced and compared to its human ortholog. The analysis showed that the *fugu* HD gene possesses a four-glutamine repeat, whereas mouse has seven and healthy humans a minimal of eight.

Because a tract of four glutamines is unlikely to form a functional site in itself, the *fugu* protein supported the view that it is only after expanding to over 37 residues in HD patients that it somehow gained a new pathogenic function. This initial sequencing project in *fugu* was followed by many more that generally showed limited long-range conservation of gene order, rarely more than four genes per conserved synteny block, thus dampening down the initial hope that this compact genome was a valuable tool to accelerate the mapping of human genes, then a priority in the Human Genome Project.

However there were still major reasons to establish complete sequences of fish genomes: (1) these early studies had shown the usefulness of comparative sequencing when constructing gene models on the human genome, (2) the initiation of large scale mutagenesis projects on zebrafish was calling for a global genome effort already in preparation with the construction of genetic maps and radiation hybrid maps.

Trawling for whole genomes

Reevaluating the Number of Human Genes

Fugu is a relatively large marine fish that contains elevated doses of tetrodotoxin causing live specimen or frozen samples to be the subject of restrictive importation laws in most countries outside of Asia and thus posing practical problems for genomic analyses. A different pufferfish, *Tetraodon nigroviridis* (Green spotted puffer, sometimes confused in the aquarium fish market with *Tetraodon fluviatilis*, which is a different species) was proposed that alleviates this restriction: *Tetraodon* also possesses a small compact genome but it is a popular aquarium fish that can live in freshwater.

In contrast to *fugu*, few specific *Tetraodon* loci were sequenced and studied in comparison to their homologs in other species. From the beginning instead, *Tetraodon* genomic DNA was exploited in large-scale comparisons between different vertebrate genomes. The initial rationale behind this second pufferfish project was to assist in the annotation of human genes, a slow and fastidious task when performed by humans, often unreliable when performed by automatic approaches, and yet one of the primary goals of the Human Genome Project. On the basis of an initial sampling of random sequences from the *Tetraodon* genome (about 30%), a tool named Exofish that was based on BLAST sequence alignments was developed to identify conserved regions in human genomic DNA that correspond to coding exons, rapidly and with high specificity.

A surprising outcome of this first example of a global sequence comparison between large samples of two vertebrate genomes is that the number of conserved sequences identified by Exofish in the human genome was not compatible with the 60,000 to 150,000 genes that it was thought to possess at the time. Indeed comparisons with the *Tetraodon* sequence sample indicated that the entire human genome would contain about 88,000 evolutionary conserved regions (termed ecores) corresponding to human exons, while known human genes possessed on average between 2.6 and 3.2 ecores.

A simple ratio between these figures yields a total of about 30,000 human genes. This new estimate, confirmed later by the initial analysis of the human genome sequence, challenged the notion that the complexity of genetic information contained in a genome is a function of the number of protein-coding genes.

Puffer Fish Genome Features in Draft Sequences

The *fugu* genome, the first vertebrate genome to be sequenced after human, was obtained using the whole-genome shotgun method. This sequence draft enabled a number of interesting observations, such as differences in specific protein families between human and *fugu*. The *Tetraodon* genome sequence was subsequently produced, also with the whole-genome shotgun method albeit with a higher redundancy in sequence reads (8.3 vs. 5.6). Both pufferfish possess about 70 different families of transposable elements against only 20 for human or mouse, but in pufferfish they comprise two to three orders of magnitude fewer copies.

Interestingly in *Tetraodon*, SINE and LINE families are distributed in opposite regions of the genome compared to human or mouse: SINEs are more abundant in G + C-rich sequences in mammals, and in A + T-rich regions in *Tetraodon*, and vice versa for LINE elements. More surprisingly, these initial studies of *Tetraodon* and *fugu* showed a number of differences in their genomes. For instance a G + C-rich region present in both *Tetraodon* and mammal genomes is absent in *fugu*. Also some gene families such as type I cytokines and their receptors, present in all vertebrates studied so far, were notably difficult to find in *fugu*, while over 30 members of the family could be identified in *Tetraodon*.

These discrepancies are most likely attributable to biases in clone libraries or differences in methodologies, and hopefully should be resolved as the genomes reach completion. When comparing fish and mammal gene catalogs, surprisingly few major differences could be documented when using the Gene Ontology classification system. More striking differences could be seen using protein domain comparisons: Proteins involved in sodium transport are more abundant in fish, which also contain an allantoin pathway for purine degradation that is absent in humans.

Neutral nucleotidic sequence evolution per year was found to be twice as fast in pufferfish as between human and mouse, and protein evolution also appears to proceed at a faster rate in fish, although the reasons for this are still unclear. It should be noted that these results depend on the dating of the divergence between *Tetraodon* and *fugu* (18–30 Mya).

Insights into Vertebrate Genome Evolution

Perhaps one of the major differences in the two pufferfish draft sequences resides in the fact that the *fugu* genome sequence was assembled purely by the whole-genome shotgun method with no physical mapping, whereas most (64%) of the *Tetraodon* genomic sequence is anchored on each of the chromosomes, providing a long-range view of gene organization in the genome. This added information primarily resolved a long-standing issue on the occurrence of a whole-genome duplication in the fish lineage.

Remarkably, the idea that an increase in chromosome numbers may be a source of phenotypic novelty is nearly a century old. In 1911, Kuwada already observed that some varieties of maize were tetraploid and suggested that this may be the source of "innumerable races". During the following 60 years, the occurrences and consequences of gene and genome duplications continued to be discussed, with for instance the proposal by Stephens in 1951 that increasing the number of genetic loci was the only path to "evolutionary progress" and his suggestion that genome duplication could be one way of achieving this. These theories reached a high point in 1970 with the publication of Susumo Ohno's book that stated several landmark notions: (1) without duplicated genes, the emergence of metazoans, vertebrates, and mammals from unicellular organisms would have been impossible, (2) this process required the creation of new loci with previously nonexistent functions, and (3) he postulated that at least one whole-genome duplication facilitated the evolution of vertebrates. While these ideas were met with mild enthusiasm at the time, it is today a widely accepted notion that to create functional novelty, gene duplications are at least as important as point mutations in individual loci.

Moreover when the duplication affects the entire genome at once, this potential for novelty is theoretically amplified by allowing the duplication and retention of partial or complete metabolic pathways.

Zebrafish and Medaka: Biological Models Comeback

The zebrafish as a model system has accompanied the development of molecular biology from the 1960s to the genomic revolution of the 1990s. Following the publication of the results from two large-scale genetic screens in 1996 in a special issue of *Development*, zebrafish was propelled at the forefront of developmental biology research. While such screens are currently being expanded to tackle a wider range of mutations in embryonic and adult stages (see below), many researchers are also studying zebrafish models of human diseases including blood diseases, heart disorders, and cancer.

As mentioned above, genome research on zebrafish started in parallel to these developments. Early efforts to map genes and genetic markers served to identify the loci of specific mutant genes, and to compare the organization of genes with other vertebrates. Microarray technology has also been applied, in combination with single nucleotide polymorphism (SNP) identification, to accelerate the linking of mutant phenotypes to their genotypes.

The zebrafish genome is about 1700 Mb, and that of the Tübingen strain is now being sequenced at the Sanger Institute using a mixed strategy where both large insert clones and the entire genome are submitted to the whole-genome shotgun method. Today the zebrafish community worldwide has expanded tremendously with over 3500 scientists registered in the ZFIN community database. The availability of the genome sequence will provide a unifying reference to integrate the wealth of functional data accumulated so far.

THE FUTURE OF FISH GENOMICS

The future of fish genomics is bright, and this prediction is sustained below by three examples where specific characteristics of fish species were successfully exploited to gain penetrating insights in a broad range of subjects.

Fish, as a highly diversified group of the vertebrate family, experience an astonishing range of environmental conditions to which their physiologies, body shapes, and lifestyles have adapted. However, a common denominator of all fish species is their aquatic habitat, meaning that water is in direct contact with several tissues and internal compartments of the animal, potentially inducing a high sensitivity to water-borne parameters such as temperature, oxygen levels, salinity, and sometimes toxic chemicals.

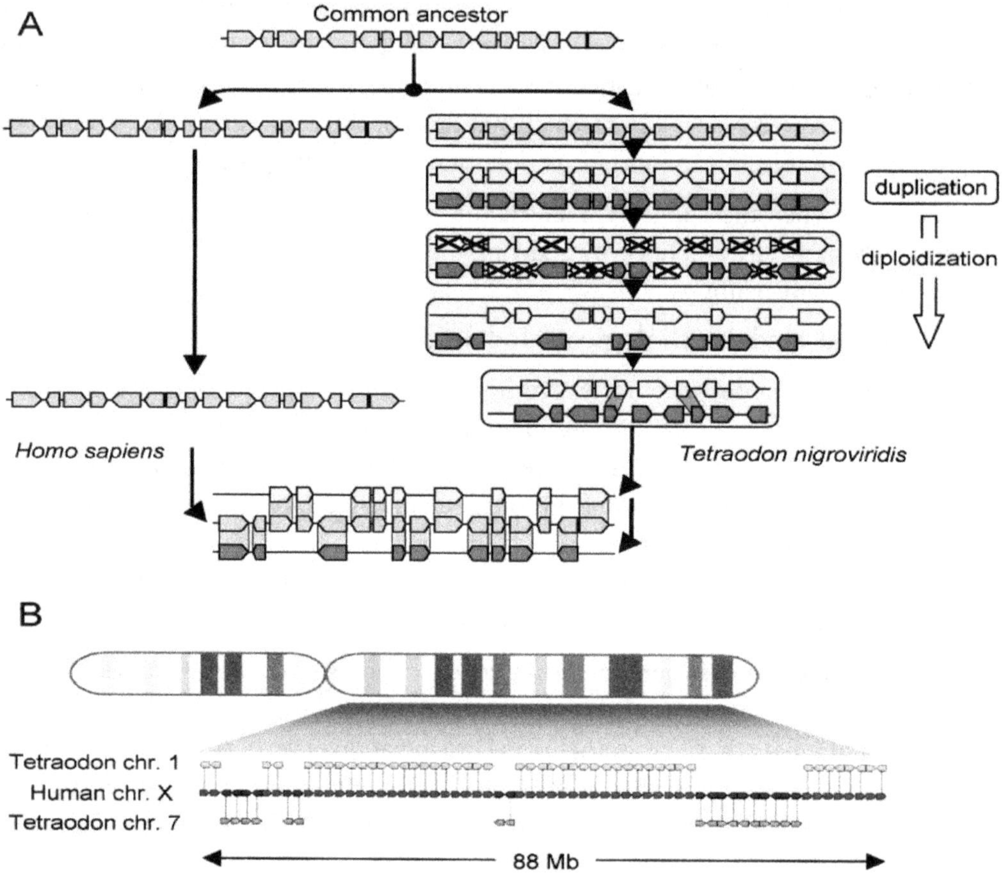

Fig. 3. (*A*) Duplication leads to double-conserved synteny. (*B*) Example of doubleconserved synteny. An 88-Mb region covering the majority of the long arm of human chromosome X contains 65 genes with orthologs in the Tetraodon genome, alternating between chromosome 1 and chromosome 7. Genes are represented by small arrows that indicate the orientation of transcription.

This intimate relationship between an organism and a wide range of different environments has recently prompted the view that fish could be used as models for "environmental genomics", in other words the study of the interface between an organism and its environment using genomic approaches. This concept had recently been illustrated by an elegant study involving the exposure of carp to increasing levels of cold, from 30°C to 10°C and measuring the change in the level of expression of several thousand genes using microarray techniques. Carp, as most fish, is poikilothermic (cold-blooded), which implies that its body temperature follows that of the water it is immersed in. In the seven tissues that were monitored, the decrease in temperature was correlated with a graded increase in expression of a core set of 252 genes predominantly from transcriptional regulation, RNA splicing and translation systems, while very few genes common to all tissues showed a decrease in expression.

Conversely, tissue specific responses showed that brain modifies its glycolytic activity while a switch to lipid metabolism is observed in liver. Interestingly this study was performed in a nonmodel species, exemplifying the use of a specific species of fish with a specific feature of interest, in this case the carp which tolerates a wide range of temperature, to investigate a whole body physiological adaptation to a change in environment. While the results are useful to better understand the molecular basis of the response to cold in carp, fish share so many aspects of their developmental pathways, physiological mechanisms and organ systems with mammals that these results are also relevant to human physiology.

The second illustration of the strength of future genomic research using fish species is its recognition by international funding agencies. One such recent example is the major impetus to zebrafish genomics provided by the European Union Framework Programme for Research. By funding the ZF-MODEL consortium, an Integrated Project devoted to the study of zebrafish models for human development and disease, the project links about 30 groups from seven European countries.

The research plan is to take advantage of the zebrafish model in combination with an array of large-scale techniques to identify zebrafish genes that replicate to some degree human pathologies, or are involved in developmental pathways similar to ours. The scale of this endeavor can be measured by the thousands of mutants already generated by the consortium members using either forward genetics (for instance with chemical mutagenesis screens), or reverse genetics and the scope of the screens used to analyze them.

Designed to identify mutants that resemble human disorders, these screens are also carried out on adult fish and encompass for instance bone malformation, skin development, eye movement or addiction behavior. Combined with enhancer detection techniques using Green Fluorescent Protein (GFP) transgenic lines and expression profiling with microarray analysis, the consortium will integrate a broad range of expertise to tackle the function of the mutated genes, which hopefully will lead to a better understanding of corresponding human disorders.

In addition to counting several member species elevated to the status of genomic models such as zebrafish or pufferfish, fish also represent a major source of food for humans. The European Union thus also recognized the need for improving aquaculture research by funding the AQUAFIRST consortium to identify genes associated with stress and disease resistance in sea bream, sea bass, and rainbow trout in order to provide a physiological and genetic basis for marker-assisted selective breeding. Genomic techniques and sequence comparisons with genomic models are cornerstones of this project, which illustrates how fish genomics may grow in coming years outside of fundamental research labs toward more applied objectives nonetheless essential to human welfare.

A final example of the use of fish genomes that may expand in the near future rests on the phylogenetic position of fish species in comparison to mammals. Today a great deal of attention is focused on the part of eukaryote genomes that do not code for proteins but are nevertheless functional. Elements that may carry out specific functions in these regions include non-coding RNA genes and regulatory control regions. One of the most powerful techniques to identify such elements is to align genomic sequences of distantly related organisms and look for regions that have remained similar during evolution, thus suggesting that a functional constraint is acting to preserve the sequence from mutations. The advantage of fish in this context is the long evolutionary distance, approximately 450 million years, since their last ancestor with mammals.

Neutral mutations have since saturated the genome to a point where any conserved region between for instance human and pufferfish is indicative of a functional constraint. This comparative approach was first applied on a genome scale to identify coding exons, and more recently to identify ultra-conserved regions (UCRs) of unknown function.

However an additional assumption that can be deduced from the discovery of UCRs conserved between such distant organisms is their fundamental importance across vertebrates. In line with this, UCRs found in this way lie in clusters around genes involved in the regulation of development. Indeed when the orthologous regions were assayed using GFP reporter constructs in zebrafish embryos, most showed significant enhancer activity in one or more tissues.

So fish–mammal sequence alignments not only provide the means to identify functional elements, they also act as a screen to select those elements essential to vertebrates. With the production of new fish genome sequences as well as new mammalian sequences, such comparative studies are likely to play an important role in guiding the identification of functional noncoding elements, and in deciphering the subtle sequence variations that might lead to phenotypic changes.

Many fish species are routinely being studied at the molecular level and even at the genomic level and have not been cited here. For instance thousands of EST sequences are available for carp, catfish, salmon, trout, killifish, stickleback, or tilapia and large insert BAC libraries are also available for several of these species, further illustrating the widespread interests in using fish for genomic research. Four fish genome sequences are or will be available soon: *fugu, Tetraodon*, medaka, and zebrafish. The stickleback genome (*Gasterosteus aculeatus*) is also well advanced, but to our knowledge no other is currently ongoing beyond these, although obviously several interest groups are actively working towards promoting certain species.

Research on salmon and trout for instance would greatly benefit from the availability of the genome sequence. Indeed more is known about the physiology and biology of rainbow trout than any other fish species, although genomic sequencing could be complicated by an additional genome duplication in the salmonid lineage some 25–100 Mya. The sarcopterygian fish coelacanth is the nearest living relative of tetrapods and thus escaped the whole-genome duplication that affected the teleosts.

This may be a serious advantage since it might provide access to a genome that resembles the early tetrapod genome, unaffected by the consequence of massive gene duplications such as gene conversion. Altogether the coelacanth would indeed be an excellent candidate for genome sequencing as it would provide a reference genome for tetrapods while allowing the identification of genomic features that differentiate them from teleosts. For similar reasons but across a wider evolutionary scale, the elephant fish (*Callorhinchus milii*, a cartilaginous fish) has recently been proposed as a good model to study the genome structure and gene content of a basal jawed vertebrate, and provide a common reference for tetrapods and ray-fined fishes.

FUNCTIONAL GENOMICS IN VACCINE DEVELOPMENT

Given that most pathogens exist as multiple strains having different antigenic proteins, the challenge in vaccine design has been to find cross protective antigens against variant strains of the same pathogen. In the case of viruses, different approaches have been used aiming at finding the most neutralizing epitopes using methods such as epitope mapping, peptide scan and reverse genetics. However, the upcoming of next generation sequencing (NGS) supported with current advances of bioinformatics tools is expected to expedite our ability to identify the most immunogenic proteins for vaccine production against viral diseases. For example, scientists used bioinformatics to identify the antigenic proteins for Singapore grouper iridovirus. They used the 162 open reading frames (ORFs) of SGIV for sequence similarity searches to identify motifs, cellular locations and other prediction domains to identify the most immunogenic epitopes required for vaccine production. They identified 13 genes that were cloned to produce DNA vaccines of which three vaccines produced relative percent survival (RPS) ranging from 58.3 to 66.7% in vaccinated grouper.

In the case of bacterial vaccines, identification of protective antigens can be a challenge given that they contain several antigenic proteins such as capsular antigens, fimbriae, pili and outer membrane proteins. Some of these proteins lead to serotype, biovar or strain differences leading to antigenic diversity within bacterial species. Hence, the challenge is to identify broad neutralizing antigens able to confer cross protection against variant bacterial strains can be a difficult task. To overcome this problem, scientists developed an *in vivo* induced antigen technology (IVIAT) that uses antibodies generated from individuals infected by the bacterial strain homologous to the vaccine strain to probe for immunogenic proteins using an *in vitro* expression system.

To do this, a genomic library is generated using DNA fragments from the bacteria strain to be used for vaccine production. The DNA fragments are digested using restriction enzymes and cloned into plasmid vectors. Induced colonies of the expression library are probed using pooled sera from bacterial infected individuals (Fig 4). Reactive clones are purified and used as vaccine candidates. This technology has been widely used to identify antigenic proteins for different bacteria species such as *Streptococcus iniae*, *Vibrio anguillarum*, *Aeromonas salmonicida*, *Edwardsiella tarda* and *Streptococcus parauberis*. Scientists used the IVIAT to identify a 510 aa peptidase protein, which they used to produce a subunit vaccine against *E. tarda* in Japanese flounder. Scientists used the IVIAT technique to identify a secretory antigen, which they designated as Sia10, and cloned it to produce a DNA vaccine against *S. iniae*. In vaccinated turbot, the Sia10 protein was detected in the muscle, liver, kidney and spleen by 7 days post vaccination (dpv) lasting until 49 dpv.

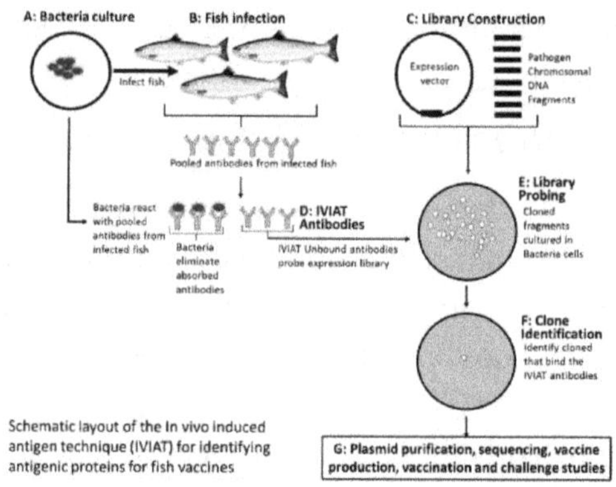

Fig. 4. Schematic layout of the IVIAT technique for the identification of bacterial antigenic proteins essential for the production of fish vaccines: A: bacteria culture. B: bacteria infection in fish and the sera from infected fish is pooled. C: library construction using chromosomal DNA fragments of the bacteria cultured in (A). D: bacteria eliminate absorbed antibodies from sera while IVIAT unbound antibodies are used to probe the library constructed in (C). E: clones from fragments of bacterial chromosomal DNA are probed with IVIAT pooled sera. F: after probing with pooled sera from infected fish, clones depicting binding capacity to IVIAT sera are sub cultured. G: the identified clones are purified, sequenced and used for subunit or DNA vaccine production followed by vaccination and challenge trials.

Post challenge RPS showed 73.9 and 92.3% in fish challenged with high and low challenge dose, respectively. In addition, the Sia10 protein produced protective antibodies in passively vaccinated fish. In another study, scientists used the IVIAT method to identify a surface antigen designated as Esa1, which they used to produce a DNA vaccine against *E. tarda* in Japanese flounder.

They showed that the pCEsa1 vaccine enhanced respiratory burst, acid phosphatase activity and bactericidal activity of headkidney macrophages. In addition, it produced RPS

= 57% in passively vaccinated fish. Overall, these studies show that genomics approaches can be used to identify the most immunogenic proteins for different bacterial strains in order to produce the most protective vaccines for use in aquaculture.

Current advances in HTS have refined genomic analyses to base pair resolution making it easier to map entire epigenomes of living organisms enabling us to identify biological markers predictive of the outcome of disease infections, reproduction, growth and adaptation to new environments. As a result of these advances, epigenetics studies in aquaculture have tremendously increased in the last decades with the view to identifying biological markers relevant for improving the production of farmed aquatic organisms.

Technologies used for epigenetics analyses in aquaculture include (i) RNA seq in Medaka and Nile tilapia; (ii) genome wide methylated DNA immunoprecipitation sequencing (MeDIP seq) in Nile tilapia and Medaka; (iii) bisulfite sequencing (BS seq) in smooth tongue sole (*Cynoglossus semilaevis*), rainbow trout and Nile tilapia; (iv) genetic linkage map analysis using simple sequence length polymorphisms (SSLPs) in medaka; (v) methylation sensitivity amplified polymorphism (MSAP) in Atlantic salmon, grass carp, brown trout, sea urchin (*Glyptocidaris crenularis*) and sea cucumber (*Apostichopus japonicas*); (vi) 5 methylcytosine immunolocation in sea lamprey (*Petromyzon marinus*); (vii) restriction endonuclease hydrolysis of DNA using methylation enzymes in Zebrafish and (viii) bisulfite sequencing PCR in Pacific Oyster (*Crassostrea gigas*) and grass carp.

11

AQUATIC GENETIC RESOURCES AND ITS CONSERVATION

INTRODUCTION

Biodiversity or Biological Diversity a sum of all the different species of animals, plants, fungi, and microbial organisms living on Earth and the variety of habitats in which they live. Each species is adapted to its unique niche in the environment, from the peaks of mountains to the depths of deep-sea hydrothermal vents, and from polar ice caps to tropical rain forests. According to the definition of the Convention on Biological Diversity, biodiversity is the variability among living organisms from all sources, including terrestrial, marine and other aquatic ecosystems and the ecological complexes of which they are part; this includes diversity within species, between species and of ecosystems.

Aquatic biodiversity can be defined as the variety of life and the ecosystems that make up the freshwater, tidal, and marine regions of the world and their interactions. Aquatic biodiversity encompasses freshwater ecosystems, including lakes, ponds, reservoirs, rivers, streams, groundwater, and wetlands. It also consists of marine ecosystems, including oceans, estuaries, salt marshes, seagrass beds, coral reefs, kelp beds, and mangrove forests. Aquatic biodiversity includes all unique species, their habitats and interaction between them. It consists of phytoplankton, zooplankton, aquatic plants, insects, fish, birds, mammals, and others.

Primary producers of the oceans are occupying the photic zones down to about 200 m from the water surface, and sea bottoms areas adjacent to the shallow coastal water. In the deep-sea zone, there is an existence of entire different life. It is a true fact that the oceans and seas are support system of an extended number of biological diverse species, which are immensely important for the ecological diversity. There are total 222,000-230,000 marine species listed, and almost 200,000 belong to Animalia. Marine habitats are fragile because they are highly reactive to the variations and transformation in physical environments.

Sensitivity of the biodiversity in fresh water resources is more than any other terrestrial ecosystems. The vulnerability of the freshwater habitat is because disproportionate numbers of plant and animal communities are growing in the water regime. As estimated by researchers freshwater bodies are enriched with more than 10,000 fish species, which comprises approximately 40% of global fish communities and one fourth of diverse vertebrate population at global scale. Combining countless number of amphibians, aquatic reptiles and mammalian populations to the total quantity of freshwater-fish clearly depicts the freshwater habitat as the only favorite biological spot of all vertebrates. Comprehensive information about the total species diversity in the freshwater resources is incomplete especially among invertebrates and microbes and in the tropical zones of the world that serves as a dwelling spot of different species of the world. From amphibian's phylum, total of 5778 species has been identified since last 10 years.

Mekong drainage in Cambodia has been identified as one of the global "hotspots" for regional river fish biodiversity; researchers stimated that Mekonge basin has variety of species richness making it globally recognized. Recent research estimation has revealed fish wealth of about 1700 species. It has been noticed that freshwater biological regions are given less interests than terrestrial zones. Documentation about invertebrate animals diversity in tropical freshwaters are not available. However, great endemism and species richness at regional habitat do exist in the groups of crustaceans, mollusks and aquatic arthropods.

Information on microbial biodiversity in the freshwater bodies is also not well recorded, not indicating the vital function of microorganisms in biogeochemical cycles operation on the earth. Most of the prokaryotic taxonomic groups are still not discovered.

Aquatic biodiversity increases with latitudes with maximum in the tropical zones. The best examples of aquatic ecosystem biodiversity can be seen in the Amazon River. The Great Barrier Reef in Australian continent is the largest coral reef ecosystem in the world, habitat of over 700 varieties of coral, and also giving shelter to diverse varieties of fish and mollusks species.

Coral reefs are the systems with extreme biodiversity of marine animals. One studies from the Red Sea region of Gulf of Aqaba, has revealed that egg releasing phase of aquatic animals are different throughout the year in this particular region of marine ecosystem. This gives the strikingly different detail from the Great Barrier Reef, where aquatic animals and coral spawning occurs at the same time.

Not only marine systems many fresh-water ecosystems are also a favorable biological spot of unique species, because freshwater habitats are isolated environment on the natural landscape, facing obstacles of distance and weather conditions which is a difficult task to tolerate. Such kind of ecological phenomena favors the evolution of new species in different fresh-water resources.

Researchers categorized a number of diverse plant species found in and around aquatic bodies: flowering plant, mosses, liverworts, species of encrusting lichens, stonewort and other enormously growing algal species. The uninterrupted water cycles in the aquatic bodies have an immense role in the arrival of minerals and nutrient components from high level to low-lying area and eventually to the extended sea and ocean water.

Aquatic bodies downstream support a rich diverse species of plants: cool hard bottom streams containing mostly bryophytes and large soft bottom rivers supporting angiosperms. Among macrophytes the plant species, which is most diverse in running water bodies, is the benthic algae. They occur on all possible surfaces along the river and are intimately associated with microbes and an extracellular organic matrix called Aufwuchs (minute plants and animals attached to rooted aquatic plants and other exposed layers, this also collects scrapers who are organisms and detritus covering rocks and plants colonizing the aquatic environment and often serve the purpose of fish food).

In aphotic zones of the rivers the diatom species abound in large numbers. Large aquatic bodies in the tropics have more species diversity than those in temperate regions; in addition species richness increases rapidly in lower latitudes than higher ones. Flow regimes of running aquatic bodies are important. It adds to the sustaining capability of Rivers and their associated flood plain. Any alteration of flowing stream often claims to be the serious and threatens wetlands and their species diversity.

Many authors have observed the distinction in plant group structure related to spatial and temporal distribution, which has influence of flood and scour, desiccation, substrate stability and localized variations in water velocity, turbulence and shear stress.

Sometimes aquatic flora has isolated or discontinuous growth and availability in the environment because of differences in distribution, frequency and intensity, establishment success and growth rates. Researchers found that increased stability of base flow and reduction of flow variability led to excessive growths of aquatic macrophytes. Similarly seedling survival and plant growth rates are affected by changes in rates of water level fluctuation and disturbance frequency and intensity. In this review information on biodiversity in aquatic habitats and their resources, in marine and fresh water ecosystems, their importance conservation and restoration mechanisms was discussed in brief.

IMPORTANCE OF AQUATIC BIODIVERSITY

Aquatic biodiversity has enormous economic and aesthetic value and is largely responsible for maintaining and supporting overall environmental health. Humans have long depended on aquatic resources for food, medicines, and materials as well as for recreational and commercial purposes such as fishing and tourism. Aquatic organisms also rely upon the great diversity of aquatic habitats and resources for food, materials, and breeding grounds.

Factors including overexploitation of species, the introduction of exotic species, pollution from urban, industrial, and agricultural areas, as well as habitat loss and alteration through damming and water diversion all contribute to the declining levels of aquatic biodiversity in both freshwater and marine environments. As a result, valuable aquatic resources are becoming increasingly susceptible to both natural and artificial environmental changes. Thus, conservation strategies to protect and conserve aquatic life are necessary to maintain the balance of nature and support the availability of resources for future generations.

Role of Biodiversity in the Ecosystem

Figure 1 represents the vast functions and services provided by the ecosystem biodiversity, which is continuously supporting the mankind and its existence.

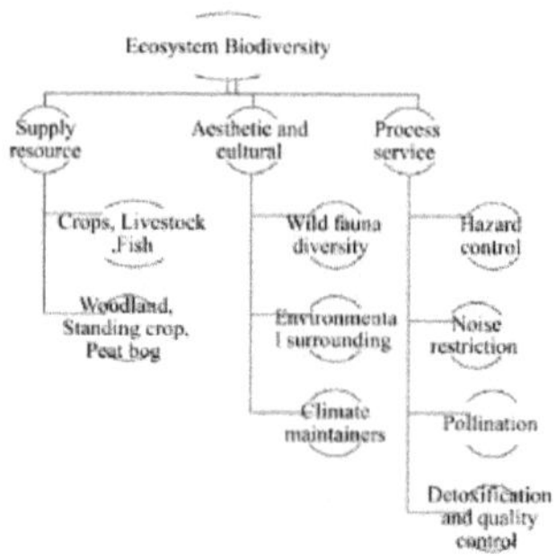

Fig. 1. Various uses of ecosystem biodiversity

THREATS TO AQUATIC BIODIVERSITY

Human activities are causing species to disappear at an alarming rate. Aquatic species are at a higher risk of extinction than mammals and birds. Losses of this magnitude impact the entire ecosystem, depriving valuable resources used to provide food, medicines, and industrial materials to human beings. Runoff from agricultural and urban areas, the invasion of exotic species, and the creation of dams and water diversion have been identified as the greatest challenges to freshwater environments. Overexploitation of aquatic organisms for various purposes is the greatest threat to marine environments, thus the need for sustainable exploitation has been identified by the Environmental Defense Fund as the key priority in preserving marine biodiversity. Other threats to aquatic biodiversity include urban development and resource-based industries, such as mining and forestry that destroy or reduce natural habitats. In addition, air and water pollution, sedimentation and erosion, and climate change also pose threats to aquatic biodiversity.

1. **Overexploitation of species —** Overexploitation of species affects the loss of genetic diversity and the loss in the relative species abundance of both individual and /or groups of interacting species. The population size gets reduced because of disturbances in age structure and sex composition. Efficient gears remove quick growing larger individuals . consequently, the proportion of slow growing ones

increases and the average size of individuals in a population decreases. Overfishing causes change in the genetic structure of fish populations due to loss of some alleles. Thus, genetic diversity gets reduced .

2. **Habitat modification** — Physical modification of habitat may lead to species extinction. This is mainly caused due to damming, deforestation, diversion of water for irrigation and conversion of marshy land and small water bodies for other purposes. Construction of dams on river impedes upstream migration of fishes and displaces populations from their normal spawning grounds and separate the popultion in two smaller groups. Deforestation leads to catchment area degradation due to soil erosion which results into sedimentation and siltation. This not only affect the breeding ground of aquatic organisms but cause gill clogging of small fishes also.

- **Pollution load** — Four forms of pollutants can be distinguished-
    - i. **Poisonous pollutants** — Agrochemicals, metals , acids and phenol cause mortality, if present in a high concentration and affect the reproductive functionality of fish.
    - ii. **Suspended solids** — it affects the respiratory processes and secration of protective mucus making the fish susceptible to infection of various pathogens.
    - iii. **Seewage and organic pollutants** — They cause deoxygenation due to eutrophication causing mortality in fishes.
    - iv. **Thermal pollution** — It cause increase in ambient temperature and reduce dissolved oxygen concentration leading to death of some sensitive species.

These factors affect the aquatic biodiversity directly or indirectly. Excessive mortality of organisms due to any of these factors may lead to two type of effects:

- Extinction of the species / populations
- Reduction of population size.

Trinidad and Tobago adequate freshwater resource ecosystems were under continuous threats and the damages to the freshwater habitats was noticeable. Important significant impacts that were causing extensive damage to the water resources and the loss of aquatic biodiversity are deforestation, use of pesticides in agriculture and water for irrigation purpose, rise in urban and industrial sector, regulation of water for dam and hydroelectricity production, mineral and petroleum extraction, Solid wastes dumps, bottom dredging and water flow channelization, fishing and aquarium market.

According to researchers Trinidad and Tobago's freshwater ecosystems has a significant local value as biologically distinguished system, but seriously at the risk of extinction in terms of conservation category. All these freshwater systems were included in the priority list of conservation at the regional basis. In Chile, freshwater resource protection was targeted towards habitat protection to conserve the exotic species of

Salmonids. This biodiversity management in turn benefited the balance of ecosystem process and preservation of indigenous fauna. The similar situation happened in the New Zealand also, where conservation of Salmonids and other valuable fish biodiversity has made great impact on the development of habitat protection.

Colonization of new species in the aquatic system through human activities, whether knowingly or unknowingly, is becoming global environmental issue. The problem of introduced species is on the rise in United States freshwater resources. In the year 1920, six exotic fishes were introduced into the freshwater; a few more had been added by 1945 and until 1980 more species had been noticed. Newly arrived species in the ecosystem creates many problems for the native species, as they start breeding and population growth in more area other than their particular spot.

Removal of chemical contaminations from the aquatic ecosystem can be eliminated at the point of source, or the habitat can be restored after pollution peak, but once the species introduction has begun, it is very difficult to reverse this task of introduction. Besides this introduced or alien species can also produce negative impacts on native organism of the water resources, such as preying of animals, competition, niche alteration, interspecific breeding, and disease and parasitic infection development. Species are introduced to the new environment to gain economic advantage or to fulfill the developmental requirements like aquaculture.

This introduction can be economically profitable initially but later it can produce disadvantages on economical and ecological terms, due to their invasive growth. Such an example is of Oysters, which has been introduced in to the global coastal waters for food. Oysters have function in estuarine ecosystem brings several complex modifications. Apart from this when oysters combine with alien oysters they get attached to their shells creating a kind of mismanagement in the aquatic system. Introduced invasive species in the marine environment are a severe threat to biodiversity of the ecosystem because of their prolific growth and other harmful effects creating ecological and economical disruption.

1. **Overexploitation of species:** over exploitation of particular species to fulfill the demand of human greed can result in the loss of genetic diversity and that lowers the relative species abundance of individual or groups of interacting species. This process can reduce the population size as a result of interruption in age structure and sex composition. Fishing exploitation disrupts the genetic constitution of fish population because of loss of some alleles from the genome, reducing the genetic diversity.

2. **Habitat modification:** structural modification in the habitat may also bring species extinction for example, dam construction, deforestation, diverting water course for agriculture supply changing marsh land and small water entities for other function. Building dams on river obstructs the migration of fishes from the

source point and cause movement of populations from their natural spawning area and divides the population in to smaller groups. Another instance is outcome of deforestation that leads to progressive degradation of catchment area due to soil erosion causing sedimentation and siltation. This process has negative impact not only on the breeding zone of aquatic organisms but also results in blockage of gills in small fishes.

3. **Organic pollutants drainage/suspended solids/sewage discharge:** this includes toxic and poisonous compounds like metals, agrochemical substances, phenolic and acidic discharge, promoting mortality rate and suppresses the reproductive function of the aquatic animals. Suspended solids create problems and suffocation in respiration physiology and production of mucus that makes the fishes prone to infections produced by aquatic pathogens. Discharge of sewage results in eutrophic aquatic system, causing deoxygenation and aggravates death rate in aquatic organism and submerged aquatic flora.
4. **Thermal pollution:** produce changes in surrounding water temperature and reduces the concentration of dissolved oxygen in the water column consequently death of small and sensitive aquatic animals. Death rate acceleration in the aquatic organisms affects the aquatic biodiversity either by promoting extinction of species or population or reduced population. Remarkable rise in ocean water temperature due to marked El Niño phenomenon has anxious the scientists who noticed the coral bleaching could extend beyond 2017. Climate change and global warming has been accepted as the main reason behind rise in ocean water temperature, which now threatening the rich but vulnerable wealth of coral reef's worldwide.

STATUS OF AQUATIC GENETIC RESOURCES

Fish genetic resources comprise the DNA; genes; gametes; wild, farmed and research populations; species; and genetically altered forms - selectively bred strains, hybrids, polyploids, and transgenes - of all exploited and potentially exploitable finfish and aquatic invertebrates.

Fish genetic resources management merits high emphasis in ecosystem approaches to the development of responsible aquaculture and ecosystem-based management of responsible capture fisheries. Fish genetic resources help determine the performance of the farmed fish and their interactions, including genetic interactions, with aquatic biodiversity. In capture fisheries, fish genetic resources help determine the productivity of fished populations and their adaptability to environmental change, including climate change.

Aquatic genetic resources encompass also the genetic diversity of farmed and harvested aquatic plants, which are plant genetic resources, but have not yet been adequately covered by the Commission and by other organizations that are participating in policymaking for and management of plant genetic resources.

The most important fish genetic resources for aquaculture and capture fisheries by species groups are:

- **For aquaculture** - carps, catfishes, milkfish, salmon, tilapias, mussels, oysters and shrimps, as well as their wild relatives;
- **For inland capture fisheries** - carps, catfishes, characins, salmonids; tilapias and other cichlids;
- **For marine capture food fisheries** - small and large pelagic fishes, reef fishes, sharks and other elasmobranchs, demersal fishes, and diadromous migratory fishes such as salmon and sturgeons;
- **For marine industrial and low value/trash fish fisheries**– small pelagic and demersal species that provide fishmeal and fish oil for farmed animal and fish feeds.

Important plant genetic resources for farmed aquatic plants include those for marine seaweeds and freshwater macrophytes.

With few exceptions, substantial domestication and genetic improvement of farmed fish is not as advanced as in the crop and livestock sectors. This is now changing for some widely farmed aquatic species, with rapid benefits to fish farmers and fish consumers.

Fish genomics is also developing rapidly and is seen as having many potential applications including marker-assisted selection for the genetic improvement of farmed fish, accurate identification of fish genetic resources for their conservation and use, and the diagnosis and prevention of fish diseases. The farming of distinct strains, hybrids, mono-sex populations, polyploids, is increasing, bringing increased needs for effective biosecurity procedures. Private sector research for the development of biotechnological products and processes in aquaculture and capture fisheries is increasing.

GENETIC VARIATION AND SUSTAINABLE AQUACULTURE

Aquaculture species have two eminent features that make them highly suitable for selective breeding: high fecundity and high genetic diversity, because most species in aquaculture are still in the early stages of domestication. High genetic variation in wild populations is an important resource for selective breeding of aquatic species. The genetic resource of wild populations can contribute in various ways to sustainable aquaculture. First, wild alleles can be reintroduced to farmed populations if inbreeding occurs due to bottleneck effects in breeding practices. Second, if there is demand for new traits, such as resistance to disease, genetic materials can be accessed from selected strains as well as the wild populations. Third, there are still many other or "novel" species, particularly native species being tested for aquaculture, such as the fat sleeper (*Dormitator latifrons*) in Latin America.

Therefore, conservation of biodiversity and genetic variation of aquatic species is important for developing and maintaining good strains for sustainable aquaculture.

Furthermore, comparison between different regional populations and farmed ones may aid understanding of genetic components that underlie aquaculture performance across different traits, such as identifying causative variants for target traits.

Conservation Status and Damaging Factors to Genetic Resources of Aquaculture Species

Currently, most attention regarding aquatic genetic resource conservation has been focused on protection of freshwater species, probably because freshwater habitats are under high pressure of degradation and have close affinity to the general public. Researchers argued that affordable and sustainable farmed aquatic foods should be focused on freshwater aquaculture. Nevertheless, many marine species also are under threat (International Union for Conservation of Nature [IUCN]). Migratory fishes are the most vulnerable group, because they often require more habitat areas to complete different life history stages. For example, some shads and anchovies, most sturgeons, and some salmonids are critically endangered.

Most conservation efforts have focused on finfish, but the characterization and protection of genetic resources of bivalves and seaweeds have also been highlighted. The major drivers for the decline of genetic resources of aquaculture species are believed to be habitat loss and degradation. Overfishing and pollution also commonly damage both freshwater and marine species. In temperate rivers, biodiversity changes are mostly due to river fragmentation and invasive species.

Strategies for Conservation of Genetic Resources

Conservation acts include setting up regional freshwater conservation reserves or MPAs, or globally, the Convention on Biological Diversity. In situ conservation zones such as small grassroots reserves in the Mekong River systems are usually localized and isolated, yet are important in protecting food webs and augmenting fisheries yields. Seasonal fishing bans are often aimed at encouraging successful breeding of target species, such as the hilsa shad (*Tenualosa ilisha*) in the Bay of Bengal and the tapertail anchovy (*Coilia nasus*) in the Yangtze River. A 10-year ban on all commercial fishing in the Yangtze River and its tributaries, including the adjoining lakes, was instituted in January 2020. A range-wide total fishing ban is an enormous task, but is crucial for restoring genetic resources of endangered species, particularly migratory ones. Spotting of rare species in historical sites of middle reach of the Yangtze River has already been reported, but the long-term effects for this 10-year ban are yet to be revealed.

Another way of protecting genetic resources is ex situ conservation. Sometimes the wild populations or habitats are too degraded for protection. Artificial breeding in captivity and conservation of endangered species is better than its disappearance entirely. A good example is the two sturgeon species in the Yangtze River, the Chinese sturgeon (*Acipenser sinensis*), and the Yangtze sturgeon (*Acipenser dabryanus*).

Both are endangered and their natural breeding activities are interrupted by the building of hydroelectric dams. Fortunately, artificial breeding has allowed healthy populations of these two species to be maintained in captivity. There is still hope that these species can be kept long enough to restore their natural habitats and establish populations in the wild. Preserving gametes as seed banks often is seen as storing genetic resources of aquaculture species, both finfish and shellfish. Restocking artificially bred seedlings, commonly used for fisheries management, is a hybrid of *ex situ* and *in situ* conservation practice, but scrutiny should be applied to monitor the genetic component of the stocked and wild individuals for protecting genetic resource of the wild populations.

Interestingly, aquaculture per se sometimes can play a positive role in conservation of genetic resources; for example, megalopa larvae of the Chinese mitten crab (*Eriocheir sinensis*) were heavily collected for aquaculture, threatening its wild populations in the 1990s. Subsequent success in commercialized artificial breeding not only supported the steady growth of aquaculture of the Chinese mitten crab but also helped in protecting its wild population and genetic resources due to lessened fishing pressure on the wild populations. Similar benefits can be found in aquaculture of the Chinese giant salamander (*Andrias davidianus*), or of Amazonian ornamental species.

New Technologies and Perspectives

The first step to initiate a conservation plan for aquaculture species is to characterize the wild populations as well as domesticated ones. Identifying population structure and genetic diversity is important for setting up management plans and sometimes can lead to discovery of cryptic species. If planning a new breeding project, characterizing the genetic resource is also fundamental work necessary for designing the best strategy.

Mitochondrial DNA loci and microsatellite markers have often been used for genetic characterization, but panels of single nucleotide polymorphism loci (SNPs) are better choices for characterizing genetic profiles of wild and domesticated species. Restriction site associated DNA sequencing (RAD-seq), genotyping by sequencing, or SNP arrays are common tools for collecting SNP data; however, the cost of these methods is prohibitive to low-budget projects. Target enrichment for multiple nuclear loci may be a good low-cost alternative, such as single-copy nuclear loci for ray-finned fishes.

Several countries have established national programs to support agriculture, beef, poultry, and pork genetic programs for those industries. Few have done it for aquaculture. Major animal breeding companies have made progress on valuable aquaculture species using those technologies, but low-profit species in developing countries have received less attention.

High-throughput phenotyping is mainly used for crop breeding and human behavior studies. Many traits of aquaculture species, such as growth, meat quality, sex-related traits, disease resistance, salinity/stress tolerance, high- and low-temperature tolerance, feed conversion rates, nutritional requirements, physiological/biochemical traits, and body colors may be efficiently evaluated using high-throughput phenotyping technology. Innovations such as automated imaging/machine vision, diode frame measurements, and deep learning networks can be further explored in collecting phenotypic data in a high-throughput manner

CONSERVATION APPROACHES FOR THE AQUATIC ECOSYSTEMS

One of the remarkable approach especially for the conservation of aquatic ecosystems was United Nations Convention on the Law of Sea (UNCLOS) which was implemented in the year 1994. This law was basically considering the promotion and support of the biodiversity along with the sustainable use of the oceanic water. Major non-government international organizations, research institutes and scientific groups can collaborate under the roof of the United Nations to work together for the assessment of global biodiversity and identification and listing of hotspots as biodiversity indicator.

However, to prevent the future losses in the biodiversity either marine or freshwater, inadequate information availability should not be considered as hindrance in the progress of conservation. Communities operating at the broader sector should not delay waiting for the right time to get all the potential facts to initiate any process. Some of the data collected in the class of fishes and marine reptiles and other taxa are sufficient and reliable to trust that aquatic biodiversity is going through the crisis and can be calamity if not prevented.

Charles Elton, the "father of ecology", once presented his idea of conservation in his words that we should be "... looking for some wise principle of co-existence between man and nature, even if it has to be a modified kind of man and a modified kind of nature. This is what I understand by conservation". Researchers also favored the idea of reconciliation ecology, suggesting different species inhabitation together with mankind to enhance species richness and positive outcome. Many wetland zones, whether fresh or brackish-water are biologically engineered by man or the environment is controlled by man.

The best example is represented at Ramsar site, which hosts a large number of water birds which are under serious threats of extinction, another wetland in Indian subcontinent known as "Kabar Taal" is a winter niche for many migratory birds. Both these examples clearly show that human alterations in the ecosystem can be compatible and cooperative in biodiversity conservation.

It has been recommended that strategies to conserve aquatic resources should comply with ecosystem conservation, which can favor the growth of biotic resources, and help to maintain the natural habitats and ecosystems (Fig. 2). In order to make aquatic biodiversity conservation effective and successful, conservation management regulations should include all broad areas.

- Aquatic regions suffering from anthropogenic destruction, degradation or habitat loss should be aimed for the restoration. The same objective should imply for the conservation of the plant or animal species, which are facing decline growth and are close to extinction from the community. Water systems and aquatic biodiversity are closely related and interdependent on each other. General assumption is that water quality and habitat quality have an effect on the species composition, diversity and physical structure of the aquatic ecosystems.

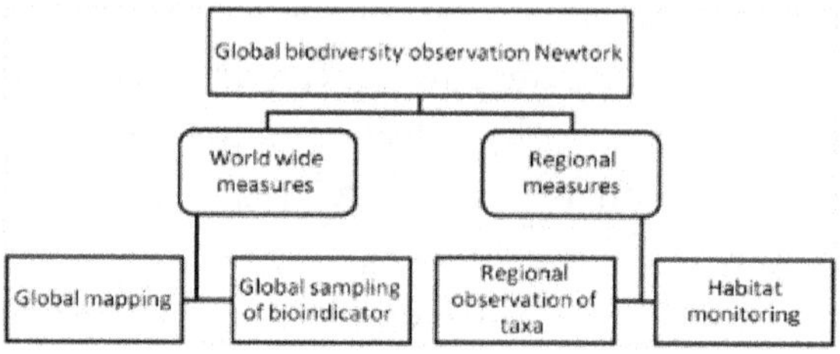

Fig. 2. Representation of global biodiversity monitoring system.

- Bio-reserve spots can be created within a water body where fishing is restricted along with the other activities to protect the plants, animals and their habitat to conserve the aquatic biodiversity. Another useful role of these bio-reserves are that they can be highly beneficial for the purpose of education, recreation, tourism as well as potential increase in fishes reserve by improving the conditions of declining fish population. These bio-reserves function as marine protected areas, fishery reserves, sanctuaries and biological parks.
- Bioregional management, can help to regulate factors influencing aquatic biodiversity, brings the balance in conservation, economy, and social demand of an area. This includes biosphere zones from small area and larger reserves. Figure 2 shows the network of global biodiversity monitoring system and regional areas protection plan.
- Red Sea Biodiversity joint project was conducted to assess the marine biodiversity of the Saudi-Arabian coast and deep water of the Red Sea and to create an information source at the King Abdul Aziz Marine Museum (KAUMM) and at Senckenberg Biodiversity Research Institute in Frankfurt am Main, Germany (SRI) in 2011. This project programme was aimed at the collection and identification of species of all marine animal groups. Marine species were preserved and catalogued as a reference source of the KAUMM and SRI.
- Watershed management has a significant role in keeping the balance and preservation of aquatic diversity. Aquatic resources are open and unprotected system, because their geographical route often crosses through many political territories. Such type of situation creates difficulty in conservation and management of resources. However, recently, this situation has improved as

regional and local watershed communities have started protection projects for lakes and streams.

- The Napo drainage depression of the Amazon basin is enormously biodiverse ecosystem located in Ecuador, has gone through the various anthropogenic threats. Researchers enlisted the freshwater vertebrates and invertebrates, and also prepared their diversity and distribution model. The study was focused on the freshwater invertebrate and vertebrate taxa. Species richness maps were prepared using compilation data for approximately 481 freshwater vertebrate species, which includes amphibians, birds, mammals, reptiles and fishes and over 54 invertebrate families recorded from the altitude of 200 - 4500 m height of the basin.
- The interest of the study was to implement conservation and management programs for the high species diversity of invertebrates in the large tropical Napo basin with the help of manpower, financial expenditure, human attempt, skilled knowledge and time.
- Afforestation around drainage or catchment area of water bodies can control soil erosion and helps to minimize sedimentation in the water systems. This can provide better survival environment to aquatic fauna.
- Construction of industeries, chemical factories and thermal power plants adjacent to the aquatic systems should be controlled as industrial effluent can damage the environment supporting aquatic ecosystem and finally results in the loss of biodiversity.
- The World Resources Institute USA, which is a global research organization, has emphasized the idea of putting particular plant or animal in threatened or endangered category can be helpful in protecting the biodiversity. Restoration of the aquatic resources can also be promoted via specialized educational programme.
- For example, USDA Forest service has begun a special course with target to protect and restore the river ecosystem and their related species. Several international universities also offer the biodiversity conservation management program, to create understanding and awareness about significance of biodiversity among upcoming generation.
- Regulation of industrial and urban wastewater discharge must be taken in to consideration to conserve the aquatic resource diversity.
- Spreading awareness at the social forum is the friendliest way to educate the community about conservation of aquatic biodiversity. This can be achieved successfully via educational programs, prize sponsored events, and volunteer promotions.

- Marine protected areas developments of Marine protected areas (MPAs) are receiving popularity as a coral reef management tools for the management of coral reef systems. Non-coral invertebrates also contribute in providing enormous biodiversity to coral reef ecology, but their role in the marine ecosystem is still not well explored. Japan has been designating Marine Protected Areas since long time.

Examples of them are:

1. Natural Parks and Natural Seashore Conservation Areas that aim to protect the natural landscape
2. Nature Conservation Areas, Wildlife Protection Areas, Natural Habitat Conservation Areas and designated areas for natural monuments for the protection of the natural environment or the habitat of organisms; and
3. Protected Water Surface zone aim towards conservation and gardening of aquatic animals and plants, coastal marine resource development areas and other various specified areas designated by different entities such as prefectural governments and fishing groups

INTERNATIONAL AGREEMENTS AND THEIR IMPACTS ON AQUATIC GENETIC RESOURCES AND ON STAKEHOLDERS

This section deals with the impacts of international agreements on AqGR of both farmed aquatic species and their wild relatives, as well as on stakeholders. Countries were asked to summarize the most important international agreements to which they subscribe that cover or impact upon AqGR of farmed species and their wild relatives. Countries were also asked to assess the impact of those agreements on AqGR and stakeholders, for example:

- Establishment and management of shared or networked aquatic protected areas;
- Aquaculture and culture-based fisheries in transboundary or shared waterbodies;
- Sharing aquatic genetic material and related information;
- Fishing rights, seasons and quotas;
- Conservation and sustainable use of shared waterbodies and watercourses; and
- Quarantine procedures for aquatic organisms and for control and notification of aquatic diseases.

Countries Take Steps Towards International Agreement to Protect Marine Biodiversity

Member States at the United Nations have agreed to develop a legally-binding instrument to conserve and sustainably use marine biological diversity of areas beyond their national borders.

The 193 member-UN General Assembly agreed to establish a preparatory committee, open to all countries, to negotiate the new instrument over 2016-2017. The committee would report back to the General Assembly on its progress at the end of 2017.

The negotiations will cover, among other issues, the sharing of benefits related to the use of marine genetic resources, marine protected areas and environmental impact assessments, as well as the transfer of marine technology.

Countries also agreed to establish a voluntary trust fund to assist developing countries – in particular least developed countries, land-locked developing countries and Small Island developing States – so they can actively take part in the process of establishing the instrument.

The instrument will fall under the UN Convention on the Law of the Sea. The Convention, which came into force on 16 November 1994, governs all aspects of ocean space – including the delimitation of maritime boundaries, exploitation of non-living resources, conservation and management of living resources, protection and preservation of the marine environment, marine scientific research and the settlement of relevant international disputes.

The Convention also provides the basis for the equitable utilization and conservation of resources in the oceans and seas.

Participation in the negotiations for the new instrument has no effect on the legal status of countries that are not party to the Convention on the Law of the Sea.

The General Assembly decision takes forward an outcome from Rio+20 that called for countries to consider whether an instrument was needed.

Secretary-General Ban Ki-moon has repeatedly stressed the crucial role of oceans in achieving sustainable development and combatting climate change.

"Given how critical oceans are to the health of our planet and the prosperity of people, they are an essential element in our emerging vision for sustainable development, including the new set of sustainable development goals now being prepared to guide the global fight against poverty for the next 15 years," Mr. Ban said in his message to commemorate this year's World Oceans Day.

"The oceans are vast – but their capacity to withstand human damage is limited. In this potentially pivotal year, we must commit to using the gifts of the oceans peacefully, equitably and sustainably for generations to come."

12

FISH SEED CERTIFICATION AND QUARANTINE

SEED PRODUCTION: AN INTRODUCTION

The production of marketable fish begins with the stocking of fry or juveniles into a rearing environment that assures optimum and rapid growth to allow harvest in the shortest possible time. The fish farmer has to obtain adequate number of young fish to meet his production goals. These fish can come from wild capture. However, there is little or no guarantee that adequate numbers can be captured and stocked in the time corresponding to optimum production conditions. The fish farmer then naturally turns to other means of obtaining his stock. By simulation of the conditions necessary for the reproduction of his fish, the farmer can spawn the fish in captivity. Successful spawning is only the beginning, however, the eggs must hatch, and these reared successfully to fry stage. These stages - spawning, hatching, and early rearing are like a steeple chase which the farmer must win. The race course is well filled with obstacles, for example physico-chemical quality of water such as available dissolved oxygen, feed of the proper nutritive composition and particle size, low resistance to diseases, and so on. A good appreciation of all these factors is needed for successful production of fish.

The ultimate goal of the fish farmer is to produce fish that meet both his needs and the market demand. Through artificial propagation, the farmer can select for desirable characteristics such as fast growth, resistance to disease, etc. By hybridization and selection, these goals can be achieved if the farmer dedicates enough time and patience.

FISH SEED IN AQUACULTURE

As stated in the introduction, fish farming begins with the stocking of fry, and these can come from the wild or be produced on the farm. Whatever their origin, they are indispensable and the means of obtaining them influences directly farm production. If supplies are erratic, there will be interruptions in other farm activities; if the supplies are regular, farm production may be maximized. The cost of the fry can vary considerably and may be an important factor in overall production costs. It any event; a good supply of fry is essential for successful fish farming.

If one looks at production of eggs, larvae, and fry that is carried out on the farm itself, the major problems are obtaining a sufficient number of eggs, a good hatching rate of these eggs, and good survival and growth of the larvae obtained. In nature, there is very high mortality at these stages, and a lot of attention and effort is needed to overcome these difficulties.

To practice reproduction and fry production, a certain investment in equipment, infrastructure (ponds, tanks, water supply), and trained labour is needed. These costs can be considered a part of overall production costs of marketable fish.

The proportion of total cost in producing saleable fish that is met by seed production should be kept always in mind, and efforts made to find new methods of seed production that are less expensive, hopefully increasing the profitability of the whole farm.

Definition of terms:

- **Larvae:** hatchlings with yolk sack until first feeding.
- **Fry:** free swimming fish - from first feeding until complete development of somatic organs.
- **Juveniles:** sometimes referred to as "fingerlings", but generally fish of a small size but adequate for stocking, not sexually mature.

METHODS OF SEED PRODUCTION

For certain farmed fish, it is not yet possible to understand and control all the stages of reproduction, and farmers depend on natural supply. In most cases, these fish are brackish water species and their biological cycle includes migrations into a different environment, either moving to the sea, or into fresh water. Thus to simulate the conditions needed for natural reproduction, this would mean a complete change in the environment. In general, fish can only be reproduced on the farm if conditions that correspond to natural spawning can be closely approximated. The cost of creating a completely marine habitat or freshwater habitat in the midst of a brackishwater situation would in most cases be prohibitive.

Collection of Seed from the Wild

Collecting eggs or fry from the wild was the first method used in obtaining stocking material. This is still occuring for species for which the spawning behavior is not controlled or not well understood, or for which the costs of artificial propagation are too high or where fry in large quantities is easily obtainable. Means of collection differ with species. Three important species of fish will be examined.

Catfish

Throughout West Africa, the catfish (*Chrysichthys* spp) is very popular and in Ivory Coast represents an important species in brackishwater lagoon aquaculture. As it is relatively difficult to reproduce this fish in captivity, eggs are collected from the wild by placing bamboo traps in areas where this fish (*C. walkeri*) spawns. The traps are placed on rocky bottoms, in deep water (4-6 m), by fishermen and the catfish spawns inside the bamboo tube. This method allowed the collection of up to 2000,000 eggs even outside of the peak spawning periods. The eggs are transported to the hatchery, and placed in artisanal Zoug jars made from empty plastic bottles. Two or three days after hatching, the larvae are transferred into tanks made of plywood and covered with polyester resin measuring 2.4 × 0.4 × 0.3 m at a. density of 5,000 to 10,000 per tank.

After about 10 days, the yolk sac is resorbed. After two and half months, they reach an average weight of 750 to 800 mg. During this time, they are fed on a mixture of cattle brain and crab eggs, finely ground which forms a colloid suspension in the water easily accessible as food for all fish. After the first month, the feed is modified and composed of ground crabs, wheat flour (15% of the weight of crab) wheat bran (5%) and fish meal (20%). The fry are then transferred after 2½ months to circular concrete tanks with a running water system. The concrete tanks are 3 m in diameter with a central drain.

After the 4th month, the fish are fed with a mixture richer in flour: crabs, fish meal 30%, wheat bran 20%, wheat flour 20%, copra cake 10%. Percentages are in relation to weight of crab. This feed is again changed during the 5th or 6th month to a feed composed of S! crab, and T! agricultural by-products (fish meal 35%, wheat bran 20%, wheat flour 25%, palm kernel cake 10%, copra cake 10%. The fish are then transferred to enclosures when they achieve an average weight of 4.5 grams.

One-fourth will have reached already a weight of 8 g, when half of the population will weight 4 grams. The fry are reared in enclosures until achieving a size of 50 g, which takes a further 6 months.

The obvious cost and time necessary for procurring catfish juveniles through this method is considerable, and further experimental work is necessary.

Larger fish are sometimes captured from the wild. In Nigeria, the catfish are captured using hook and line with soap as bait. The catfish are injured, and care must be taken to avoid mortalities. It is easy to capture large numbers of catfish using traps to collect fish attracted to tilapia cages.

Milkfish

The milkfish (*Chanos chanos*) is found in warm waters of the Red Sea, and the Indian and Pacific oceans and is one of the most important brackish water species cultured particularly in S. E. Asia. In Africa, it is only found on the east coast. The fish reproduces once or twice a year in coastal marine waters of 25 m in depth. Each female spawns 1.5 to 1.7 million eggs, and sometimes up to 5 million.

The larvae drift to the coastal waters near the estuaries with a temperature of about 23°C, the eggs taking about 24 hours to hatch. Salinity isn't too important (10 to 32 ppt) and the fish are captured where phytoplankton is abondant. When the fry have achieved 1.3 m in length, the yolk sac is completely absorbed and they begin looking for food.

At this life stage, their diet is a mixture of blue-green algaes and the associated organisms (bacteria, protozoans). The diet will change later on. Artificial reproduction has been achieved in the laboratory, but not yet on a large scale, and thus almost all of the milkfish larvae are captured from the wild. For example, in 1970, in Taiwan only, 207 million fry were captured from the wild.

According to the regions, capture techniques and material can differ. In Java, the best collection zones are on the north side of the island, and the fry are abundant 7 months out of the year. The best time for capture is at the spring tide, and the three days before and following the full moon. Sandy beaches, a slight slope, and clear, calm water are ideal.

The fish are often concentrated in the zone near estuaries where the influence of freshwater is high. They will often take refuge leeward of sand bars. If no sand bar exists, the fishermen will build a rock wall perpendicular to the beach into the ocean, or build artificial shelters using palm branches.

As the fry are collected, they are transferred into basins or similar containers, and the sea water is immediately diluted with freshwater. This will kill off many of the other organisms inadvertantly collected, and also acclimate the fry to brackishwater. Other species of fish are also captured, and these must be sorted out. This needs a certain level of expertise and experience on the part of the fisherman.

In the Philippines, the fry are captured by using seine nets or traps. In Mauritius, the fry are captured usually in July using a fine mesh seine net, and are transferred into acclimatization tanks. The mortality is usually high in the tanks.

Mullet

Mullet species are cultivated around the world, and have a large distribution. Mullets belong to the genera of Mugil or Liza, many species are known. The artificial reproduction is possible, however there are still many instances where fry available in important quantities is collected from the wild. The mullets reproduce in the sea, and will thrive well in the sea, in estuaries, in brackishwater, and sometimes migrate to freshwater. When they are sexually mature, they form schools, and the reproductive products are released into the sea where the fertilization occurs.

The eggs are pelagic, and hatch in the following two days. After hatching, the larvae begin to swim near the coast where they usually arrive in mass after 2 months. At this point, they have reached a size of 25 mm. On the coast, they form small schools and begin to move towards the estuaries. The capture is usually done as they enter the estuary, at low or rising tide, using seine nets, deep nets, etc. The fish are euryhaline and have a good resistance to abrupt changes in salinity.

Shrimps

Among the shrimps used in aquaculture, there are two main groups, the Penaeidae and the Caridea. Their biology and life cycles are different. For an example of the Penaidae, Penaeus notialis is widespread and often farmed. Spawning is carried out in the sea and the female is recognized by well developed ovaries and coloration. She molts shortly before spawning. The male places a spermatophore on the female theca which is kept until egg deposition.

The various stages of post egg development occur in the sea. After roughly 3 weeks, the young, termed post-larvae, begin to migrate to estuaries. The larvae are attracted by light, and are often collected by seine nets and deep nets at night. In other instances, the farm is filled with water known to contain large numbers of postlarvae. Larvae can also be collected by using bunches of palm or coconut branches lashed together and placed on the bottom of the lagoon. After a few days, the traps are surrounded by a fine mesh net and the branches removed.

For the group of Caridea, the principal species raised is Macrobrachium rosenbergii. This species spends its life in brackish-water or freshwater, and migrates to the estuaries or lagoons for spawning. Either gravid females or post-larvae are captured. Again traps, deep nets, etc. are used for capture.

Pawning

The reproduction of a species has to assure the survival of the species. The development of the oocytes is a long process that occurs over a relatively long period of time, and spawning occurs when the moment is the best for the survival of the fry, abundant food, shelter etc. Reproduction is controlled by internal factors (intrinsic) and environmental factors (extrinsic). When the fish is about to breed, proper environmental conditions are not enough to get the process of spawning started.

There is need for the female to encounter a male in the place suited for laying the eggs. Often the behaviour of male and female has to be coordinated in such a way that male and female of the same species come close to assure fertilization of eggs by the male's milt. The adjustment of the respective movements to this effect is fixed in the specific pattern of the breeding behaviour. Sometimes this behaviour is quite elaborate, as for tilapias, and is extended beyond the time of laying the eggs to the time the fry is able to sustain on their own and provide to their own needs. After spawning occurs, the eggs may be left or abandoned, or there may be some degree of parental care provided for by the male and/ or female

Spawning Techniques

Common carp

The carp will reproduce in ponds when conditions are favourable: water temperature at 18-22 degrees, and up to 25 degrees, submerged aquatic plants covering the bottom, clean water, rich in dissolved oxygen and little turbidity, and stable meterologic conditions. Increasing the water depth acts as a stimulant. Even under these conditions, spawning is not controlled and erratic. For this reason, farmers turn to semi-artificial or artificial reproduction. Spawning ponds are small, 100 to 1000 m^2, shallow, and one places "kakabans" or bunches of grasses to collect the eggs at 20 to 30 cm from the surface.

One needs 1 to 1.5 m^2 of kakabans per female. The pond is filled with clean water at the correct temperature the same day as the fish are introduced. the broodstock receive an injection of pituitary extract at a dose of 2 to 3.5 mg/kg using an intramuscular injection. As soon as the eggs are seen on the kakabans, they can be picked up and transferred to a hatching pond or first stage larval pond.

Artificial reproduction of common carp is also carried out using pituitary extract. Generally two doses of extract are given at 14 to 24 hour intervals, the initial dose is 0.3 to 0.5 mg/kg, and the second dose of 3.5 to 6 mg/kg for the females, the males receive one single dose (0.5 to 2.5 mg/kg) at the same time as the second dose for the females. The genital papilla of the female is sutured to avoid a premature release of eggs. The female is ready for egg extraction 9 to 13 hours after the last injection (9 hours if the temperature is 26 degrees and 13 h if the temperature is 19°C).

If the male is kept with the female, he will indicate the right time for the mating; he will start following and courting the female. At this point the broostocks are captured and sexual products harvested. The eggs and milt are collected dry. Eggs are obtained as soon as the suture is removed using slight abdominal pressure.

Heterotis Niloticus

Heterotis niloticus was the object of a large amount of research in the 1960's, and the species was reproduced naturally in ponds. Reproduction is possible without hormones, but by simulating natural spawning conditions in the ponds. It isn't possible to distinguish the sexes from the exterior, which is a major inconvenience. The fish are sexually mature when they have a length of 400 mm, and weight around 600 g, and generally fish that weight 800 g to 1 kg are mature. This is around 20 to 24 months old.

Heterotis has only one gonad, located on the left side. The spawning behavior is complex. It begins with the spawning couple looking for a nest site in shallow water filled with weeds. The fish pulls out the weeds with his teeth and then swims in a circle to push out weeds and debris. He then digs a nest, the top of which is just about at the surface of the water. The fish can enter and leave through a passage way that is left.

The nest is usually 1 to 1.5 m in diameter, and surrounded by a band or vegetation, the depth is 30 to 50 cm. Both parents participate in the nest bulding. Spawning occurs inside the nest. After spawning, the brooders will guard the eggs and fry, the fry eventually group into school and will remain in this formation in a length of time. Spawning can occur several times in a season. Reproduction is stimulated by three principal changes in the environment; increase in photoperiod, beginning of the rainly season, general lowering of the pH.

Mortality of the fry is very high and continues until the fish reach 50 g. Heterotis can reproduce without difficulty in larg ponds (more than 1 ha) with a lot of peripheral aquatic vegetation. They can also spawn in smaller ponds of 200 to 1000 m^2. A slight slope is advantegeous (2 to 2.5%), and the pond should be desilted before the rainy season to avoid dissolved oxygen depletion. The pond is cleaned of all weeds except for a few square meters in the corners of the ponds.

These weedy area are isolated from the rest of the pond by constructing a picket barrier. These areas are the spawning sites, and as the rest of the pond is free of weeds, there is less change of predation on the fry. Enough brood stock should be introduced to assure that the two sexes are present. The fry can be removed as soon as parental care has stopped.

OBJECTIVE OF SEED CERTIFICATION

The main objective of the Seed Certification is to ensure the acceptable standards of seed viability, vigour, purity and seed health. A well organized seed certification should help in accomplishing the following three primary objectives.

- The systematic increase of superior varieties;
- The identification of new varieties and their rapid increase under appropriate and generally accepted names.
- Provision for continuous supply of comparable material by careful maintenance.

CERTIFICATION AGENCY

Certification shall be conducted by the Certification Agency notified under Section 8 of the Seeds Act, 1966.

Eligibility Requirements for Certification

Seed of only those varieties which are notified under Section 5 of the Seeds Act, 1966 shall be eligible for certification. Any variety to become eligible for seed certification should meet the following requirement:

- **General requirements** - Should be a notified variety under Section-5 of the Indian Seed Act, 1966. Should be in the production chain and its pedigree should be traceable.

- **Field standards** - Field standards include the selection of site, isolation requirements, spacing, planting ratio, border rows etc.
- **Specific requirements** - Presence of off-types in any seed crop, pollen-shedders in Sorghum, Bajra, Sunflower etc., Shedding tassels in maize crosses, disease affected plants, objectionable weed plants etc., should be within the maximum permissible levels for certification.
- **Seed standards** - Minimum seed certification standards have been evolved crop-wise.

PROCESS FOLLOWED IN SEED CERTIFICATION

- **An Administrative check on the origin of the propagating material:** Source seed verification is the first step in Seed Certification Programme. Unless the seed is from approved source and of designated class certification agency will not accept the seed field for certification, thereby ensuring the use of high quality true to type seed for sowing of seed crops.
- **Field Inspection :** Evaluation of the growing crop in the field for varietal purity, isolation of seed crop is to prevent out-cross, physical admixtures, disease dissemination and also ensure crop condition as regards to the spread of designated diseases and the presence of objectionable weed plants etc.
- **Sample inspection:** assessing the planting value of the seeds by laboratory tests. Certification agency draws representative samples from the seeds produced under certification programme and subjects them to germination and other purity tests required for conforming to varietal purity. Bulk Inspection: Under certification programme provision has been made for bulk inspection. Hence, the evaluation of the lot for the purpose of checking homogeneity of the bulk seed produced as compared with the standard sample is carried out. This gives an idea about the genuinity of lot and sample.
- **Control plot testing:** Here the samples drawn from the source and final seed produced are grown side by side along with the standard samples of the variety in question. By comparison it can be determined whether the varietal purity and health of the produced seed are equal to the results based on field inspection.
- **Grow-out test:** Evaluation of the seeds for their genuineness to species or varieties or seed borne infection. Here the samples drawn from the lots are grown in the field along with the standard checks. Growing plants are observed for the varietal purity. Grow-out test helps in the elimination of the sub-standard seed lots.

PHASES OF SEED CERTIFICATION

Seed Certification is carried out in six broad phases listed as under:

- Receipt and scrutiny of application.

- Verification of seed source, class and other requirements of the seed used for raising the seed crop.
- Inspection of the seed crop in the field to verify its conformity to the prescribed field standards.
- Supervision at post-harvest stages including processing and packing.
- Drawing of samples and arranging for analysis to verify conformity to the seed standards; and
- Grant of certificate, issue of certification tags, labelling, sealing etc.

VALIDITY PERIOD OF THE CERTIFICATE

The validity period shall be nine months from the date of test at the time of initial certification. The validity period could be further extended for six months provided on retesting seed conforms to the prescribed standards in respect of physical purity, germination and insect damage for all seeds except vegetatively propagating material for which lot shall be re-examined for seed standards specified for respective crop. A seed lot will be eligible for extension of the validity period as long as it conforms to the prescribed standards.

FISH SEED CERTIFICATION

A proper fish seed certification process does not exist in many countries. Table shows the kind of certification existing in six countries. In Bangladesh, the Ministry of Fisheries and Livestock has drafted the Law for Fish and Shrimp Hatchery 2005 known as "Matsha and Chingri Hatchery Ayen, 2005", which includes requirements for registration of hatcheries and the rules for fish and shrimp hatchery operation such as physical infrastructure/ facilities of hatchery, ponds, selection of brood fish for breeding, source of selected brood fish, environment, etc. Similar certification process on hatchery management and operation exists in Indonesia. These initiatives will help improved the quality of fish seed produced.

The focus of certification process of NCCA-WBA in China and BFAR in the Philippines is on the genetically improved varieties. NCCA-WBA, which is operational at national and some province in China, certifies genetically improved varieties. BFAR has a certification process for tilapia seed to ensure good quality seed of tilapia will be disseminated to fishpond operators. BFAR distributes its certified seed of GET EXCEL strain and broodstock to BFAR multiplier stations in order to disseminate among private hatcheries encouraged them to breed their own fish and obtain feedback and superior breeding material.

Raising awareness among farmers to use the suitable strains and their identification will help the farmers to select the suitable strains. In practice, most Nile tilapia strains are difficult to determine at field level and are only distinguishable through DNA and other biochemical markers.

While the framed Fisheries Acts and Laws exist, it is a known fact that many countries are faced with difficulties in regulating aquaculture activities. These difficulties are attributed to the numerous interests involved, the diversity of the natural resources, the variety of institutions involved, the tangled web of laws and regulations and related difficulties in enforcement as well as the forces driving global and regional markets towards environmentally and socially sustainable practices.

Difficulties in regulating aquaculture activities, urged a self-regulation approach in the forms of codes of conduct and good or best management practices. A few countries have the interest in codes of conducts and good/best management practices, but largely on shrimp aquaculture.

Alternative to fish seed certification, voluntary codes and guideline are implemented in Thailand from hatchery management and operation to out-grower and processor. The Department of Fisheries certifies Code of Conduct products from hatcheries to the processor to enhance customer/consumer confidence on quality, safety and environmental friendliness.

The guidelines on Good Aquaculture Practices help the hatcheries and the farms focus on farms/hatcheries, farm/hatchery management, feed, chemical input, animal health, farm/hatchery sanitation, post harvest and data collection. The Department of Fisheries should have a monitoring mechanism to carryout regular monitoring for hygiene and good aquaculture practices of hatcheries/farms to be certified for Code of Conduct and Good Aquaculture Practices. This monitoring is, however, emphasized on the use of feed and therapeutants.

In a situation where large quantities of fish seed are imported and distributed throughout the country needs, a fish seed certification and quarantine process to preclude transboundary movement of aquatic animal diseases is necessary. To start with, fish seed quality certification process may be initiated for the fish seed importers to observe.

Table 1.Modes of fish seed certification in asian countries

Country	Committee	Role
Bangladesh	National Committee	Preparing a policy for seed certification
India	NBFGR, Lucknow, Uttar Pradesh	Recognized nodal agency for formulating legislation on aquatic animal health certification and quarantine
China	National CertificationCommittee	Certification of genetically improved varieties on Aquatic Wild and Bred Varieties (NCCA-WBV)
Philippines	BFAR	Certifies and distributes improved tilapia strains

INTRODUCTION TO QUARANTINE

"Quarantine" has been defined in many ways, ranging from the very broad definition of "quarantine measures" as encompassing all activities related to the prevention of the international and domestic spread of serious animal diseases (what is now usually referred to as "biosecurity"), or rearing of animals under conditions which prevent their escape or the escape of organisms and potential disease agents infecting or associated with them into the natural environment.

According to the World Organisation for Animal Health, "Quarantine means maintaining a group of aquatic animals in isolation with no direct or indirect contact with other aquatic animals, in order to undergo observation for a specified length of time and, if appropriate, testing and treatment, including proper treatment of the effluent waters."

At the international level, the primary purpose of quarantine is to minimize the risk of introducing pathogens into the territory of the importing country and their transmission to susceptible species.

GLOSSARY

Allele: Alternative form of a given gene or DNA sequence.

Ammonia: Tne chemical compound NH3 that is produced as a waste-product of metabolism. Ammonia is either excreted through the gills or through the gills and urine after being converted to urea in the liver.

Ampullae of Lorenzini: Electromagnetic field receptors on sharks, primitive fishes and some teleosts. It appears as pores on the snouts of sharks and some fishe, and all over the body on others. The ampullae are tubes filled with gelatinous material that lead to eletroreceptors at their base. Some fishes have electrorecptors that are placed directly at the skin surface rather than embedded below.

Anadromous: Fish that live most of their adult life in saltwater but spawn in freshwater, such as salmon.

Anterior: On a fish, the side of the fish that contains the head, opposite the anterior side with the tail.

Aquatic: An adjective that describes a habitat that is surrounded by water. The term describes both freshwater and marine habitats.

Amphidromous fish: migratory fish that live in both fresh and salt water, independent of breeding.

Anadromous fish: migratory fish that live in salt water but breed in fresh water.

Androgenesis: asexual reproduction of males, where offspring bear nuclear genes from their father only. Androgenesis can occur via the fertilization of enucleate eggs, or by active elimination of the maternal nuclear genome soon after fertilization.

Aneuploidy: a deviation from the canonical number of chromosomes in a species, typically caused by non-disjunction at meiosis. Aneuploidy is usually lethal or strongly deleterious because it modifies gene dosage.

Aquaculture: also known as aquafarming, it is the controlled growth of aquatic species.

Bony Fish: Ray-finned and lobe-finned fishes (Classes Actinopterygii and Sarcopterygii) that have skeletons made primarily of bone rather than cartilage (such as cartilaginous fishes in the Class Chondrichthyes.

Brackish: An adjective that describes a habitat that is surrounded by water that is saltier than freshwater, but less saline than normal seawater. Brackish habitats include estuaries where rivers meet the sea and freshwater mixes with seawater.

Bootstrapping Pseudorandom resampling of data with replacement to estimate the error of estimates of parameters of interest.

Breed: to produce offspring by hatching (or gestation).

Brood stock: fish of any particular species which are raised for reproduction purposes.

Brood: guard and groom eggs until they hatch.

Cartilaginous Fish: Fishes such as sharks, rays, skates and chimaeras that have cartilaginous rather than bony skeletons.

Catadromous: Fish that live most of their adult life in freshwater but spawn in saltwater, such as anguillid eels.

Clade: A group of species, living or exting, consisting of a single common ancestor and all its descendants.

Cosmopolitan: Species that lives in many places around the world rather than in a very specific and restricted location.

CT scan: Cross-sectional images produced by computer-processing of component x-ray images taken from several angles. Computer tomography or CAT scan.

Catadromous fish: migratory fish that live in fresh water but breed in salt water.

Contaminants: Both natural and artificial materials not typically found or found in unusually high concentrations which can be detrimental to the health of wild or farmed aquatic species

Credibility Interval In Bayesian statistics, a credibility interval is a posterior probability interval. Credibility intervals differ from the confidence intervals in frequentist statistics in that they do not assume an underlying distribution (e.g., normal distribution) and are direct statements of probability: i.e., a %90 credibility interval has a %90 chance of containing the true answer.

Culture tanks: Artificial holding areas used to rear fish farm stocks.

Demersal: Fishes and other aquatic organisms that live on or in close association with the ground whether it's the bottom of the sea or a lake.

Derived Characteristic: A characteristic that is shared with an ancestor (i.e. inherited) that has been modified in a novel way, such as the dorsal fin of a remora that has the shape of a sucking disc.

Diadromous: A fish that migrates between freshwater and saltwater, regardless of when it does so in its life history, how many times or which direction it migrates.

Diel migration: The daily vertical migration of organisms into the photic zone at night and back down below the photic zone during the day. The diel migration in the oceans is the greatest migration of biomass on Earth. It involves selected organisms and may be dependent on other environmental factors in detail.

Dorsal: On a fish, the side of the fish that contains the back fin, opposite the ventral side with the belly.

Ectothermy: Organisms that control body temperature by external means such as basking in sunlight.

Endemic: Species associated with a specific and restricted location rather than in many places.

Endothermy: Organisms that control body temperature through internal means such as regulating their metabolism to maintain homeostasis.

Estuary: Place where a river meets the sea and freshwater and seawater mix.

Effective Population Size (Ne). The size of an ideal population that would be affected by genetic drift at the same rate as the actual population. This idealized population has discrete generations, an even sex ratio, constant size, random union of gametes and random survivorship of offspring.

Extensive aquaculture: managed aquaculture dependent on the local natural setting, such as a pond or coastal sea area.

Fertilization: fusion of eggs and sperm (creating diploids from haploids); mating and fertilization need not occur simultaneously (e.g., surfperches and rockfishes).

Fingerling: Development stage of fish following the fry stage and continuing into the first three to four months of life.

Fish farm: locations used to grow populations of aquatic organisms, primarily fish.

Fish finder: A sonar-based device used to determine location and depth of fishes.

Fish meal: Ground dried fish used as fertilizer and as feed for domestic livestock.

Fish Hatchery: one form of a fish farm, managed with the intent of resupplying native wild populations in natural environments.

Fish husbandry: the breeding and rearing of fish for a variety of reasons.

Fish kill: a description of the number of fish stock to die in a specific amount of time.

Fish meal: commercially processed food source used in fish farming as a source of protein for the fish stock.

Fry: Development stage of fish immediately after the larvae stage, at an age of less than a week.

F-statistics (FIS, FST, FIT). Measures used to hierarchically partition genetic diversity within and among populations. Common measures include: 1) FIS, which is the average departure of genotype frequencies from Hardy-Weinberg expectations within a population; 2) FST, which is the proportion of the variation due to allele-frequency differences among populations; and 3) FIT, which is the departure of genotype frequencies from Hardy-Weinberg expectations relative to the entire collection of populations. In this common hierarchy, the subscripts refer to comparisons between levels in the hierarchy: IS refers to individuals within populations, ST to subpopulations within the total population, and IT to individuals within the total population. Hierarchies and subscript notation can be extended to any level to accommodate different study designs.

Gadiform: Fishes such as cod.

Gigantothermy: An animal that is usually ectothermic that uses its large size to control body temperature to a signicant degree. Large size slows the rate of heat loss to the surrounding environment all other things being equal.

Game fish: Fish, such as trout, raised to become replenishment stock for natural sport fisheries.

Gametic Disequilibrium (or Linkage Disequilibrium) A state that exists in a population when alleles at different loci are not distributed independently of each other in a population's gamete pool, often because the loci are physically linked on the same fragment of DNA.

Genetic Drift Chance changes in allele frequencies that result from the sampling of gametes from generation to generation in a finite population. The magnitude of these changes is inversely related to twice effective population size (2Ne), because each diploid fish carries two sets of genes.

Genetic Marker (Marker) A known DNA sequence that can be identified by a simple assay. For the purpose of this document, a marker can be a single SNP or a combination of SNPs.

Genotype The set of alleles at one or more loci for an individual.

Genotyping: determining the genetic makeup of a fish or stock or the purpose of managing the efficiency of aquaculture production.

Hardy-Weinberg Expectations (HWE). Genotypic frequencies expected from given allele frequencies, assuming random mating, no mutation (the alleles do not change), no migration or emigration (no exchange of alleles between populations), infinitely large population size, and no selective pressure for or against any of the genotypes.

Hatching: when the larva frees itself from the egg.

Heterosis: also termed hybrid vigor; a fitness advantage gained by hybrids over their parental species. This is generally due to heterozygosity advantages, where the fitness of heterozygotes is higher than that of both homozygotes.

Heterozygosity The proportion of individuals in a population that carry two different alleles at a particular marker. This statistics can be used as a measure of genetic variability in a population.

High-resolution: A subjective term that describes the quality of an image that shows less than normal detail. The term describes an aspect of images that is different than focus. A high-resolution image can still be very out of focus and fuzzy. A low-resolution image is fuzzy becuase the image was not collected or saved in enough detail.

Homeotherm: An animal that maintains a relatively constant body temperature independent of the environmental temperature. Homeotherms expend energy to maintain body temperature by reguating metabolic processes.

Hypersaline: An adjective that describes water that has significantly high salinity. Term is not used to describe seawater. Term applies to some land-locked bodies of water that accumulate salt and have unusually high levels of salinity.

In vitro fertilization: artificially fertilizing fish eggs in a laboratory setting.

Incubation time: time from egg fertilization to hatching. Gestation applies only to live-bearing fishes; it is the time young stay within the female.

Intensive aquaculture: managed aquaculture controlled through human engineered means, such as managing water quality and sources of food.

Lobe-finned: Fishes in the class Sarcopterygii such as coelacanths and lungfishes.

Low-resolution: A subjective term that describes the quality of an image that shows less than normal detail. The term describes an aspect of images that is different than focus. A high-resolution image can still be very out of focus and fuzzy. A low-resolution image is fuzzy becuase the image was not collected or saved in enough detail.

Locus :(plural Loci). A fixed position or region on a chromosome that may contain more than one genetic marker.

Mating: pairing (one-on-one) for the purpose of fertilizing eggs; copulatory organ present.

Microsatellites: DNA sequences containing short (5–2 base pairs) tandem repeats of nucleotides (e.g., GTGTGTGT).

Migration: The periodic movement of groups of animals from one region to another for feeding or breeding. Fishes migrate daily, seasonally and in synchrony with other periodic climatic events.

MRI: Acronym for Magnetic Resonance Imaging or Nuclear Magnetic Reseonace Imaging. The term is used for the process of taking the image as well as for images produced by this method.

Myotome: Group of tissues (usually muscle) served by a single nerve root from the spinal cord. The bundle of muscle served by a single nerve root that we associate with a flake of cooked fish.

Mitochondrial DNA (mtDNA). Circular, maternally inherited DNA located outside the nucleus in organelles in a cell called mitochondria.

Mixed Stock Analysis (MSA). Method using allele frequencies from populations and genotypes from mixture samples to estimate stock compositions.

Muller's ratchet: the inevitable accumulation of mutations as a consequence of drift in small asexual populations. Because genomes with the lowest number of mutations can be lost by drift and mutation-free genomes cannot be recovered in the absence of recombination, the average number of mutations per genome is expected to increase over time, in a way that evokes the unidirectional notches of a ratchet.

Nictating membrane: The third eyelid possessed by some sharks, particularly Carcharhinids and their allies. The nictating membrane is made up of an extra fold under the lower eyelid that can move upward and protect the eye.

Oceanodromous fish: migratory fish that move within salt water only.

Ornamental fish: are small fish, such as koi, which are typically grown to be kept in aquariums or small landscape ponds.

Overfishing: detrimental practice of removing more of a natural aquatic species than what natural reproduction can support.

Parthenogenesis: from the Greek parthenos, virgin, and genesis, creation, meaning reproduction without fertilization.

Pectoral Girdle: A skeletal support to which the arms, forelimbs, or pectoral fins of vertebrates are attached.

Pelvic Girdle: A skeletal support to which the legs, hind limbs, or pelvic fins of vertebrates are attached.

Perciform: Perch-like fish.

Poikilotherm: An animal whose body temperature varies with the temperature of its surroundings. Not all ectotherms are poikilotherms.

Posterior: On a fish, the side of the fish that contains the tail, opposite the anterior side with the head.

Primitive Fish: Fishes that retain characteristics in common with early ancestors, or lack characteristics of more recent ancestors.

Pisciculture: aquaculture practice involving finned fish.

Polymerase Chain Reaction (PCR). The enzymatic amplification a single or few copies of a DNA fragment several orders of magnitude, generating millions of copies.

Pond acreage: area committed to extensive aquaculture holdings at a fish farm or company.

Population A randomly mating group of fish that are largely reproductively isolated from other populations.

Potamodromous fish: migratory fish that move within fresh water only.

Ray-finned: Fishes in the class Actinopterygii that have fins supported by internal skeletons that fan out from a point of attachment to the body (as opposed to from an intervening stalk like the Sarcopterygians.)

Reporting Group A group of populations in a genetic baseline to which portions of a mixture are allocated during mixed stock analyses. In this study, reporting groups were constructed based on stakeholder needs, on the extent of genetic distinction among populations, and on advice from the WASSIP Technical Committee and Advisory Panel.

Segmentation: The delineation of anatomical structures that are visible on an MRI or other image. This process can be done manually based on visual inspection or by an automated digital process.

Spiracle: A small opening behind the eye used by bottom-dwelling rays and some benthic sharks to bring respiratory current to the gills. The spiracle is a remnant gill slit found in rays, sharks and some primitive bony fishes such as bichirs, sturgeons and paddlefishes.

Single nucleotide polymorphism (SNP). DNA sequence variation occurring when a single nucleotide (A, T, C, or G) differs among individuals or within an individual between paired chromosomes.

Spawning: release of unfertilized eggs into the environment or release of larvae into the environment; mating and spawning need not occur simultaneously (e.g., surfperches). Spawning can occur without true mating (e.g., herring, which are broadcast spawners).

Transgressive phenotype: the appearance of extreme phenotypes in hybrids as compared to their parental species. This can be caused by overdominance, where the trait value in heterozygotes falls outside the range of both homozygotes. Adaptive transgressive phenotypes are thought to be the major force driving hybrid speciation.

Ventral: On a fish, the side of the fish that contains the belly, opposite the dorsal side with the back fin.

Water Quality: the characteristics of water, such as water temperature and contamination, which define its ability to sustain life and its purity from chemicals.

Wild fishing harvest: the commercial and personal fishing consumption in a specific area and over a specific time frame.

Yearling: Development stage of fish following the fingerling stage and lasting until approximately one year of age.

BIBLIOGRAPHY

Adams, M.D., Kelley, J.M., Gocayne, J.D., Dubnick, M., Polymeropoulos, M.H., Xiao, H., Merril, C.R., Wu, A., Olde, B., Moreno, R.F., Kerlavage, A.R., McCombie, W.R. and Venter, J.C. (1991) Complementary DNA sequencing: Expressed sequence tags and human genome project. Science, 252(5013), 1651-1656.

Alder, T. 1995. Fishy sex: Uncovering the wild ways of fish. Science News 148: 266-7.

Allen, S. K Jr., and Stanley, J. G. (1978). Reproductive sterility in polyploid brook trout, *Salvelinus fontinalis*. *Trans. Am. Fish. Soc.* 107, 473–478. doi: 10.1577/1548-8659(1978) 107<473: RSIPBT>2.0.CO;2

Asensio, L., Gonzalez, I., Fernandez, A., Rodriguez, M.A., Lobo, E., Hernandez, P.E., Garcia, T. and Martin, R. (2002) Application of random amplified polymorphic DNA (RAPD) analysis for identification of grouper (Epinephelus guaza), wreckfish (Polyprion americanus), and nile perch (Lates niloticus) fillets. Journal of Food Product, 65(2), 432-435.

Avise, J.C. (1994) Molecular Markers, Natural History and Evolution. Chapman and Hall, New York, London. [Citation Time(s): 2]

Avise, J.C., Helfman, G.S., Saunders, N.C. and Hales, L.S. (1986) Mitochondrial DNA differentiation in North Atlantic eels: Population genetic consequences of an unusual life history pattern. Proceeding of the National Academy Science (USA), 83(12), 4350-4354.

Avise, J.C., Walker, D. and Johns, G.C. (1998) Speciation durations and Pleistocene effects on vertebrate phylogeography. Proceedings of the Royal Society of London Series B, 265(1407), 1707-1712.

Backer, J., Bentzen, P. and Moran, P. (2002) Molecular markers distinguish coastal cutthroat trout from coastal rainbow trout/steelhead and their hybrids. Transaction of American Fisheries Society, 131(3), 404-417.

Baker, A.J. and Marshall, H.D. (1997) Molecular evolution of the mitochondrial genome. In: Mindell, D.P. Ed., Avian Molecular Evolution and Systematics, Academic Press, San Diego, 51-82.

Bartly, D.M. and Gall, G.A.E. (1991) Genetic identification of native cutthroat trout (Onchorynchus clarki) and introgressive hybridization with introduced rainbow trout (O. mykiss) in streams associated with the Alvord basin, Oregon and Nevada. Copeia, 3, 854-859.

Biradar, D., and Rayburn, A. L. (1993). Heterosis and nuclear DNA content in maize. *Heredity* 71: 300. doi: 10.1038/hdy.1993.139

Birky, C.W., Fuerst, P. and Maruyama, T. (1989) Organelle gene diversity under migration, mutation, and drift: equilibrium expectations, approach to equilibrium, effect of heteroplasmic cells, and comparison to nuclear genes. Genetics, 121(3), 613-627. [Citation Time(s): 2]

Boguski, M.S. and Schuler, G.D. (1995) Establishing a human transcript map. Nature Genetics, 10(4), 369-371.

Brown, B. and Epifanio, J. (2003) Nuclear DNA. In: Hallermann, E.M. Ed., Population Genetics: Principles and Applications for Fisheries Scientists, American Fisheries Society, Bethesda, 458-472.

Brown, J.R., Bechenbach, A.T. and Smith, M.J. (1993) Intraspecific DNA sequence variation of the mitochondrial control region of white sturgeon (Acipenser transmontanus). Molecular Biology Evolution, 10(2), 326-341.

Brown, W.M. (1985) The mitochondrial genome of animals. In: MacIntyre, R.J. Ed., Molecular Evolutionary Genetics, Plenum, New York, 95-130.

Bullini, L. (1994). Origin and evolution of animal hybrid species. *Trends Ecol. Evol.* 9, 422–426. doi: 10.1016/0169-5347(94) 90124-4

Cal, R., Vidal, S., Camacho, T., Piferrer, F., and Guitian, F. (2005). Effect of triploidy on turbot haematology. *Comp. Biochem. Physiol. Part A Mol. Integr. Physiol.* 141, 35–41. doi: 10.1016/j.cbpb.2005.03.007

Carmichael, G.J., Hanson, J.N., Schmidt, M.E. and Morizot, D.C. (1993) Introgression among apache, cutthroat and rainbow trout in Ariozona. Transaction of American Fisheries Society, 122, 121-130

Carrasco, L. A., Doroshov, S., Penman, D. J., and Bromage, N. (1998). Long-term, quantitative analysis of gametogenesis in autotriploid rainbow trout, *Oncorhynchus mykiss*. *J. Reprod. Fertil.* 113: 197. doi: 10.1530/jrf.0.1130197

Chen, S., Wang, J., Liu, S., Qin, Q., Xiao, J., Duan, W., et al. (2009). Biological characteristics of an improved triploid crucian carp. *Sci. China C Life Sci.* 52, 733–738. doi: 10.1007/s11427-009-0079-3

Chinnery PF, Howell N, Lightowlers RN et al: Genetic counseling and prenatal diagnosis for mtDNA disease. Am J Hum Genet 63: 1908–1911, 1998

Chourrout, D. & I. Nakayama, 1997. Chromosome studies of progenies of tetraploid female rainbow trout. Theoret. Appl. Genet. 74: 687–692.

Chourrout, D. (1984). Pressure-induced retention of second polar body and suppression of first cleavage in rainbow trout: production of all-triploids, all-tetraploids, and heterozygous and homozygous diploid gynogenetics. *Aquaculture* 36, 111–126. doi: 10.1016/0044-8486(84) 90058-9

Chourrout, D. (1988). Induction of gynogenesis, triploidy, and tetraploidy in fish. *Isi. Atlas Sci. Anim. Plant Sci.* 65–70.

Chow, S. and Kishino, H. (1995) Phylogenetic relationships between tuna species of the genus Thunnus (Scombriidae: Teleosrei): Inconsistent implications from morphology, nuclear and mitochondrial genomes. Journal of Molecular Evolution, 41, 741-748.

Chow, S., Clarke, M.E. and Walsh, P.J. (1993) PCR-RFLP analysis of thirteen western Atlantic snappers (subfamily Lutjaninae): A simple method for species and stock identification. Fisheries Bulletin (US), 91, 619-627.

Chow, S., Okamoto, H., Uozumi, Y., Takeuchi, Y. and Takeyama, H. (1997) Genetic stock structure of the swordfish (Xiphias gladius) inferred by PCR-RFLP analysis of the mitochondrial DNA control region. Marine Biology, 127(3), 359-367.

Colas, P., and Dubé, F. (1998). Meiotic maturation in mollusc oocytes. *Sem. Cell Dev. Biol.* 9, 539–548. doi: 10.1006/scdb.1998.0248

Compton, D. and Utter, F.M. (1985) Natural hybridization between steelhead trout (Salmo gairdneri) and coastal cutthroat trout (Salmo clarki clarki) in two Puget sound streams. Canadian Journal of Fisheries and Aquatic Science, 42, 110-119.

Cooney CA, Dave AA, Wolff GL: Maternal methyl supplements in mice affect epigenetic variation and DNA methylation of offspring. J Nutr 132: 2393S–2400S, 2002

Cox, D.R., Burmeister, M., Price, E., Kim, S. and Myers, R.M. (1990) Radiation hybrid mapping: a somatic cell genetic method for constructing high-resolution map of mammalian chromosomes. Science, 250(4978), 245-250.

Crozier, R.H. (1997) Preserving the information content of species: Genetic diversity, phylogeny, and conservation worth. Annual Review of Ecology and Systematics, 28(1), 243-268.

Cuñado, N., Terrones, J., Sánchez, L., Martínez, P., and Santos, J. L. (2002). Sex-dependent synaptic behaviour in triploid turbot, *Scophthalmus maximus* (pisces, *Scophthalmidae*). *Heredity* 89: 460. doi: 10.1038/sj.hdy.6800165

Danan C, Sternberg D, Van Steirteghem A et al: Evaluation of parental mitochondrial inheritance in neonates born after intracytoplasmic sperm injection. Am J Hum Genet 65: 463–473, 1999

Daugherty, C.H., Cree, A., Hay, J.M. and Thompson, M.B. (1990) Neglected taxonomy and continuing extinctions of tuatara (Sphenodon). Nature, 347(6289), 177- 179.

Debacker K, Kooy RF: Fragile sites and human disease. Hum Mol Genet 16 Spec No. 2: R150-8, 2007

DeBaun MR, Niemitz EL, Feinberg AP: Association of in vitro fertilization with Beckwith-Wiedemann syndrome and epigenetic alterations of LIT1 and H19. Am J Hum Genet 72: 156–160, 2003

Del Bosco, S. F., Siragusa, M., Abbate, L., Lucretti, S., and Tusa, N. (2007). Production and characterization of new triploid seedless progenies for mandarin improvement. *Sci. Hortic.* 114, 258–262. doi: 10.1016/j.scienta.2007.06.014

Delomas, T., Gomelsky, B., Anil, A., Schneider, K., and Warner, J. (2017). Spontaneous polyploidy, gynogenesis and androgenesis in second generation (F2) koi *Cyprinus carpio* × goldfish *Carassius auratus* hybrids. *J. Fish Biol.* 90, 80–92. doi: 10.1111/jfb.13157

Dobson, A.P., Rodriguez, J.P., Roberts, W.M. and Wilcove, D.S. (1997) Geographic distribution of endangered species in the United States. Science, 275(5299), 550-555.

Donaldson, E.M. & R.H. Devlin, 1996. Uses of biotechnology to enhance production, pp. 969–1020 in Principles of Salmonid Culture. Developments in Aquaculture and Fisheries Science, Vol. 29, edited by W. Pennell & B. Barton. Elsevier Science Publishers, Amsterdam, The Netherlands.

Doornbos ME, Maas SM, McDonnell J et al: Infertility, assisted reproduction technologies and imprinting disturbances: a Dutch study. Hum Reprod 22: 2476–2480, 2007

Faith, D.P. (1994) Genetic diversity and taxonomic priorities for conservation. Biological Conservation, 68(1), 69-74.

Falls JG, Pulford DJ, Wylie AA et al: Genomic imprinting: Implications for human disease. Am J Pathol 154: 635–647, 1999

Ferguson, A., Taggart, J.B., Prodohl, P.A., McMeel, O., Thompson, C., Stone, C., McGinnity, P. and Hynes, R.A. (1995) The application of molecular markers to the study and conservation of fish populations with special reference to Salmo. Journal of Fish Biology, 47(A), 103-126.

Ferreira, D. N. N., de Siqueira-Silva, D. H., Pereira-Santos, M., Fujimoto, T., Senhorini, J. A., Lso, N., et al. (2017). Stereological analysis of gonads from diploid and triploid fish yellowtail tetra *Astyanax altiparanae* (Garutti & Britski) in laboratory conditions. *Zygote* 55, 537–544. doi: 10.1017/S0967199417000399

Fisher, R.A. (1930) The Genetical Theory of Natural Selection. Oxford University Press, UK.

Geifman-Holtzman O, Fay K: Prenatal diagnosis of congenital myotonic dystrophy and counseling of the pregnant mother: Case report and literature review. Am J Med Genet 78: 250–253, 1998

Giovannucci E, Ogino S: DNA methylation, field effects, and colorectal cancer. J Natl Cancer Inst 97: 1317–1319, 2005

Gold, J.R. Sun, F. and Richardson, L.R. (1997) Population structure of red snapper from the Gulf of Mexico as inferred from analysis of mitochondrial DNA. Transaction of American Fisheries Society, 126(3), 386-396.

Gold, J.R., Richardson, L.R., Furman, C. and King, T.L. (1993) Mitochondrial DNA differentiation and population structure in red drum (Sciaenops ocellatus) from the Gulf of Mexico and Atlantic Ocean. Marine Biology, 116(2), 175-185.

Gomelsky, B., Delomas, T. A., and Warner, J. L. (2016). Ploidy variation and viability of aneuploid ornamental koi carp obtained by crossing triploid females with diploid males. *North Am. J. Aquacult.* 78, 218–223. doi: 10.1080/15222055.2016.1150923

Graves, J.E., McDowell, J.R. and Jones, M.L. (1992) A genetic analysis of weakfish Cynoscion regalis stock structure along the mid-Atlantic coast. Fisheries Bulletin, 90, 469-475.

Greenberg, R., Cordero, P.J., Droege, S. and Fleischer, R.C. (1998) Morphological adaptation with no mitochondrial DNA differentiation in the coastal plain swamp sparrow. Journal of American Ornithologists' Union, 115(3), 706-712.

Gregory, T. R., and Mable, B. K. (2005). "Polyploidy in animals, " in *The Evolution of the Genome*, ed T. Gregory (Amsterdam: Elsevier), 427–517. doi: 10.1016/b978-012301463-4/50010-3

Haffray, P., Aubin, J., Houis, V., Labbe, L., and Jalabert, B. (2007). Comparison of pressure or thermal treatments on triploid yields and malformations up to swim up stage in rainbow trout (*Oncorhynchus mykiss*). *Aquaculture* 272, S265–S265.

Haig, S.M. (1998) Molecular contributions to conservation. Ecology, 79(2), 413-425.

Han, Y., Liu, M., Zhang, L. L., Simpson, B., and Zhang, G. X. (2010). Comparison of reproductive development in triploid and diploid female rainbow trout *Oncorhynchus mykiss*. *J. Fish Biol.* 76: 1742. doi: 10.1111/j.1095-8649.2010.02613.x

Haymer, D.S. (1994) Random amplified polymorphic DNAs and microsatellites: What are they, and can they tell us anything we don't already know? Annals of Entomological Society of American, 87, 717-722.

He, W., Xie, L., Li, T., Liu, S., Xiao, J., Hu, J., et al. (2013). The formation of diploid and triploid hybrids of female grass carp × male blunt snout bream and their 5S rDNA analysis. *BMC genetics* 14: 110. doi: 10.1186/1471-2156-14-110

Hebert, P.D.N., Ratnasingham, S. and deWaard, J.R. (2003) Barcoding animal life: Cytochrome c oxidase subunit 1 divergences among closely related species. Royal Society London, 270 (Suppl 1), S96-S99.

Hebert, P.D.N., Stoeckle, M.Y., Zemlak, T.S. and Francis, C.M., (2004) Identification of birds through DNA barcodes. PLoS Biology, 2(10), 312-316.

Hecker, K.H., Taylor, P.D. and Gjerde, D.T. (1999) Mutation detection by denaturing DNA chromatography using fluorescently labeled polymerase chain reaction products. Analytical Biochemistry, 272(2), 156-164.

Heist, E.J. and Gold, J.R. (1999) Microsatellite DNA variation in sandbar sharks (Carcharhinus plumbeus) from the Gulf of Mexico and mid-Atlantic bight. Copeia, 1, 182-186.

Hillis, D.M., Mable, B.K. and Moritz, C. (1996) Applications of molecular systematics: The state of the field and a look to the future. In: Hillis, D.M., Moritz, C. and Mable, B.K. Eds., Molecular systematics, Sinauer Associates, Massachusetts, 515-543.

Horsthemke B, Ludwig M: Assited reproduction: the epigenetic perspective. Hum Reprod Update 11: 473–480, 2005

Hu, F., Wu, C., Zhou, Y., Cao, L., Xiao, J., Wang, S., et al. (2018). Production of androgenetic, triploid and tetraploid hybrids from the interspecific hybridization of female Japanese crucian carp and male blunt snout bream. *Aquaculture* 491, 50–58. doi: 10.1016/j.aquaculture.2018.03.014

Hu, J., Liu, S., Xiao, J., Zhou, Y., You, C., He, W., et al. (2012). Characteristics of diploid and triploid hybrids derived from female *Megalobrama amblycephala* Yih × male *Xenocypris davidi Bleeker*. *Aquacultures* 364–365, 157–164. doi: 10.1016/j.aquaculture.2012.08.025

Iafrate AJ, Feuk L, Rivera MN et al: Detection of large-scale variation in the human genome. Nat Genet 36(9): 949-51, 2004

Jacquemont S, Hagerman RJ, Leehey MA et al: Penetrance of the fragile X-associated tremor/ataxia syndrome in a premutation carrier population. JAMA 291: 460–9, 2004b

Jehle, R. and Arntzen, J.W. (2002) Microsatellite markers in amphibian conservation genetics. Herpetological Jour nal, 12, 1-9.

Jiang Y, Lev-Lehman E, Bressler J et al: Genetics of Angelman syndrome. Am J Hum Genet 65: 1–6, 1999

Jin P, Warren ST: Understanding the molecular basis of fragile X syndrome. Hum Molec Genet 9: 901–908, 2000

Johnstone, R., H.A. McLay & M.V. Walsingham, 1991. Production and performance of triploid Atlantic salmon in Scotland. Canadian Technical Report Fisheries and Aquatic Sciences 1789, pp. 15–36.

Kaul, M., and Murthy, T. (1985). Mutant genes affecting higher plant meiosis. *Theor. Appl. Genet.* 70, 449–466. doi: 10.1007/BF00305977

Kavumpurath, S., and Pandian, T. J. (2010). Effects of induced triploidy on aggressive display in the fighting fish, Betta splendens Regan. *Aquacult. Res.* 23, 281–290. doi: 10.1111/j.1365-2109.1992.tb00771.x

Kocabas, A.M., Kucuktas, H., Dunham, R.A., Liu, Z. (2002) Molecular characterization and differential expression of the myostatin gene in channel catfish (*Ictaluruspunctatus*). *Biochimicaetbiophysicaacta* **1575,** 99-107.

Koga, A., Suzuki, M., Inagaki, H., Bessho, Y., Hori, H. (1996) Transposable element in fish.Nature **383,** 1- 30.

Korwin-Kossakowska, A., Reed, K.M., Pelak, C., Krause, E., Morrison, L. and Alexander, L.J. (2002) Radiation hybrid mapping of 118 new porcine microsatellites. Animal Genetics, 33(3), 224-227.

Krasnov, A., Agren, J.J., Pitaknen, T.I., Molsa, H. (1999) Transfer of growth hormone (GH) transgenes into Arctic charr. (*Salvelinusalpinus* L.) II. Nutrient partitioning in rapidly growing fish. Genetic analysis: biomolecular engineering**15,** 99-105.

Krisfalusi, M., Wheeler, P. A., Thorgaard, G. H., and Cloud, J. G. (2000). Gonadal morphology of female diploid gynogenetic and triploid rainbow trout. *J. Exp. Zool.* 286, 505–512. doi: 10.1002/(SICI) 1097-010X(20000401) 286: 5<505:: AID-JEZ8>3.0.CO;2-3

Lange, W., and Wagenvoort, M. (1973). Meiosis in triploid *Solanum tuberosum* L. *Euphytica* 22, 8–18. doi: 10.1007/bf00021550

Langenau, D.M., Traver, D., Ferrando, A.A., Kutok, J.L., Aster, J.C. et al. (2003) Myc-induced T cell leukemia in transgenic zebrafish. Science **299,** 887-90.

Latham KE: Epigenetic modification and imprinting of the mammalian genome during development. Curr Top Dev Biol 43: 1–49, 1999

Lawrence, C.S. & N.M. Morrissy, 2000. Genetic improvement of marron Cherax tenuimanus Smith and yabbies Cherax spp. In Western Australia. Aquacult. Res. 31: 69–82.

Lee, O., Green, J.M., Tyler, C.R. (2015) Transgenic fish systems and their application in ecotoxicology. Critical reviews in toxicology **45,** 124-141.

Lee, S.C., Tosi, S.C.M., Cheng, H.L. and Chang, J.T. (1997) Identification of Anguilla Japonica and A. marmorata elvers by allozyme electrophoresis. Journal of Fish Biology, 51(1), 208-210.

Levesque, H.M., Shears, M.A., Fletcher, G.L., Moon, T.W. (2008) Myogenesis and muscle metabolism in juvenile Atlantic salmon (Salmosalar) made transgenic for growth hormone. The Journal of experimental biology **211,** 128-137.

Li, D., Hu, W., Wang, Y., Zhu, Z., Fu, C. (2009) Reduced swimming abilities in fast-growing transgenic common carp Cyprinuscarpio associated with their morphological variations. Journal of fish biology **74,** 186-97.

Lian, H., Hu, W., Huang, R., Du, F., Liao, L. et al. (2013) Transgenic common carp do not have the ability to expand populations. PloS one**8,** e65506.

Lightowlers RN, Chinnery PF, Turnbull DM et al: Mammalian mitochondrial genetics: heredity, heteroplasmy and disease. Trends Genet 13: 450–455, 1997

Lin, C.H., John, J.A., Ou, L.W., Chen, J.C., Lin, C.H. et al. (2004) Cloning and characterization of metallothionein gene in ayu *Plecoglossusaltivelis.* Aquatic toxicology **66,** 111-124.

Linda, K.P. and Paul, M. (1995) Developments in molecular genetic techniques in fisheries. In: G.R. Carvalho and T.J. Pitcher, Eds., Molecular Genetics in Fisheries, Chapman and hall, London, 1-28.

Litt, M. and Luty, J.A. (1989) A hypervariable microsatellite revealed by in-vitro amplification of dinucleotide repeat within the cardiac muscle actin gene. American of Journal of Human Genetics, 44(3), 397-401.

Liu, S. (2010). Distant hybridization leads to different ploidy fishes. *Sci. China Life Sci.* 53, 416–425. doi: 10.1007/s11427-010-0057-9

Liu, S., Hu, F., Zhou, G., Zhang, X., He, X., Feng, H., et al. (2000). Gonadal structure of triploid crucian carp produced by crossing allotetraploid hybrids of *Carassium auratus* red var.(@&) × *Cyprinus carpio* () with Japanese crucian carp (*Carassius auratus* cavieri T. et S). *Acta Hydrobiol. Sin.* 24, 301–306.

Liu, S., Liu, Y., Zhou, G., Zhang, X., Luo, C., Feng, H., et al. (2001). The formation of tetraploid stocks of red crucian carp × common carp hybrids as an effect of interspecific hybridization. *Aquaculture* 192, 171–186. doi: 10.1016/s0044-8486(00) 00451-8

Liu, S., Qin, Q., Xiao, J., Lu, W., Shen, J., Li, W., et al. (2007). The formation of the polyploid hybrids from different subfamily fish crossings and its evolutionary significance. *Genetics* 176, 1023–1034. doi: 10.1534/genetics.107.071373

Liu, T., Liu, L., Wei, Q., Hong, Y. (2011) Sperm nuclear transfer and transgenic production in the fish medaka.International journal of biological sciences **7,** 469-475.

Liu, W.Y., Wang, Y., Qin, Y., Wang, Y.P., Zhu, Z.Y. (2007) Site-directed gene integration in transgenic zebrafish mediated by crerecombinase using a combination of mutant lox sites. Marine biotechnology **9,** 420-428.

Liu, Z.J. and Cordes, J.F. (2004) DNA marker technologies and their applications in aquaculture genetics. Aquaculture, 238, 1-37.

Liu, Z.J., Karsi, A. and Dunham, R.A. (1999) Development of polymorphic EST markers suitable for geneticlinkage mapping of catfish. Marine Biotechnology, 1(5), 437-447.

Liu, Z.J., Li, P., Kocabas, A., Ju, Z., Karsi, A., Cao, D. and Patterson, A. (2001) Microsatellite-containing genes from the channel catfish brain: evidence of trinucleotide repeat expansion in the coding region of nucleotide excision repair gene RAD23B. Biochemical Biophysical Research Communication, 289(2), 317-324.

Liu, Z.J., Moav, B., Faras, A.J., Guise, K.S., Kapuscinski, A.R. (1990) Development of expression vectors for transgenic fish. Bio/technology **8,** 1268-1272.

Lohmus, M., Bjorklund, M., Sundstrom, L.F., Devlin, R.H. (2010) Effects of temperature and growth hormone on individual growth trajectories of wild-type and transgenic coho salmon *Oncorhynchuskisutch*. Journal of fish biology**76,** 641-654.

Lohmus, M., Raven, P.A., Sundstrom, L.F., Devlin, R.H. (2008) Disruption of seasonality in growth hormone-transgenic coho salmon (*Oncorhynchuskisutch*) and the role of cholecystokinin in seasonal feeding behavior. Hormones and behavior **54,** 506-513.

Lupski JR: Genomic rearrangements and sporadic disease.Nat Genet 39(7 Suppl): S43-7, 2007

Lutz RE: Trinucleotide repeat disorders. Semin Pediatr Neurol 14(1): 26-33, 2007

Mable, B. (2004). 'Why polyploidy is rarer in animals than in plants': myths and mechanisms. *Biol. J. Linnean Soc.* 82, 453–466. doi: 10.1159/000351998

Maceira, N., Haan, A. D., Lumaret, R., Billon, M., and Delay, J. (1992). Production of 2n gametes in diploid subspecies of *Dactylis glomerata* L. 1. Occurrence and frequency of 2n pollen. *Ann. Bot.* 69, 335–343. doi: 10.1093/oxfordjournals.aob.a088350

Maclea, K.S., Abuhagr, A.M., Pitts, N.L., Covi, J.A., Bader, B.D. et al. (2012) Rheb, an activator of target of rapamycin, in the blackback land crab, Gecarcinuslateralis: cloning and effects of molting and unweighting on expression in skeletal muscle. The Journal of experimental biology **215,** 590-604.

Maclean, N., Iyengar, A., Rahman, A., Sulaiman, Z., Penman, D. (1992) Transgene transmission and expression in rainbow trout and tilapia. Molecular marine biology and biotechnology **1,** 355-365.

Maher ER, Brueton LA, Bowdin SC et al: Beckwith-Wiedemann syndrome and assisted reproduction technology (ART). J Med Genet 40: 62–64, 2003

Mali, P., Yang, L., Esvelt, K.M., Aach, J., Guell, M. et al. (2013) RNA-guided human genome engineering via Cas9. Science **339,** 823-826.

Mallet, J. (2007). Hybrid speciation. *Nature* 446, 279–283. doi: 10.1038/nature05706

Mamuris, Z., Sfougaris, A.I., Stamatis, C. and Suchentrunk, F. (2002) Assessment of genetic structure of Greek Brown Hare (Lepus europeaus) populations based on variation in Random Amplified Polymorphic DNA (RAPD). Biochemical Genetics, 40(9-10), 323- 338.

Mao, H., Wang, D.H., Yang, W.X. (2012) The involvement of metallothionein in the development of aquatic invertebrate. Aquatic toxicology **110,** 208-213.

Marchington DR, Macaulay V, Hartshorne GM et al: Evidence from human oocytes for a genetic bottleneck in an mtDNA disease. Am J Hum Genet 63: 769–775, 1998

Markert, C.L. and Moller, F. (1959) Multiple forms of enzymes: Tissue, ontogenetic and species-specific patterns. Proceedings of the Naionall Academy of Science (USA), 45(5), 753-763.

Marris, E. (2010) Transgenic fish go large. Nature **467,** 259.

Martinez, R., Juncal, J., Zaldivar, C., Arenal, A., Guillen, I. (2000) Growth efficiency in transgenic tilapia (*Oreochromis* sp.) carrying a single copy of an homologous cDNA growth hormone. Biochemical and biophysical research communications **267,** 466-472.

Maxime, V. (2008). The physiology of triploid fish: current knowledge and comparisons with diploid fish. *Fish Fish.* 9, 67–78. doi: 10.1111/j.1467-2979.2007.00269.x

McAndrew, B.J. and Majumdar, K.C. (1983) Tilapia stock identification using electrophoretic markers. Aquaculture, 30(1-4), 249-261.

McCoard, S.A., Fahrenkrug, S.C., Alexander, L.J., Freking, B.A., Rohrer, G.A., Wise, T.H. and Ford, J.J. (2002) An integrated comparative map of the porcine X chromosome. Animal Genetics, 33(3), 178-185.

McKinney, M.L. (1999) High rates of extinction and threat in poorly studied taxa. Conservation Biology, 13(6), 1273-1281.

Menezes, M.R., Naik, S. and Martins, M. (1993) Genetic characterization in four sciaenid species from the Arabian Sea. Journal of Fish Biology, 43(1), 61-67.

Meng, X., Noyes, M.B., Zhu, L.J., Lawson, N.D., Wolfe, S.A. (2008) Targeted gene inactivation in zebrafish using engineered zinc-finger nucleases. Nature biotechnology **26,** 695-701

Menotti-Raymond, M. and O'Brien, S.J. (1993) Dating the genetic bottleneck of the African cheetah. Proceedings of the National Academy of Science (USA.), 90(8), 3172-3176.

Meyer, A. (1993) Evolution of mitochondrial DNA in fishes. In: Mochachka, P.W. and Mommsen, T.P. Eds., Biochemistry and molecular biology of fishes, Elsevier Press Amsterdam, New York, 1-38. [Citation Time(s): 3]

Meyer, B.M., Froehlich, J.M., Galt, N.J., Biga, P.R. (2013) Inbred strains of zebrafish exhibit variation in growth performance and myostatin expression following fasting. Comparative biochemistry and physiology Part A, Molecular and integrative physiology **164,** 1-9.

Mila, B., Girman, D.J., Kimura, M. and Smith, T.B. (2000) Genetic evidence for the effect of a postglacial population expansion on the phylogeography of a North American songbird. Proceedings of the Royal Society London Series B, 267(1447), 1033-1040

Miller, J.C., Holmes, M.C., Wang, J., Guschin, D.Y., Lee, Y.L. et al. (2007) An improved zinc-finger nuclease architecture for highly specific genome editing. Nature biotechnology **25,** 778-785.

Moav, B., Z.J. Liu, Y. Groll & P.B. Hackett, 1992. Selection of promoters for gene transfer into fish. Molec. Mar. Biol. And Biotechnol. 1: 338–345.

Mohanta, R., Jayasankar, P., Das Mahapatra, K., Saha, J.N., Barman, H.K. (2014) Molecular cloning, characterization and functional assessment of the myosin light polypeptide chain 2 (mylz2) promoter of farmed carp, Labeorohita. Transgenic research **23,** 601-607.

Moreau, D.T., Conway, C., Fleming, I.A. (2011) Reproductive performance of alternative male phenotypes of growth hormone transgenic Atlantic salmon (Salmosalar). Evolutionary applications **4,** 736-748.

Moreau, D.T., Gamperl, A.K., Fletcher, G.L., Fleming, I.A. (2014) Delayed phenotypic expression of growth hormone transgenesis during early ontogeny in Atlantic salmon (Salmosalar) PloS one **9,** e95853.

Morello R, Bertin TK, Chen Y et al: CRTAP is required for prolyl 3- hydroxylation and mutations cause recessiveosteogenesis imperfecta. Cell 127(2): 291-304, 2006

Morin, P.A., Luikart, G., Wayne, R.K. and the SNP working group, SNPs in ecology, evolution and conservation. Trends in Ecology and Evolution, 19(4), 208-216.

Morison IM, Reeve AE: A catalogue of imprinted genes and parent-of-origin effects in humans and animals. Hum Mol Genet 7: 1599–1609, 1998

Moritz, C. (2002) Strategies to protect diversity and the evolutionary processes that sustain it. Systematic Biology, 51(2), 238-254.

Mosher, D.S., Quignon, P., Bustamante, C.D., Sutter, N.B., Mellersh, C.S. (2007) A mutation in the myostatin gene increases muscle mass and enhances racing performance in heterozygote dogs. PLoS genetics **3,** e79.

Mueller RF, Cook J: Mendelian inheritance. In Rimoin DL, Connor JM, Pyeritz RE (eds): Emery and Rimoin's Principles and Practice of Medical Genetics, pp 87–102. Vol 1, 3rd ed. New York, Churchill Livingstone, 1997

Muir, W.M., Howard, R.D. (2002) Assessment of possible ecological risks and hazards of transgenic fish with implications for other sexually reproducing organisms. Transgenic research**11,** 101-114.

Murakami, Y, Motohashi, K, Yano, K, Ikebukuro, K, Yokoyama, K, Tamiya, EKarube, I (1994) Micromachined electroporation system for transgenic fish. Journal of biotechnology **34,** 35-42.

Murphy, R.W., Sites, J.J.W., Buth, D.G. and Haufler, C.H. (1996) Proteins I: Isozyme electrophoresis. In: Hillis, D.M., Moritz, C. and Mable, B.K. Eds., Molecular Systematics, Sinauer Associates, Sunderland, 51-132.

Nadjar-Boger, E., Hinits, Y., Funkenstein, B. (2012) Structural and functional analysis of myostatin-2 promoter alleles from the marine fish Sparusaurata: evidence for strong muscle-specific promoter activity and post-transcriptional regulation. Molecular and cellular endocrinology **361,** 51-68.

Nam, Y.K., Cho, Y.S., Kim, D.S. (2000) Isogenic transgenic homozygous fish induced by artificial parthenogenesis. Transgenic research**9,** 463-9.

Nam, Y.K., Noh, J.K., Cho, Y.S., Cho, H.J., Cho, K.N. (2001) Dramatically accelerated growth and extraordinary gigantism of transgenic mud loach Misgurnusmizolepis. Transgenic research **10,** 353-362.

Neff, B.D. and Gross, M.R. (2001) Microsatellite evolution in vertebrates: Inference from AC dinucleotide repeats. Evolution, 55(9), 1717-1733.

Novotny, V., Basset, Y., Miller, S.E., Weiblen, G.D., Bremer, B., Cizek, L. and Drozd, P. (2002) Low host specificity of herbivorous insects in a tropical forest. Nature, 416, 841-844.

O'Brien, S.J. (1991) Molecular genome mapping: lessons and prospects. Current Opinion in Genetic Development, 1(1), 105-111.

Olivar, M.P., Moser, H.G. and Beckley, L.E. (1999) Lantern fish larvae from the Agulhas current (SW Indian Ocean). Science of Marine, 63, 101-120.

Otto, S. P. (2007). The evolutionary consequences of polyploidy. *Cell* 131, 452–462. doi: 10.1016/j.cell.2007.10.022

Overturf, K. (2009) Molecular research in aquaculture, Ames, Iowa: Wiley-Blackwell.

Ozato, K., Kondoh, H., Inohara, H., Iwamatsu, T., Wakamatsu, Y. et al. (1986) Production of transgenic fish: introduction and expression of chicken delta-crystallin gene in medaka embryos.Cell differentiation **19,** 237-244.

Pai, R., Soreghan, B., Szabo, I.L., Pavelka, M., Baatar, D. (2002) Prostaglandin E2 transactivates EGF receptor: a novel mechanism for promoting colon cancer growth and gastrointestinal hypertrophy. Nature medicine **8;** 289-293.

Pan, X., Zhan, H., Gong, Z. (2008) Ornamental expression of red fluorescent protein in transgenic founders of white skirt tetra (*Gymnocorymbusternetzi*). Marine Biotechnology (New York) **10,** 497-501.

Peter, J. G. S., Nagy, A., Horvath, L., and Csanyi, V. (2010). Induced triploidy in carp, *Cyprinus carpio* L. *J. Fish Biol.* 17, 667–671.

Phelps, M.P., Jaffe, I.M., Bradley, T.M. (2013) Muscle growth in teleost fish is regulated by factors utilizing the activin II B receptor. The Journal of experimental biology **216,** 3742-3750.

Pie, M.R., Alvares, L.E. (2006) Evolution of myostatin in vertebrates: is there evidence for positive selection? Molecular phylogenetics and evolution **41,** 730-734.

Piferrer, F., Beaumont, A., Falguière, J. C., Flajšhans, M., Haffray, P., and Colombo, L. (2009). Polyploid fish and shellfish: production, biology and applications to aquaculture for performance improvement and genetic containment. *Aquaculture* 293, 125–156.

Poirier LA: The effects of diet, genetics and chemicals on toxicity and aberrant DNA methylation: an introduction. J Nutr 132: 2336S–9S, 2002

Poirier LA, Vlasova TI: The prospective role of abnormal methyl metabolism in cadmium toxicity. Environ Health Perspect 110(Suppl 5): 793–795, 2002

Poontawee, K., Werner, C., Müller-Belecke, A., Hörstgen-Schwark, G., and Wicke, M. (2007). Flesh qualities and muscle fiber characteristics in triploid and diploid rainbow trout. *J. Appl. Ichthyol.* 23, 273–275. doi: 10.1111/j.1439-0426.2007.00843.x

Poulton J, Marchington DR: Prospects for DNA-based prenatal diagnosis of mitochondrial disorders. Prenat Diagn 16: 1247–1256, 1996

Powers, D.A., Gonzalez-Villasenor, L.I., Zhang, P.J., Chen, T.T., Dunham, R.A. (1991) Studies on transgenic fish: gene transfer, expression, and inheritance. Biotechnology **16,** 307-324.

Qin, Q., Wang, Y., Wang, J., Dai, J., Xiao, J., Hu, F., et al. (2014). The autotetraploid fish derived from hybridization of *Carassius auratus* red var.(female) × *Megalobrama amblycephala* (male). *Biol. Reprod.* 91: 93.

Rahman, M.A. & N. Maclean, 1999. Growth performance of transgenic tilapia containing an exogenous piscine growth hormone gene. Aquaculture 173: 333–346.

Rahman, M.A., A. Iyengar & N. Maclean, 1997. Co-injection strategy improves integration efficiency of a growth hormone gene construct, resulting in lines of transgenic tilapia (Oreochromis niloticus) expressing an exogenous growth hormone gene. Transgenic Res. 6: 369–378.

Rahman, M.A., Mak, R., Ayad, H., Smith, A., Maclean, N. (1998) Expression of a novel piscine growth hormone gene results in growth enhancement in transgenic tilapia (*Oreochromisniloticus*). Transgenic research **7,** 357-369.

Rajesh, R., Majumdar, K.C. (2005) Transgene integration - an analysis in autotransgenic Labeorohita Hamilton (Pisces: Cyprinidae). Fish physiology and biochemistry **31,** 281-287.

Rajesh, R., Majumdar, K.C. (2008) The growth hormone-encoding gene isolated and characterized from Labeorohita Hamilton is expressed in CHO cells under the control of constitutive promoters in 'autotransgene' constructs. Fish physiology and biochemistry **34,** 413-436.

Ramseyer, L.J. (2002) Transgenic fish: a boon or threat? Science **298,** 1715.

Rao, N.M., Rambabu, K.M., Rao, S.H. (2008) Electroporation of adult zebrafish.Methods in molecular biology **423,** 289-298.

Rasmussen, C., Ostberg, C.O., Clifton, D.R., Holloway, J.L. and Rodriguez, R.J. (2003) Identification of a genetic marker that discriminates ocean-type and stream-type Chinook salmon in the Columbia River Basin. Transaction American Fisheries Society, 132(1), 131-142.

Raven, P.A., Sakhrani, D., Beckman, B., Neregard, L., Sundstrom, L.F. (2012) Growth and endocrine effects of recombinant bovine growth hormone treatment in non-transgenic and growth hormone transgenic coho salmon. General and comparative endocrinology **177,** 143-152.

Rembold, M., Lahiri, K., Foulkes, N.S., Wittbrodt, J. (2006) Transgenesis in fish: efficient selection of transgenic fish by co-injection with a fluorescent reporter construct. Nature protocols **1,** 1133-1139.

Ren, H., Xu, M., He, P., Muto, N., Itoh, N. et al. (2006) Cloning of crucian carp (Carassiuscuvieri) metallothionein-II gene and characterization of its gene promoter region.Biochemical and biophysical research communications **342,** 1297-304.

Riberon, A., Sotiriou, E., Miaud, C., Andreone, F. and Taberlet, P. (2002) Lack of genetic diversity in Salamandra lanzai revealed by cytochrome b gene sequences. Copeia, 2002, 229-232.

Robertson, D. R. 1972. Social control of sex reversal in a coral-reef fish. Science, 1007-9.

Rocha-Olivares, A., Moser, H.G. and Stannard, J. (2000) Molecular identification and description of pelagic young of the rockfishes Sebastes constellatus and Sebastes ensifer. Fisheries Bulletin, 98, 353-363.

Rosen, J.N., Sweeney, M.F., Mably, J.D. (2009) Microinjection of zebrafish embryos to analyze gene function. Journal of visualized experiments: JoVE.

Rossi, A.R., Capula, M., Crosetti, D., Sola, L. and Campton, D.E. (1998) Allozyme variation in global populations of striped mullet, Mugil cephalus (Pisces: Mugilidae). Marine Biololgy, 131(2), 203-212.

Sanematsu, F., Takami, Y., Barman, H.K., Fukagawa, T., Ono, T.(2006) Asf1 is required for viability and chromatin assembly during DNA replication in vertebrate cells. The Journal of biological chemistry, **281,** 13817-13827.

Sanjana, N.E., Cong, L., Zhou, Y., Cunniff, M.M., Feng, G. (2012) A transcription activator-like effector toolbox for genome engineering. Nature protocols **7,** 171-192.

Sarmasik, A., Chun, C.Z., Jang, I.K., Lu, J.K., Chen, T.T. (2001a) Production of transgenic live-bearing fish and crustaceans with replication-defective pantropic retroviral vectors. Marine biotechnology **3,** S177-S184.

Sarmasik, A., Jang, I.K., Chun, C.Z., Lu, J.K., Chen, T.T. (2001b) Transgenic live-bearing fish and crustaceans produced by transforming immature gonads with replication-defective pantropic retroviral vectors. Marine biotechnology **3,** 470-477.

Sebat J, Lakshmi B, Malhotra D et al: Strong association of de novo copy number mutations with autism. Science 316(5823): 445-9, 2007

Serb, J.M., Phillips, C.A. and Iverson, J.B. (2001) Molecular phylogeny and biogeography of Kinosternon flavenscens based on complete mitochondrial control region sequences. Molecular Phylogenetics and Evolution, 18(1), 149-162.

Shaffer LG, McCaskill C, Adkins K et al: Systematic search for uniparental disomy in early fetal losses: The results and a review of the literature. Am J Med Genet 79: 366–372, 1998

Shanker, K., Ramadevi, J., Choudhaury, B.C., Singh, L. and Aggarawal, R.K. (2004) Phylogeny of olive ridley turtles (Lepidochelys olivacea) on the east coast of India: implications for conservation theory. Moelcular Ecology, 13(7), 1899-1909.

Shimizu, A., Shimizu, N. (2013) Dual promoter expression system with insulator ensures a stringent tissue-specific regulation of two reporter genes in the transgenic fish. Transgenic research **22,** 435-444.

Sin, F.Y., Walker, S.P., Symonds, J.E., Mukherjee, U.K., Khoo, J.G. et al. (2000) Electroporation of salmon sperm for gene transfer: efficiency, reliability, and fate of transgene. Molecular reproduction and development **56,** 285-288.

Soltis, P.S. and Gitzendanner, M.A. (1999) Molecular systematics and the conservation of rare species. Conservation Biology, 13(3), 471-483.

Starck, M.G. (1998) Isozymes in Molecular tools for screening biodiversity. In: Angela, K., Peter, G.I. and. David, S.I. Eds., Chapmann and Hall, London, 75-80.

Stinckens, A., Luyten, T., Bijttebier, J., Van Den Maagdenberg, K., Dieltiens, D. et al. (2008) Characterization of the complete porcine MSTN gene and expression levels in pig breeds differing in muscularity. Animal genetics **39,** 586-596.

Stoeckle, M. (2003) Taxonomy, DNA, and the bar code of life. BioScience, 53(9), 2-3.

Stoeckle, M., Janzen, D., Hallwachs, W., Hanken, J. and Baker, J. (2003) Taxonomy, DNA, and the barcode of life, " Draft conference report. Rubinoff DNA Barcodes and Conservation Barcode Conference, The Rockefeller University, New York, http: // phe.rockefeller.edu/Barcode Confeence/docs/B2summary.doc

Stokstad, E. (2002) Transgenic species. Engineered fish: friend or foe of the environment? Science, **297,** 1797-1799.

Stuppia L, Calabrese G, Franchi PG et al: Widening of a Y-chromosome interval-6 deletion transmitted from a father to his infertile son accounts for an oligozoospermia critical region distal to the RBM1 and DAZ genes (letter). Am J Hum Genet 59: 1393–1395, 1996

Su, B., Shang, M., Li, C., Perera, D.A., Pinkert, C.A. (2015) Effects of transgenic sterilization constructs and their repressor compounds on hatch, developmental rate and early survival of electroporated channel catfish embryos and fry. Transgenic research **24,** 333-352.

Sun, Y.-D., Liu, S.-J., Zhang, C., Li, J.-Z., Huang, W.-R., Zhang, J., et al. (2003). The chromosome number and gonadal structure of F9-F11 allotetraploid crucian-carp. *Acta Genet. Sin.* 30, 414–418.

Sundstrom, L.F., Lohmus, M., Devlin, R.H. (2005) Selection on increased intrinsic growth rates in coho salmon, Oncorhynchuskisutch. Evolution; international journal of organic evolution **59,** 1560-1569.

Sundstrom, L.F., Lohmus, M., Devlin, R.H. (2010) Migration and growth potential of Coho salmon smolts: implications for ecological impacts from growth-enhanced fish. Ecological applications: a publication of the Ecological Society of America **20,** 1372-1383.

Sundstrom, L.F., Lohmus, M., Johnsson, J.I., Devlin, R.H. (2004) Growth hormone transgenic salmon pay for growth potential with increased predation mortality. Proceedings Biological sciences/The Royal Society**271,** S350-S352.

Suneetha, B.K. (2000) Interspecific and inter specific genetic variation in selected mesopelagic fishes with emphasis on microgeographic variation and species characterization. Dr. Scient. Dissertation, Department of Fisheries and Marine Biology, University of Bergen, Bergen, Norway.

Tautz, D. (1989) Hypervariability of simple sequences as a general source for polymorphic DNA markers. Nucleic Acids Research, 17(16), 6463-6471.

Taylor, A.C., Sherwin, W.B. and Wayne, R.K. (1994) Genetic variation of microsatellite loci in a bottlenecked species: The northern hairy-nosed wombat Lasiorhinus krefftii. Molecular Ecology, 3(4), 277-290.

Thayanithy, V., Anathy, V., Kirankumar, S., Pandian, T.J. (2004) Growth enhancement and food conversion efficiency of transgenic fish Labeorohita. Journal of experimental zoology Part A, Comparative experimental biology **301,** 477-90.

Tilghman SM: The sins of the fathers and mothers: Genomic imprinting in mammalian development. Cell 96: 185–193, 1999

Tsou JA, Hagen JA, Carpenter CL et al: DNA methylation analysis: a powerful new tool for lung cancer diagnosis.Oncogene 21: 5450-5461, 2002

Venugopal, T., Mathavan, S., Pandian, T.J. (2002) Molecular cloning of growth hormone encoding cDNA of Indian major carps by a modified rapid amplification of cDNA ends strategy. Journal of biosciences **27,** 261-272.

Vick, B.M., Pollak, A., Welsh, C., Liang, J.O. (2012) Learning the scientific method using GloFish. Zebrafish **9,** 226-241.

Vrijenhoek, R. C. (1994). Unisexual fish: model systems for studying ecology and evolution. *Ann. Rev. Ecol. Syst.* 25, 71–96. doi: 10.1146/annurev.ecolsys.25.1.71

Vrijenhoek, R. C. (2006). Polyploid hybrids: multiple origins of a treefrog species. *Curr. Biol.* 16, 245–247.

Walsh T, McClellan JM, McCarthy SE et al: Rare structural variants disrupt multiple genes in neurodevelopmental pathways in schizophrenia. Science 320(5875): 539-43, 2008

Wang, K., Gan, L., Jeffry, E., Gayle, M., Gown, A.M., Skelly, M., Nelson, P.S., Ng, W.V., Schummer, M., Hood, L. and Mulligan, J. (1999) Monotoring gene expression profile changes in ovarian carcinomas using cDNA microarray. Gene, 229(1-2), 101-108.

Wang-Hurng, Y., Lee, S.C. and Yu, M.J. (1997) Genetic evidence to clarify the systematic status of the genra Zacco and Candidia (Cypriniformes: Cprinidae). Zoological Studies, 36(3), 170-177.

Warmbrodt, R.D., Stone, V. (1993) National Agricultural Library (U.S.) Transgenic fish research: a bibliography: a selected bibliography of research in the field of molecular biology and genetic engineering using fresh water fish, Beltsville, Md.: National Agricultural Library.

Wasko, A.P., Martins, C., Oliveira, C. and Foresti, F. (2003) Non-destructive genetic sampling in fish. An improved method for DNA extraction from fish fins and scales. Hereditas, 138(3), 161-165.

Weaver, D. (2002) Dealing with the risks of transgenic fish. Science **298,** 1715.

Welsh, J. and McClelland, M. (1990) Fingerprinting genomes using PCR with arbitrary primers. Nucleic Acids Research, 18(24), 7213-7218.

Werner, C., Poontawee, K., Mueller-Belecke, A., Hoerstgen-Schwark, G., and Wicke, M. (2008). Flesh characteristics of pan-size triploid and diploid rainbow trout (*Oncorhynchus mykiss*) reared in a commercial fish farm. *Arch. Anim. Breed.* 51, 71–83. doi: 10.5194/aab-51-71-2008

Wertheim, B., Beukeboom, L., and Van de Zande, L. (2013). Polyploidy in animals: effects of gene expression on sex determination, evolution and ecology. *Cytogenet. Genome Res.* 140, 256–269. doi: 10.1159/000351998

Widschwendter M, Jones PA: DNA methylation and breast carcinogenesis. Oncogene 21: 5462-5482, 2002

Williams, J.G.K., Kubelik, A.R., Livak, K.J., Rafalski, J.A. and Tingey, S.V. (1990) DNA polymorphisms amplified by arbitrary primers are useful as genetic markers. Nucleic Acids Research, 18(22), 6531-6535.

Wilmot GR, Warren ST: A new mutational basis for disease. In Wells RD, Warren ST (eds): Genetic Instabilities and Hereditary Neurological Diseases, pp 3–12. San Diego, Academic Press, 1998

Wilson, A.C., Cann, R.L., Carr, S.M., George, M., Gyllensten, U.B., Helm-Bychowski, K.M., Higuchi, R.G., Palumbi, S.R. and Prager, E.M. (1985) Mitochondrial DNA and two perspectives on evolutionary genetics. Biological Journal of Linnean Society, 26(4), 375-400.

Wirgin, I.I. and Waldman, J.R. (1994) What DNA can do for you? Fisheries, 19, 16-27.

Wooten, M.C. and Lydeard, C. (1990) Allozyme variation in a natural contact zone between Gambusia affinis and Gambusia holbrooki. Biochemical Systematics and Ecology, 18(2-3), 169-173.

Wright, J.M. (1993) DNA fingerprinting in fishes. In: W. Hochachka, P. and Mommsen, T. Eds., Biochemistry and Molecular Biology of Fishes, Elsevier, Amsterdam, 58-91.

Wu, B., Sun, Y.H. Wang, Y.W., Wang, Y.P., Zhu, Z.Y. (2005) Characterization of transgene integration pattern in F4 hGH-transgenic common carp (Cyprinuscarpio L.). Cell research **15,** 447-454.

Wu, B., Sun, Y.H., Wang, Y.P., Wang, Y.W., Zhu, Z.Y. (2004) Sequences of transgene insertion sites in transgenic F4 common carp. Transgenic research **13,** 95-96.

Wu, B., Zhu, Z.Y. (2003) [Integration sites of transgenes in transgenic animals]. Yi chuan = Hereditas / Zhongguoyichuanxuehuibianji **25,** 77-80.

Xiao, J., Kang, X., Xie, L., Qin, Q., He, Z., Hu, F., et al. (2014). The fertility of the hybrid lineage derived from female *Megalobrama amblycephala* × male *Culter alburnus. Anim. Reprod. Sci.* 151, 61–70. doi: 10.1016/j.anireprosci.2014.09.012

Xu, J., Huang, W., Zhong, C., Luo, D., Li, S. (2011) Defining global gene expression changes of the hypothalamic-pituitary-gonadal axis in female sGnRH-antisense transgenic common carp (Cyprinuscarpio). PloS one **6**, e21057.

Xu, J., You, F., Wu, X., Zhang, P., Lin, Y., Jiang, H., et al. (2010). Induction of triploidy in large yellow crocker *Pseudosciaena crocea* (Richardson, 1846): effects of pressure shocks and growth performance in the first rearing year. *Aquacult. Res.* 39, 1369–1376. doi: 10.1111/j.1365-2109.2008.02005.x

Xu, Q., Feng, C.Y., Hori, T.S., louffe, D.A., Buchanan, J.T. (2013) Family-specific differences in growth rate and hepatic gene expression in juvenile triploid growth hormone (GH) transgenic Atlantic salmon (Salmosalar). Comparative biochemistry and physiology Part D, Genomics and proteomics **8,** 317-333.

Xu, Q., Stemple, D., Joubin, K. (2008) Microinjection and cell transplantation in zebrafish embryos.Methods in molecular biology **461,** 513-520.

Yazawa, R., Hirono, I., Aoki, T. (2005) Characterization of promoter activities of four different Japanese flounder promoters in transgenic zebrafish. Marine Biotechnology (New York) **7,** 625-633.

Ye, X., Tian, Y.Y., Gao, F.Y. (2011) [Progress in transgenic fish techniques and application]. Yi chuan = Hereditas / *Zhongguoyichuanxuehuibianji* **33,** 494-503.

Yin, H., Sun, Z., Pan, W., and Chen, H. (2000). Study on gonadal development of triploid catfish (*Silurus asotus* L.). *Oceanol. Limnol. Sin.* 31, 123–129.

Yoshikawa, H., Morishima, K., Fujimoto, T., Arias-Rodriguez, L., Yamaha, E., and Arai, K. (2008). Ploidy manipulation using diploid sperm in the loach, *Misgurnus anguillicaudatus*: a review. *J. Appl. Ichthyol.* 24, 410–414.

Zbikowska, H.M. (2003) Fish can be first—advances in fish transgenesis for commercial applications. Transgenic research **12,** 379-89.

Zeng, Z., Liu, X., Seebah, S., Gong, Z. (2005a) Faithful expression of living color reporter genes in transgenic medaka under two tissue-specific zebrafish promoters. Developmental Dynamics **234,** 387-392.

Zeng, Z., Shan, T., Tong, Y., Lam, S.H., Gong, Z. (2005b) Development of estrogen-responsive transgenic medaka for environmental monitoring of endocrine disrupters. Environmental science andtechnology **39,** 9001-9008.

Zhang, P.J., Hayat, M., Joyce, C., Gonzalez-Villasenor, L.I., Lin, C.M. (1990) Gene transfer, expression and inheritance of pRSV-rainbow trout-GH cDNA in the common carp, Cyprinuscarpio (Linnaeus). Molecular reproduction and development**25,** 3-13.

Zhang, Q., Arai, K., and Yamashita, M. (1998). Cytogenetic mechanisms for triploid and haploid egg formation in the triploid loach *Misgurnus anguillicaudatus. J. Exp. Zool. Part A Ecol. Genet. Physiol.* 281, 608–619. doi: 10.1002/(SICI) 1097-010X(19980815) 281: 6<608:: AID-JEZ9>3.0.CO;2-R

Zhang, X., Yang, X. (2004) [Safety evaluation of food from transgenic fish and the molecular biological mechanism]. Wei sheng yanjiu = Journal of hygiene research **33,** 233-236.

Zhong, C., Song, Y., Wang, Y., Zhang, T., Duan, M. (2013) Increased food intake in growth hormone-transgenic common carp (Cyprinuscarpio L.) may be mediated by upregulating Agouti-related protein (AgRP). General and comparative endocrinology **192,** 81-88.

Zhong, Q.W., Fan, T.J. (2002) Advances in fish antifreeze protein research. Sheng wuhuaxueyu sheng wuwu li xuebaoActabiochimicaetbiophysicaSinica **34,** 124-130.

Zhu, Z., He, L., Chen, S. (1985) Novel gene transfer into the fertilized eggs of gold fish (*Carassiusauratus* L. 1758).Journal of Applied Ichthyology **1,** 31-34.

Zimmers, T.A., Davies, M.V., Koniaris, L.G., Haynes, P., Esquela, A.F. (2002) Induction of cachexia in mice by systemically administered myostatin. Science **296,** 1486-1488.

Zink, R.M., Barrowclough, G.F., Atwood, J.L. and Blackwell-Rago, R.C. (2000) Genetics, taxonomy, and conservation of the threatened California gnatcatcher. Conservation Biology, 14(5), 1394-1405.

Zoghbi HY, Orr HT: The spinocerebellar ataxias: In Scriver CR, Beaudet AL, Sly WS, Valle D (eds): The Metabolic and Molecular Bases of Inherited Disease, 8th ed. New York, McGraw Hill, 1995

INDEX

www.ingramcontent.com/pod-product-compliance
Lightning Source LLC
LaVergne TN
LVHW062116200926
848484LV00007B/1687